AF761977

Biological Nitrogen Fixation and Beneficial Plant–Microbe Interactions

Fernando González-Andrés
Euan James
Editors

Biological Nitrogen Fixation and Beneficial Plant–Microbe Interactions

Springer

Editors
Fernando González-Andrés
Institute of Environment, Natural Resources and Biodiversity, University of León
León
Spain

Euan James
The James Hutton Institute
Dundee, Huntingdonshire
UK

All the chapters included in this book have been peer-reviewed with a positive result. They were selected among the works presented at the XV SEFIN National Meeting of the Spanish Society of Nitrogen Fixation and the IV Portuguese-Spanish Congress on Nitrogen Fixation, held on June 16–18, 2015 in León, Spain.

ISBN 978-3-319-32526-2 ISBN 978-3-319-32528-6 (eBook)
DOI 10.1007/978-3-319-32528-6

Library of Congress Control Number: 2016939923

© Springer International Publishing Switzerland 2016
This work is subject to copyright. All rights are reserved by the Publisher, whether the whole or part of the material is concerned, specifically the rights of translation, reprinting, reuse of illustrations, recitation, broadcasting, reproduction on microfilms or in any other physical way, and transmission or information storage and retrieval, electronic adaptation, computer software, or by similar or dissimilar methodology now known or hereafter developed.
The use of general descriptive names, registered names, trademarks, service marks, etc. in this publication does not imply, even in the absence of a specific statement, that such names are exempt from the relevant protective laws and regulations and therefore free for general use.
The publisher, the authors and the editors are safe to assume that the advice and information in this book are believed to be true and accurate at the date of publication. Neither the publisher nor the authors or the editors give a warranty, express or implied, with respect to the material contained herein or for any errors or omissions that may have been made.

Printed on acid-free paper

This Springer imprint is published by Springer Nature
The registered company is Springer International Publishing AG Switzerland

Preface

The book entitled Biological Nitrogen Fixation and Beneficial Plant–microbe Interactions includes 20 chapters derived from the same number of selected contributions to the XV SEFIN, the National Meeting of the Spanish Society of Nitrogen Fixation (SEFIN in the Spanish acronym) from June 16th to 18th, 2015, with the slogan "The bridge between biotechnology, agriculture and environment." This event was celebrated together with the 4th Portuguese–Spanish Congress on Nitrogen Fixation and contained more than 100 original scientific contributions. This international event brought together more than 110 scientists from four continents, Europe, America, Asia, and Africa, and was held in the historical city of León in the interior Northwestern Spain, with almost 20 centuries of history.

The book consists of five different parts: The first being about "Ecology, Diversity, and Evolution of Plant Probiotic Microorganisms (PPM)" with five contributions; the second on "Genetics, Genomics and Proteomics of PPM and Their Associated Plants" with two contributions; the third on "Plant–Microorganisms Interactions" with five contributions; the fourth on "Physiology and Biochemistry of Beneficial Microorganisms and Associated Plants" with one contribution; and the final one on "Inoculants For Agriculture and Environmental Science" with seven contributions.

The use of microorganisms for agriculture and environmental applications is gaining importance worldwide to improve crop performance, but also for other environmental applications, such as bioremediation in chemically polluted soils. This book covers the most recent advances in all the topics with which researchers and professionals need to be familiar in order to obtain a better understanding of, and to better exploit, beneficial plant–microbe interactions. The search for an equilibrium between fundamental and applied aspects makes this book useful for professionals at various levels in the value chain of the "microbial biofertilizers." Microbial biofertilizers contain a completely known microbial community which has been selected based on its functional efficiency and its safety for human health and the environment. For this reason, there are several chapters focused on approaches to assessing bacterial biodiversity associated with plants and on techniques for a quick and precise identification of microbes. Microbial biofertilizers present clear advantages compared with other kinds of organic fertilizers. The latter contain unregulated bacterial populations/communities, whereas biofertilizers have a constant, known, and safe bacterial population, and these are the reasons for their gaining importance in agricultural and environmental applications. This book combines a solid scientific basis at an undergraduate level, together with a practical approach for scientists and professionals interested in applying knowledge about biofertilizers.

The editors acknowledge the SEFIN executive board and the staff at the University of León for their participation in the organization of the event, as well as the sponsors of the Congress, including the enterprises Agrocode Bioscience, Agrogenia, AMC Chemical, Biomasa Peninsular, BYDT, and Laboratorios Microkit, for their support.

Fernando González-Andrés
Euan James

Contents

Part I
Ecology, Diversity, and Evolution of Plant Probiotic Microorganisms (PPM)

Chapter 1
Identification of Human Pathogenic Bacteria in Plant Roots by Using MALDI-TOF MS Methodology

Encarna Velázquez, Esther Menéndez, Fernando Sánchez-Juanes, Noemí Valencia Daza, Juan Pérez Yépez, Milagros León Barrios, Ricardo Pérez Galdona, Ana Garrido and José Manuel González Buitrago

Abstract Bacterial endophytes are a good source of plant growth promoting bacteria to design biofertilizers, but they can include pathogenic species. Since matrix-assisted laser desorption ionization-time-of-flight mass spectrometry (MALDI-TOF MS) is a reliable method for bacterial identification and all cultivable bacterial species causing human diseases are included in the Biotyper 3.0 database, we used this methodology to analyze a collection of endophytic bacteria isolated from maize roots in Canary Islands. We identified several human pathogens such as *Bacillus cereus*, *Enterobacter cloacae*, *Kosakonia cowanii* (formerly *Entrobacter cowanii*), *Klebsiella pneumoniae*, *Pantoea agglomerans*, *Pseudomonas aeruginosa*, *Pseudomonas fulva*, *Pseudomonas mendocina*, *Pseudomonas mosselii*, and *Staphylococcus epidermidis*. These results showed that MALDI-TOF MS is a useful technique in the process of characterization of endophytic bacteria allowing to discard strains belonging to pathogenic species that cannot be included in biofertilizers.

Keywords MALDI-TOF MS · Bacterial endophytes · Maize · Identification · Pathogenic bacteria

---

E. Velázquez (✉) · E. Menéndez
Departamento de Microbiología y Genética, Universidad de Salamanca, Salamanca, Spain
e-mail: evp@usal.es

F. Sánchez-Juanes · J.M. González Buitrago
Instituto de Investigación Biomédica de Salamanca (IBSAL), Salamanca, Spain

N. Valencia Daza · J. Pérez Yépez · M. León Barrios · R. Pérez Galdona
Departamento de Bioquímica, Microbiología, Biología Celular y Genética, Universidad de La Laguna, Tenerife, Islas Canarias, Spain

A. Garrido
Granja Agrícola Experimental, Área de Agricultura y Ganadería, Cabildo de Lanzarote, Lanzarote, Islas Canarias, Spain

© Springer International Publishing Switzerland 2016

F. González-Andrés and E. James (eds.), *Biological Nitrogen Fixation and Beneficial Plant–Microbe Interactions*, DOI 10.1007/978-3-319-32528-6_1

1.1 Introduction

The endophytic microorganisms take part of the plant microbiome; they live in the inner plant tissues without cause damage for the plants and can be commensal or beneficial when they present different mechanisms of plant growth (Gaiero et al. 2013). Microbial endophytes are the subset of the microbiome that have a most direct influence in the growth and health of plants due their location inside the plants. Therefore, they are the best candidates for design biofertilizers; nevertheless, it is necessary to take into account that bacterial endophytes can include pathogenic species (Rosenblueth and Martínez-Romero 2006) making necessary to identify the bacterial pathogens for plants, animals, and humans among endophytic bacteria in order to remove them from biofertilization schemes, since they must include only non-pathogenic microorganisms (García-Fraile et al. 2012).

Matrix-assisted laser desorption ionization-time-of-flight mass spectrometry (MALDI-TOF MS) is a reliable method for bacterial identification mainly applied to date to isolates from clinical samples (Nomura 2015; Patel 2015; van Belkum et al. 2015). Although some attempts for in situ identification of bacterial strains and bacterial microbiomes in plant tissues have been performed (Ziegler et al. 2012; Stets et al. 2013), this methodology has been scarcely used to identify plant pathogenic bacteria (Zaluga et al. 2011). Nevertheless, the Biotyper databases contain the cultivable bacterial species causing disease in humans, animals, or plants, and then, MALDI-TOF MS analysis is an excellent tool for the recognition of pathogenic bacteria inhabiting plant tissues.

In this work, we use MALDI-TOF MS to analyze a collection of endophytic bacteria isolated from maize roots in Canary Islands in order to identify the pathogenic species present in its microbiome.

1.2 Materials and Methods

1.2.1 Strains Isolation

The strains were isolated from roots of maize cultivated in a soil from Guatiza (Tenerife, Canary Islands, Spain). The maize plants were harvested directly in the soil six months after the sowing. The roots were surface-disinfected with 70 % ethanol for 1 min, 2 % NaClO for 3 min, and 70 % ethanol for 30 s. Then, roots were rinsed five times with sterile distilled water and crushed into a tube containing 10 ml of PBS pH 7–7.4. Samples were incubated at room temperature and 160 rpm during 1 h. 100 μ/plate of serial dilutions was spread on TSA (Difco) plates and evaluated daily for strain selection.

1.2.2 Sample Preparation for MALDI-TOF MS

Strains were cultivated on TY medium (Beringer 1974) for 24 h. As indicate the manufacturer we used amounts of biomass between 5 and 100 mg to be transferred from the plate to a 1.5-ml tube (Eppendorf, Germany) with a pipette tip and mixed thoroughly in 300 μl of water to resuspend the bacterial cells. Then, 900 μl of absolute ethanol was added and the mixture was centrifuged at 15,500 *g* for 2 min, and the supernatant was discarded. The pellet was air-dried at room temperature for 1 h. Subsequently, 50 μl of formic acid (70 % v/v) was added to the pellet and mixed thoroughly before the addition of 50 μl acetonitrile to the mixture. The mixture was centrifuged again at 15,500 *g* for 2 min. One microliter of the supernatant was placed onto a spot of the steel target and air-dried at room temperature. Each sample was overlaid with 1 μl of matrix consisting of a saturated solution of alpha-cyano-4-hydroxycinnamic acid in 50 % acetonitrile and 2.5 % trifluoroacetic acid, and air-dried.

1.2.3 MALDI-TOF MS Performing and Data Analysis

The sample preparation and the MALDI-TOF MS performing were carried out as previously published (Ferreira et al. 2011). The calibration masses were the Bruker Bacterial Test Standards (BTS) which were as follows (masses as averages): RL36, 4365.3 Da; RS22, 5096.8 Da; RL34, 5381.4 Da; RL33meth, 6255.4 Da; RL29, 7274.5 Da; RS19, 10,300.1 Da; RNase A, 13,683.2 Da; and myoglobin, 16,952.3 Da.

The identification of strains was performed by using the Biotyper 3.0 database, and the results were classified using the score values proposed by the manufacturer: a score value between 2.300 and 3.000 indicated species identification; a score value between 2.000 and 2.299 indicated genus identification and possible species identification; a score value between 1.700 and 1.999 indicated genus identification; and a score value <1.7 indicated no identification.

1.3 Results and Discussion

The identification of bacterial endophytes is commonly carried out by 16S rRNA gene sequencing because it is the only gene currently available in databanks for all bacterial species (Hardoim et al. 2015). MALDI-TOF MS is a technology mainly used for clinical isolates (Nomura 2015; Patel 2015; van Belkum et al. 2015), although the Biotyper databases contain all known species of bacterial pathogens for humans and plants. Considering that in plant sources the presence of pathogenic bacteria has been reported (Rosenblueth and Martínez-Romero 2006), we applied by first time MALDI-TOF MS to the identification of a collection of endophytes from roots of maize plants growing in Canary Islands.

The results showed a surprising high number of pathogenic species (Table 1.1) many of them correctly identified using this methodology with score values higher than 2.3, which indicate a good identification at species level.

According to the results obtained, several strains belong to the phylum Firmicutes containing Gram-positive bacteria and concretely to the families *Bacillaceae* and *Staphylococcaceae* (Table 1.1). Within the family *Bacillaceae,* one strain was identified as *Bacillus cereus*, a sporulating bacteria which promote plant growth (Niu et al. 2011), but also causes infections in the man (Veysseyre et al. 2015) producing emetic and diarrheal toxins (Berthold-Pluta et al. 2015). Within the family *Staphylococcaceae,* several strains were identified as *Staphylococcus epidermidis*, a commensal organism able to produce biofilms (Büttner et al. 2015; Calà et al. 2015), which is found ubiquitously on human skin and mucous membranes and can cause several infections (Rogers et al. 2009) including bacteremia (Kleinschmidt et al. 2015).

One strain was identified as *Micrococcus luteus* belonging to the family *Micrococcaceae* within the phylum Actinobacteria also including Gram-positive bacteria (Table 1.1). This species, ubiquitous in the nature, promote the maize growth (Raza and Faisal 2013) and is commonly considered as human commensal because it is present in the skin of healthy individuals, but it has been found in immunosuppressed patients such as those affected by VIH (Smith et al. 1999) and in endocarditis or bacteriemia of susceptible patients (von Eiff et al. 1996; Miltiadous and Elisaf 2011). The spectra of representative strains from different species of phyla Firmicutes and Actinobacteria identified in this study are shown in Fig. 1.1.

Nevertheless, the most abundant pathogenic bacteria found in this work are Gram-negative rods belonging to the families *Enterobacteriaceae* and *Pseudomonadaceae* within the phylum Proteobacteria (Table 1.1). Within the family *Enterobacteriaceae* several strains were identified as *Klebsiella pneumoniae*, which causes diverse nosocomial infections (Holt et al. 2015) and *Pantoea agglomerans*, which causes infections in immnunocompromissed patients (Shubov et al. 2011). Other strains were classified within the genus *Enterobacter*, and concretely, they were identified as *Enterobacter cloacae* an opportunistic human pathogen with multiple resistance to antibiotics (Davin-Regli and Pagès 2015) and *Enterobacter cowanii*, a species originally isolated from clinical specimens (Inoue et al. 2000), recently reclassified into *Kosakonia cowanii* (Brady et al. 2013). The spectra of representative strains from different species of family *Enterobacteriaceae* identified in this study are shown in Fig. 1.2.

Within the family *Pseudomonadaceae* (Table 1.1), two strains were identified as *Pseudomonas aeruginosa*, which although has been reported as plant growth promoter (Adesemoye et al. 2008), is an opportunist pathogen that commonly infects the lungs of patients with chronic obstructive pulmonary disease and cystic fibrosis (Rada and Leto, 2013) and also can cause pneumonia in healthy individuals (Hatchette et al. 2000). Two strains were identified as *Pseudomonas mosselii*, a species isolated from clinical specimens (Dabboussi et al. 2002) with ability to

Table 1.1 Results of the identification using MALDI-TOF MS and the database Biotyper

Strain	ID organism best match	Score value[a]
Phylum Firmicutes, Family *Bacillaceae*		
GZME01	*Bacillus cereus* ATCC 14579^{T}	2.095
GZME14	*Bacillus megaterium* IAM13418^{T}	2.369
GZME22	*Bacillus mycoides* ATCC6462^{T}	2.165
Phylum Firmicutes, Family *Staphylococcaceae*		
GZME02	*Staphylococcus epidermidis* ATCC 14990^{T}	2.292
GZME03	*Staphylococcus epidermidis* ATCC 14990^{T}	2.302
GZME06	*Staphylococcus epidermidis* ATCC 14990^{T}	2.226
GZME07	*Staphylococcus epidermidis* ATCC 14990^{T}	2.199
GZME08	*Staphylococcus epidermidis* ATCC 14990^{T}	2.213
Phylum Firmicutes, Family *Paenibacillaceae*		
GZME10	*Paenibacillus pabuli* ATCC 43899^{T}	2.179
Phylum Actinobacteria, Family *Micrococcaceae*		
GZME09	*Micrococcus luteus* DSM 20030^{T}	2.381
Phylum Proteobacteria, Class alpha-Proteobaceria, Family *Brucellaceae*		
GZME27	*Ochrobactrum grignonense* OgA9a^{T}	2.243
Phylum Proteobacteria, Class gamma-Proteobaceria, Family *Enterobacteriaceae*		
GZME04	*Pantoea agglomerans* DSM 3493^{T}	2.066
GZME11	*Klebsiella pneumoniae* ATCC 13883^{T}	2.391
GZME15	*Klebsiella pneumoniae* ATCC 13883^{T}	2.378
GZME17	*Pantoea agglomerans* DSM 3493^{T}	2.206
GZME18	*Klebsiella pneumoniae* ATCC 13883^{T}	2.244
GZME20	*Kosakonia cowanii* CIP107300^{T}	2.113
GZME21	*Enterobacter cloacae* DSM30054^{T}	2.306
GZME24	*Klebsiella pneumoniae* ATCC 13883^{T}	2.407
GZME25	*Klebsiella pneumoniae* ATCC 13883^{T}	2.396
GZME28	*Klebsiella pneumoniae* ATCC 13883^{T}	2.314
GZME29	*Kosakonia cowanii* CIP 107300^{T}	1.891
Phylum Proteobacteria, Class gamma-Proteobaceria, Family *Pseudomonadaceae*		
GZME05	*Pseudomonas mendocina* ATCC 25411^{T}	2.292
GZME23	*Pseudomonas aeruginosa* DSM 50071^{T}	2.333
GZME19	*Pseudomonas aeruginosa* DSM 50071^{T}	2.345
GZME12	*Pseudomonas mosselii* CIP 105259^{T}	2.068
GZME13	*Pseudomonas mosselii* CIP 105259^{T}	2.005
GZME16	*Pseudomonas fulva* NRIC0180^{T}	2.282

[a]No reliable identification (0–1.699), probable genus identification (1.700–1.999), probable species identification 2.000-2.299, secure species identification (2.300–3.000)

solubilize phosphate, as occurs in the case of *P. aeruginosa* (Jha et al. 2009). One strain was identified as *Pseudomonas mendocina*, which causes sepsis (Nseir et al. 2011) and endocarditis (Johansen et al. 2001), and another strain as *Pseudomonas fulva*, which has been found in several human infections (Almuzara et al. 2010;

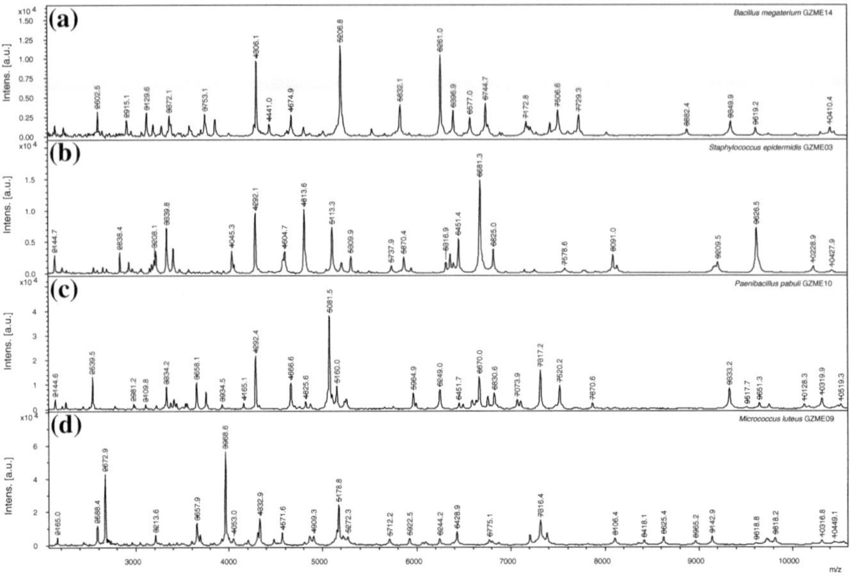

Fig. 1.1 MALDI-TOF MS spectra of whole-cell extracts obtained from a representative strain of different species identified within phyla Firmicutes and Actinobacteria: *Bacillus megaterium* GZME14 (**a**), *Staphylococcus epidermidis* GZME03 (**b**), *Paenibacillus pabuli* GZME10 (**c**), and *Micrococcus luteus* GZME09 (**d**)

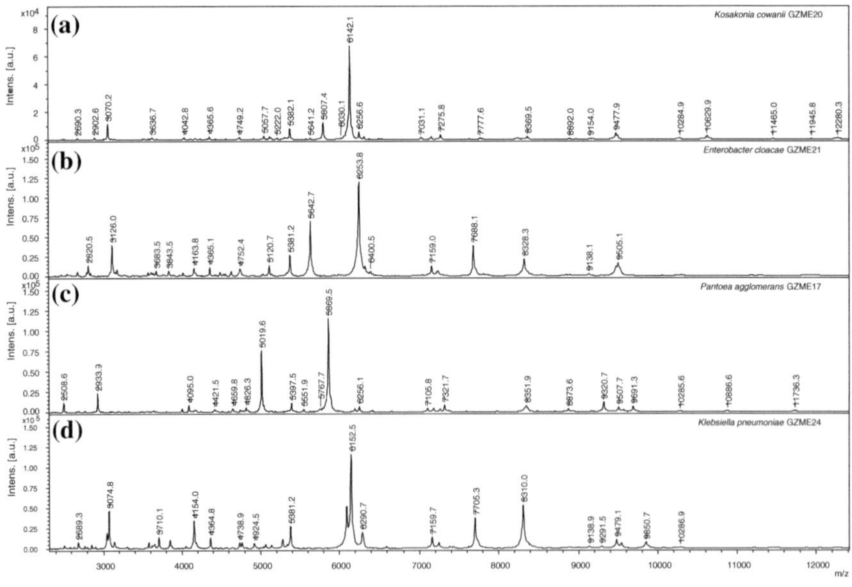

Fig. 1.2 MALDI-TOF MS spectra of whole-cell extracts obtained from a representative strain of different species identified within the family *Enterobacteriaceae* from the phylum Proteobacteria: *Kosakonia cowanii* GZME20 (**a**), *Enterobacter cloacae* GZME21 (**b**), *Pantoea agglomerans* GZME17 (**c**), and *Klebsiella pneumoniae* GZME24 (**d**)

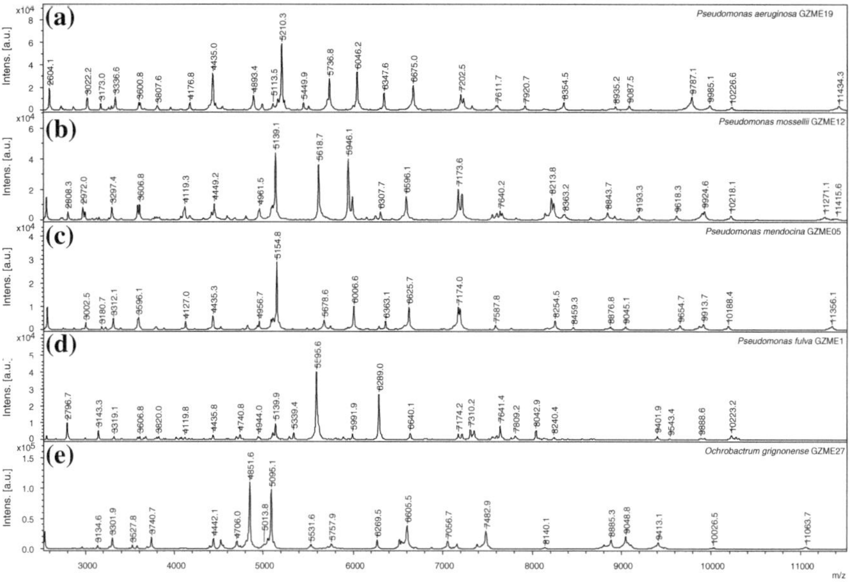

Fig. 1.3 MALDI-TOF MS spectra of whole-cell extracts obtained from a representative strain of different species identified within the families *Pseudomonadaceae* and *Brucellaceae* from the phylum Proteobacteria: *Pseudomonas aeruginosa* GZME19 (**a**), *Pseudomonas mossellii* GZME12 (**b**), *Pseudomonas mendocina* GZME05 (**c**) *Pseudomonas fulva* GZME1 (**d**), and *Ochrobactrum grignonense* GZME27 (**e**)

Rebolledo et al. 2014). The spectra of representative strains from different species of family *Pseudomonadaceae* identified in this study are shown in Fig. 1.3.

These results showed that MALDI-TOF MS is a good approach to identify the human pathogenic species present in plant endophytic bacterial populations which is a prerequisite for biofertilizers formulation in order to avoid health problems during their fabrication and application to the crops (García-Fraile et al. 2012).

Moreover, this technology allowed the identification of other species because in Biotyper databases are included both pathogenic and non-pathogenic species from the same genus, as occurs in the case of genera *Ochrobactrum* and the already mentioned *Bacillus*. The first genus containing pathogenic species such as *Ochrobactrum anthropi* (Quirino et al. 2014), and rhizospheric ones such as *Ochrobactrum grignonense* (Lebuhn et al. 2000), that was present among the endophytic bacteria isolated from maize roots in this work (Table 1.1; Fig. 1.3). The genus *Bacillus* contains pathogenic species such as *B. cereus,* and environmental ones such as *B. megaterium* and *B. mycoides* that have been not associated with foodborne illness (Logan 2012) and also were identified in maize roots (Table 1.1; Fig. 1.1). Also the Biotyper databases are increasing with the inclusion of some environmental species such as those from genus *Paenibacillus*, such as *Paenibacillus pabuli*, which was identified in maize roots growing in Canary Islands (Table 1.1; Fig. 1.1).

Therefore, we can conclude that MALDI-TOF MS allowed a rapid and reliable identification at species level of human pathogenic bacteria present in plant tissues in order to remove them from the biofertilizers design schemes. Since this methodology also allowed the identification of several non-pathogenic species we can affirm that MALDI-TOF MS is the most suitable methodology for a rapid and reliable identification in a short time for the analysis of plant endophytic bacteria, that only should be completed with 16S rRNA gene sequencing for those species absent in Biotyper databases.

Acknowledgements This work was supported by the Grant AGL2013-48098-P from the MINECO (Spanish Government).

References

Adesemoye AO, Obini M, Ugoji EO (2008) Comparison of plant growth-promotion with *Pseudomonas aeruginosa* and *Bacillus subtilis* in three vegetables Braz. J Microbiol 39:423–426

Almuzara MN, Vazquez M, Tanaka N et al (2010) First case of human infection due to *Pseudomonas fulva*, an environmental bacterium isolated from cerebrospinal fluid. J Clin Microbiol 48:660–664

Beringer JE (1974) R factors transfer in *Rhizobium leguminosarum*. J Gen Microbiol 84:188–198

Berthold-Pluta A, Pluta A, Garbowska M (2015) The effect of selected factors on the survival of *Bacillus cereus* in the human gastrointestinal tract. Microb Pathog 82:7–14

Brady C, Cleenwerck I, Venter S et al (2013) Taxonomic evaluation of the genus *Enterobacter* based on multilocus sequence analysis (MLSA): proposal to reclassify *E. nimipressuralis* and *E. amnigenus* into *Lelliottia* gen. nov. as *Lelliottia nimipressuralis* comb. nov. and *Lelliottia amnigena* comb. nov., respectively, *E. gergoviae* and *E. pyrinus* into *Pluralibacter* gen. nov. as *Pluralibacter gergoviae* comb. nov. and *Pluralibacter pyrinus* comb. nov., respectively, *E. cowanii*, *E. radicincitans*, *E. oryzae* and *E. arachidis* into *Kosakonia* gen. nov. as *Kosakonia cowanii* comb. nov., *Kosakonia radicincitans* comb. nov., *Kosakonia oryzae* comb. nov. and *Kosakonia arachidis* comb. nov., respectively, and *E. turicensis*, *E. helveticus* and *E. pulveris* into *Cronobacter* as *Cronobacter zurichensis* nom. nov., *Cronobacter helveticus* comb. nov. and *Cronobacter pulveris* comb. nov., respectively, and emended description of the genera *Enterobacter* and *Cronobacter*. Syst Appl Microbiol 36:309–319

Büttner H, Mack D, Rohde H (2015) Structural basis of *Staphylococcus epidermidis* biofilm formation: mechanisms and molecular interactions. Front Cell Infect Microbiol 5:14

Calà C, Amodio E, Di Carlo E et al (2015) Biofilm production in *Staphylococcus epidermidis* strains, isolated from the skin of hospitalized patients: genetic and phenotypic characteristics. New Microbiol (in press)

Dabboussi F, Hamze M, Singer E et al (2002) *Pseudomonas mosselii* sp. nov., a novel species isolated from clinical specimens. Int J Syst Evol Microbiol 52:363–376

Davin-Regli A, Pagès JM (2015) *Enterobacter aerogenes* and *Enterobacter cloacae*; versatile bacterial pathogens confronting antibiotic treatment. Front Microbiol 6:392

Ferreira L, Sánchez-Juanes F, García-Fraile P et al (2011) MALDI-TOF mass spectrometry is a fast and reliable platform for identification and ecological studies of species from family *Rhizobiaceae*. PLoS ONE 6:e20223

Gaiero JR, McCall CA, Thompson KA, Day NJ, Best AS, Dunfield KE (2013) Inside the root microbiome: bacterial root endophytes and plant growth promotion. Am J Botany 100:1738–1750

García-Fraile P, Carro L, Robledo M et al (2012) *Rhizobium* promotes non-legumes growth and quality in several production steps: towards a biofertilization of edible raw vegetables healthy for humans. PLoS ONE 7:e38122

Hardoim PR, van Overbeek LS, Berg G et al (2015) The hidden world within plants: ecological and evolutionary considerations for defining functioning of microbial endophytes. Microbiol Mol Biol Rev 79:293–320

Hatchette TF, Gupta R, Marrie TJ (2000) *Pseudomonas aeruginosa* community-acquired pneumonia in previously healthy adults: case report and review of the literature. Clin Infect Dis 31:1349–1356

Holt KE, Wertheim H, Zadoks RN et al (2015) Genomic analysis of diversity, population structure, virulence, and antimicrobial resistance in *Klebsiella pneumoniae*, an urgent threat to public health. Proc Natl Acad Sci USA 112:E3574–E3581

Inoue K, Sugiyama K, Kosako Y et al (2000) *Enterobacter cowanii* sp. nov., a new species of the family *Enterobacteriaceae*. Curr Microbiol 41:417–420

Jha BK, Pragash MG, Cletus J et al (2009) Simultaneous phosphate solubilization potential and antifungal activity of new fluorescent pseudomonad strains, *Pseudomonas aeruginosa*, *P. plecoglossicida* and *P. mosselii*. World J Microbiol Biotechnol 25:573–581

Johansen HK, Kjeldsen K, Høiby N (2001) *Pseudomonas mendocina* as a cause of chronic infective endocarditis in a patient with situs inversus. Clin Microbiol Infect 7:650–652

Kleinschmidt S, Huygens F, Faoagali J et al (2015) *Staphylococcus epidermidis* as a cause of bacteremia. Future Microbiol (in press)

Logan NA (2012) *Bacillus* and relatives in foodborne illness. J Appl Microbiol 112:417–429

Lebuhn M, Achouak W, Schloter M et al (2000) Taxonomic characterization of *Ochrobactrum* sp. Isolates from soil samples and wheat roots, and description of *Ochrobactrum tritici* sp. nov. and *Ochrobactrum grignonense* sp. nov. Int J Syst Evol Microbiol 50:2207–2223

Miltiadous G, Elisaf M (2011) Native valve endocarditis due to *Micrococcus luteus*: a case report and review of the literature. J Med Case Rep 5:251

Niu DD, Liu HX, Jiang CH et al (2011) The plant growth-promoting rhizobacterium *Bacillus cereus* AR156 induces systemic resistance in *Arabidopsis thaliana* by simultaneously activating salicylate- and jasmonate/ethylene-dependent signaling pathways. Mol Plant Microbe Interact 24:533–542

Nomura F (2015) Proteome-based bacterial identification using matrix-assisted laser desorption ionization-time of flight mass spectrometry (MALDI-TOF MS): a revolutionary shift in clinical diagnostic microbiology. Biochim Biophys Acta 1854:528–537

Nseir W, Taha H, Abid A et al (2011) *Pseudomonas mendocina* sepsis in a healthy man. Isr Med Assoc J 13:375–376

Patel R (2015) MALDI-TOF MS for the diagnosis of infectious diseases. Clin Chem 61:100–111

Quirino A, Pulcrano G, Rametti L et al (2014) Typing of *Ochrobactrum anthropi* clinical isolates using automated repetitive extragenic palindromic-polymerase chain reaction DNA fingerprinting and matrix-assisted laser desorption/ionization-time-of-flight mass spectrometry. BMC Microbiol 14:74

Rada B, Leto TL (2013) Pyocyanin effects on respiratory epithelium: relevance in *Pseudomonas aeruginosa* airway infections. Trends Microbiol 21:73–81

Raza FA, Faisal M (2013) Growth promotion of maize by desiccation tolerant *Micrococcus luteus*-chp37 isolated from Cholistan desert, Pakistan. Austr J Crop Sci 7:1693–1698

Rebolledo PA, Vu CC, Carlson RD et al (2014) Polymicrobial ventriculitis involving *Pseudomonas fulva*. J Clin Microbiol 52:2239–2241

Rogers KL, Fey PD, Rupp ME (2009) Coagulase-negative staphylococcal infections. Infect Dis Clin North Am 23:73–98

Rosenblueth M, Martínez-Romero E (2006) Bacterial endophytes and their interactions with hosts. Mol Plant Microbe Interact 19:827–837

Smith K, Neafie R, Yeager J et al (1999) *Micrococcus* folliculitis in HIV-1 disease. Br J Dermatol 141:558–561

Stets MI, Soares Pinto A, Fernandes Huergo L Jr et al (2013) Rapid identification of bacterial isolates from wheat roots by high resolution whole cell MALDI-TOF MS analysis. J Biotechnol 165:167–174

Shubov A, Jagannathan P, Chin-Hong PV (2011) *Pantoea agglomerans* pneumonia in a heart-lung transplant recipient: case report and a review of an emerging pathogen in immunocompromised hosts. Transpl Infect Dis 13:536–539

van Belkum A, Chatellier S, Girard V et al (2015) Progress in proteomics for clinical microbiology: MALDI-TOF MS for microbial species identification and more. Expert Rev Proteomics 15:1–11

Veysseyre F, Fourcade C, Lavigne JP et al (2015) *Bacillus cereus* infection: 57 case patients and a literature review. Med Mal Infect. doi:10.1016/j.medmal.2015.09.011

von Eiff C, Kuhn N, Herrmann M et al (1996) *Micrococcus luteus* as a cause of recurrent bacteremia. Pediatr Infect Dis J 15:711–713

Zaluga J, Heylen K, Van Hoorde K et al (2011) *GyrB* sequence analysis and MALDI-TOF MS as identification tools for plant pathogenic *Clavibacter*. Syst Appl Microbiol 34:400–407

Ziegler D, Mariotti A, Pflüger V et al (2012) In situ identification of plant-invasive bacteria with MALDI-TOF mass spectrometry. PLoS ONE 7(5):e37189. doi:10.1371/journal.pone.0037189

Chapter 2
Rhizobial Biofertilizers for Ornamental Plants

Esther Menéndez, Rocío Escribano-Viana, José D. Flores-Félix, Pedro F. Mateos and Raúl Rivas

Abstract Plant growth-promoting rhizobacteria (PGPR) are one of the most important groups of rhizospheric microorganisms, such as free-living bacteria, endophytes, and endosymbiotic bacteria. PGPRs colonize plant roots, conferring several benefits to the plants through different mechanisms. Despite its classical involvement in root nodule symbiosis with legumes, recent studies revealed the beneficial role of the genus *Rhizobium* as PGPR for several non-legume crops, supporting its inclusion in biofertilization schemes. However, studies based in the application of PGPRs as inoculants for ornamental crops are limited. Ornamental crops have a high economic importance for several regions, being Spain among the major producers of cut flowers. Due to their importance, we selected *Dianthus caryophyllus* (carnation) and a rhizobial strain for the evaluation of the PGPR effects in this plant. Selected strain was identified as *Rhizobium* sp. and presented PGPR mechanisms; phosphate solubilization, siderophores production, and IAA and precursors biosynthesis were positive. In vitro plant assays showed an increase in carnation development in early steps of plant development, when plants were inoculated respect to uninoculated plants. Moreover, effective root colonization was shown. Interestingly, an increase in root hairs redirections was observed in inoculated plants. To the best of our knowledge, this is the first study reporting the effects of *Rhizobium* in *D. caryophyllus*, showing effective root colonization and an improvement in early stages of plant development.

Keywords *Dianthus caryophyllus* · *Rhizobium* · Root colonization · Plant growth promotion · Biofertilizer

---

E. Menéndez (✉) · R. Escribano-Viana · J.D. Flores-Félix · P.F. Mateos · R. Rivas
Department of Microbiology and Genetics, University of Salamanca, Salamanca, Spain
e-mail: esthermenendez@usal.es

P.F. Mateos · R. Rivas
Associated Unit I+D University of Salamanca—CSIC 'Plant-Microbe Interactions Group', Salamanca, Spain

© Springer International Publishing Switzerland 2016

F. González-Andrés and E. James (eds.), *Biological Nitrogen Fixation and Beneficial Plant–Microbe Interactions*, DOI 10.1007/978-3-319-32528-6_2

2.1 Introduction

Many rhizobacteria can promote plant growth through a wide range of mechanisms. These rhizobacteria are commonly called plant growth-promoting rhizobacteria (PGPR). Despite its classic role in legume symbiosis, the genus *Rhizobium* has increased its importance as PGPR, showing the ability of rhizobial strains to promote plant growth and development in several non-legume crops, such as lettuce, carrots, tomato, or pepper (Santillana et al. 2005; García-Fraile et al. 2012; Flores-Félix et al. 2013), and also increase yields in cereal crops, such as rice and maize (Yanni et al. 2001; Gutiérrez-Zamora and Martínez-Romero 2001).

Interestingly, some studies suggest that rhizobacteria can be implicated in the increase of other interesting compounds, which are beneficial not only for plant development but also for human health. For example, *Phyllobacterium endophyticum* is involved in the increase of vitamin C in strawberries (Flores-Felix et al. 2015). Moreover, other studies showed that *Rhizobium* species might promote the ripeness stage in fruits of inoculated pepper plants (Silva et al. 2014a, b).

Nevertheless, the application of rhizobial strains in other plants, such as ornamental plants, is still not common. Only a few examples of application of PGPR bacteria in ornamental plants have been described, such as the effects of some members of the genus *Pseudomonas* and *Bacillus* in *Euphorbia*, *Dahlia*, *Dianthus*, *Chrysantemum,* and *Pelargionium* (Göre and Altin 2006; Sharma and Kaur 2010; Zulueta-Rodriguez et al. 2014) and of some *Azotobacter* and *Azospirillum* strains in *Gladiolus* or *Matthiola* (Srivastava and Govil 2007; Eid et al. 2009). However, there are no evidences of studies involving the application of Rhizobia in ornamental plants.

For that reason in this study, we will show the effects of a rhizobial strain in *Dianthus caryophyllus* (carnation), one of the most important ornamental plant in Spain, being the second cut flower in economical importance, after roses.

2.2 Materials and Methods

2.2.1 Bacterial Strain and Plant Material

Strain PEPV13 was isolated from an effective nodule of *Phaseolus vulgaris*, growing in a soil from Salamanca (NW, Spain) as described (Vincent 1970). As plant material, commercial variety of red carnation (*D. caryophyllus*, Vilmorin®) was selected among ornamentals for this study.

2.2.2 16S rRNA Gene Identification of Bacterial Strain

DNA extraction, PCR reactions, and PCR conditions were performed according to REDExtract-N-Amp Plant Kit (Sigma), with minor modifications. For 16S rRNA gene amplification, primers 27F and 1522R were used (Rivas et al. 2002). PCR product was purified using GeneJet Purification kit (Thermo-Fisher) according to the manufacturer′s specifications. Partial sequencing was performed on an ABI PRISM sequencer (Applied Biosystem®, EEUU) using a "Dye Terminator Cycle Sequencing Ready Reaction Kit" (Sequencing Service, University of Salamanca). Sequence was compared with type strain by using EzBiocloud Server (Chun et al. 2007).

2.2.3 *Determination of Plant Growth Promotion Mechanisms (PGPM)*

To determine PEPV13 strain PGPR abilities, three mechanisms were analyzed: (i) siderophores production, (ii) indol acetic acid (IAA) biosynthesis, and (iii) phosphate solubilization. Siderophores production was tested using M9-CAS agar medium (Alexander and Zuberer 1991). IAA and precursors biosynthesis was determined using modified JMM medium (O'Hara et al. 1989; Khalid et al. 2004), measuring the absorbance at 550 nm after the addition of Salkowski reagent and incubation during 15 min. Pikovskaya medium (Pikovskaya 1948), containing $CaHPO_4$ as source of inorganic *P*, was used for phosphate solubilization determination.

2.2.4 *Plant Assays in Axenic Conditions*

Red carnation seeds were surface-disinfected with 70 % ethanol for 30 s and 2 % NaClO for 5 min. Subsequently, several washes with sterile distilled water were performed. Seeds were placed in 1 % inverted agar plates and incubated in dark for four days at room temperature. Then, seedlings were transferred to 1 % agar square plates overlaid with sterile filter paper (five seedlings per plate). Seedlings were inoculated with 250 μl $plant^{-1}$ of a bacterial suspension with a turbidity of 6 in McFarland Standards (1.5×10^6 CFU $ml^{-1)}$. Plates were incubated vertically in a growth chamber and observed at 5, 7, and 12 days post-inoculation. An uninoculated treatment was included as negative control. The root length was measured.

2.2.5 *Colonization Assays*

For colonization events observation, a GFP-tagged derivative of PEPV13 strain was obtained by biparental mating with *Escherichia coli* S17.1, carrying pHC60 plasmid (Cheng and Walker 1998) as described in García-Fraile et al. (2012). Tetracycline was added to the media when required (10 μg/mL).

The colonization assays were performed in axenic conditions, set similarly to the assays described in Sect. 2.2.4. Colonization events and root hair deformations were observed at 5, 7, and 12 days post-inoculation under NIKON Eclipse 80i epifluorescence microscope.

2.2.6 *Glasshouse Assays*

For the evaluation of plant development, microcosms assays were performed under glasshouse conditions. Red carnation seeds were surface-disinfected as described in Sect. 2.2.4. Then, four-day-old seedlings were transferred to pots (1 seedling per pot) with a mix of non-sterile peat:vermiculite (3:1) soil. Seven days after the sow, each plant was inoculated with 5 ml of a PEPV13 strain suspension with a turbidity of 6 in McFarland Standards (1.5×10^6 CFU ml^{-1}). Uninoculated plants were also included as negative control. Red carnation plants were harvested 73 days post-inoculation. Shoot and root length and weight (fresh weight and dry weight) were measured in each case.

2.2.7 *Statistical Analysis*

Data obtained from each experiment were analyzed with one-way ANOVA using a post hoc test (Fisher's PSLD $p \geq 0.05$). Statview 5.0 software (SAS Institute Inc.) was used for these analyses.

2.3 Results and Discussion

2.3.1 *Identification of PEPV13 Strain*

We obtained a partial 16S rRNA gene sequence of the strain PEPV13 (up to 400 nt). This strain was classified into the genus *Rhizobium*, showing 100 % of identity with two *Rhizobium* type strains, *R. leguminosarum* and *R. laguerreae*. Further sequencing reactions will be necessary in order to confirm the species; however, strain PEPV13, isolated from common bean nodules, belongs to the genus *Rhizobium*, which is considered a safe microorganism for agriculture and human health (García-Fraile et al. 2012).

2.3.2 *In Vitro Plant Growth Promotion Assays*

In first place, we analyzed the in vitro PGPR ability of PEPV13 strain. This strain was able to produce siderophores, as revealed by the appearance of a yellow orange halo surrounding the colony. Moreover, this strain was able to solubilize phosphate, producing a clear halo around the colony. IAA and precursors biosynthesis was also evaluated. PEPV13 strain showed a moderate production of this auxin (0,077 mg l^{-1}). Several studies reported that strains of *Rhizobium*, mostly *R. leguminosarum*, produce siderophores such as cathecolate, vivibactin, or schizokinen (Patel et al. 1988; Dilworth et al. 1998; Storey et al. 2006). Moreover, this genus was described to produce IAA (Spaepen and Vanderleyden 2011). Regarding to phosphate solubilization, the genus *Rhizobium* is shown to be a phosphate solubilizer bacteria (Sharma et al. 2013), being in agreement with our results.

Then, we preformed an in vitro assay to test the potentiality of this strain to promote plant growth in early stages of development. We measured root length in *D. caryophyllus* plants at 5, 7, and 12 days after inoculation (Table 2.1). Our results showed that PEPV13 inoculated plants presented significative differences in root length respect to uninoculated plants.

These findings suggest that this strain is a good candidate as PGPR for in vitro culture of *D. caryophyllus* and with potentiality to promote growth in natural conditions.

2.3.3 *Colonization of* Dianthus caryophyllus *Roots*

Due to the positive interaction observed before, next step was to test colonization ability of the selected strain. For that, we obtained a GFP-labeled transformant of PEPV13 and inoculated carnation (Fig. 2.1). Root colonization was positive and bacterial cells were disposed along the root surfaces. Moreover, typical 3D structures or biofilms were observed. Also, bacterial cells formed aggregates in the root emergence zones and occupying intercellular spaces. This behavior is typical of rhizobial strains, which colonize legumes (Fujishige et al. 2006) and non-legumes different than ornamentals (Chabot et al. 1996; Antoun et al.1998; García-Fraile et al. 2012; Flores-Félix et al. 2013).

Table 2.1 Effect of PEPV13 strain in plant root length (mm) in in vitro conditions

	5 DAI	7 DAI	12 DAI
Uninoculated	4.20 ± 0.69 a	10.00 ± 0.49 a	17.90 ± 0.55 a
PEPV13	16.14 ± 1.84 b	16.66 ± 0.50 a	22.00 ± 0.65 b

Each value represent the mean (±S.E.) of at least 12 plants of three different assays. Values followed by the same letter in each treatment are not significantly different ($p \geq 0.05$) according to Fisher's Protected LSD test (Least Significant Differences). *DAI* days after inoculation

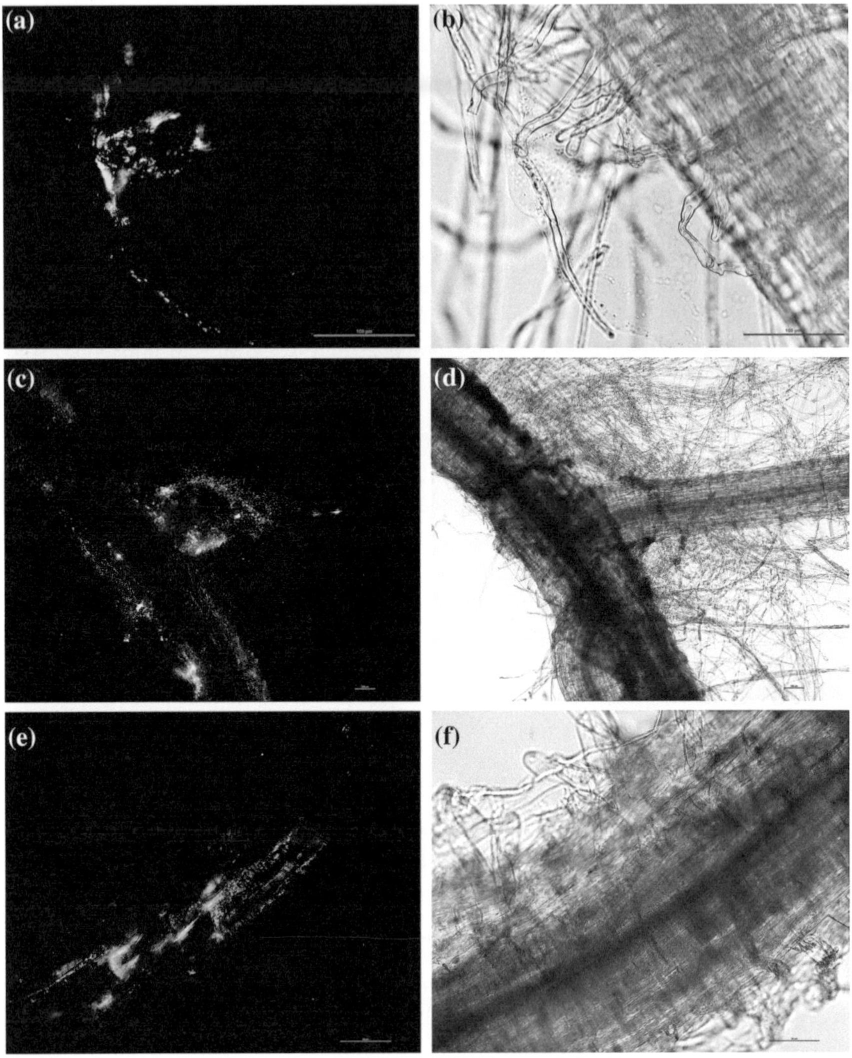

Fig. 2.1 Scanning micrographs of colonization events in carnation seedling roots 12 days after inoculation with GFP-tagged PEPV13 strain. Images **a**, **c**, and **e** were obtained by epifluorescence microscopy. Images **b**, **d**, and **f** were obtained by bright field microscopy, Bar: 100 μm

Interestingly, we observed an increased number of root hair deformations, and redirection in PEPV13 inoculated roots, in contraposition to the root hairs of uninoculated plants. This effect is in concordance with the effects observed in rice, when is inoculated with a rhizobial strain (Perrine-Walker et al. 2007). Also, similar observed deformations and polar redirections in root hairs of legumes inoculated with cellulase-overproducing strains were described before (Robledo et al. 2011). However, these phenotypes should be analyzed in depth in the near future.

2.3.4 Plant Assays: Analysis of Plant Growth Parameters Under Glasshouse Conditions

Finally, microcosms assays were performed under glasshouse conditions. Plants were harvested at 63 days after inoculation and several measures were obtained, such as shoot length and fresh and dry weight of the aerial parts. The results were not statistically significative in any of the obtained data, being PEPV13 inoculated plants slightly less developed than plants of the uninoculated treatment (data not shown). Fresh and dry weight of inoculated plants also presented lower values than non-inoculated ones.

Several studies addressed the competitiveness events among soil microorganisms. In legumes, competition for nodulation determines the ability of *Rhizobium* strains to nodulate a legume host in competition with other rhizospheric strains close to those legume roots (Dowling and Broughton 1986). Specifically, Duodu et al. (2009) reported the difference in the competitiveness of two *R. leguminosarum* strains to nodulate white clover, showing that they were able to nodulate, despite of their limited survival rates in soil. In non-legumes, such as cereals and horticultural crops, rhizobial strains can compete for root colonization in different soils (Flores-Félix et al. 2013). However, plants can also secrete antimicrobial molecules and other compounds or have other antibacterial mechanisms (De Coninck et al. 2015), affecting bacterial colonization and competitiveness. Rosenblueth and Martinez-Romero (2004) described that a highly competitive rhizobial strain belonging to *Rhizobium etli* has resistance to antimicrobial molecules synthesized by maize roots. Moreover, other soil microorganisms have antimicrobial activity, which might make difficult the survival of other rhizobacteria (Sharma and Kaur 2010). Despite the promising PGPR results that strain PEPV13 showed, our result showed that PEPV13 strain is not able to compete with other soil microorganisms by itself, suggesting the occurrence of any of the events cited above that may affect rhizobial competitiveness. Further assays should be performed in order to elucidate the cause of this result.

2.4 Conclusion

Despite the results of PEPV13 strain inoculation in microcosm assays, our conclusion is that this strain might be considered as PGPR for the initial steps of plant development.

References

Alexander DB, Zuberer DA (1991) Use of chrome azurol S reagents to evaluate siderophore production by rhizosphere bacteria. Biol Fertil Soils 12:39–45

Antoun H, Beauchamp CJ, Goussard N et al (1998) Potential of *Rhizobium* and *Bradyrhizobium* species as growth promoting rhizobacteria on non-legumes: effect on radishes (*Raphanus sativus* L.). Plant Soil 204:57–67

Chabot R, Antoun H, Cescas MP (1996) Growth promotion of maize and lettuce by phosphate-solubilizing *Rhizobium leguminosarum* biovar phaseoli. Plant Soil 184:311–321

Cheng HP, Walker GC (1998) Succinoglycan is required for initiation and elongation of infection threads during nodulation of alfalfa by *Rhizobium meliloti*. J Bacteriol 180:5183–5191

Chun J, Lee JH, Jung Y et al (2007) EzTaxon: a webbased tool for the identification of prokaryotes based on 16S ribosomal RNA gene sequences. Int J Syst Evol Microbiol 57:2259–2261

De Coninck B, Timmermans P, Vos C (2015) What lies beneath: below ground defense strategies in plants. Trends Plant Sci 20(2):91–101

Dilworth MJ, Carson KC, Giles RG, Byrne LT, Glenn AR. (1998) *Rhizobium leguminosarum* bv. viciae produces a novel cyclic trihydroxamate siderophore, vicibactin. Microbiology, 144(3): 781–791

Dowling DN, Broughton WJ (1986) Competition for nodulation of legumes. Annu Rev Microbiol 40:131–157

Duodu S, Brophy C, Connolly J et al (2009) Competitiveness of a native *Rhizobium leguminosarum* biovar trifolii strain for nodule occupancy is manifested during infection. Plant Soil 318(1):117–126

Eid AR, Awad MN, Hamouda HA (2009) Evaluate effectiveness of bio and mineral fertilization on the growth parameters and marketable cut flowers of *Matthiola incana* L. Am-Eur J Agr Environ Sci 5(4):509–518

Flores-Félix JD, Menéndez E, Rivera LP et al (2013) Use of *Rhizobium leguminosarum* as a potential biofertilizer for *Lactuca sativa* and *Daucus carota* crops. J Plant Nutr Soil Sci 176 (6):876–882

Flores-Felix JD, Silva LR, Rivera LP et al (2015) Plants probiotics as a tool to produce highly functional fruits: the case of *Phyllobacterium* and vitamin C in strawberries. PLoS ONE 10(4): e0122281

Fujishige NA, Kapadia NN, De Hoff PL et al (2006) Investigations of *Rhizobium* biofilm formation. FEMS Microbiol Ecol 56(2):195–206

García-Fraile P, Carro L, Robledo M et al (2012) *Rhizobium* promotes non-legumes growth and quality in several production steps: towards a biofertilization of edible raw vegetables healthy for humans. PLoS ONE 7(5):e38122

Göre ME, Altin N (2006) Growth promoting of some ornamental plants by root treatment with specific fluorescent pseudomonads. J Biol Sci 6(3):610–615

Gutiérrez-Zamora M, Martínez-Romero E (2001) Natural endophytic association between *Rhizobium etli* and maize (*Zea mays* L.). J Biotechnol 91:117–126

Khalid A, Arshad M, Zahir ZA (2004) Screening plant growth-promoting rhizobacteria for improving growth and yield of wheat. J Appl Microbiol 96(3):473–480

O'Hara GW, Goss TJ, Dilworth MJ et al (1989) Maintenance of intracellular pH and acid tolerance in *Rhizobium meliloti*. Appl Environ Microbiol 55(8):1870–1876

Patel HN, Chakraborty RN, Desai SB (1988) Isolation and partial characterization of phenolate siderophore from *Rhizobium leguminosarum* IARI 102. FEMS Microbiol Lett, 56(2): 131–134

Perrine-Walker FM, Prayitno J, Rolfe BG et al (2007) Infection process and the interaction of rice roots with rhizobia. J Exp Bot 58(12):3343–3350

Pikovskaya RI (1948) Mobilization of phosphorus in soil in connection with the vital activity of some microbial species. Mikrobiologiya 17:362–370

Rivas R, Velázquez E, Palomo JL et al (2002) Rapid identification of *Clavibacter michiganensis* subspecies *sepedonicus* using two primers random amplified polymorphic DNA (TP-RAPD) fingerprints. Eur J Plant pathology 108(2):179–184

Robledo M, Jiménez-Zurdo JI, Soto MJ et al (2011) Development of functional symbiotic white clover root hairs and nodules requires tightly regulated production of rhizobial cellulase CelC2. Mol Plant-Microbe Interact 24:798–807

Rosenblueth M, Martinez-Romero E (2004) *Rhizobium etli* maize populations and their competitiveness for root colonization. Arch Microbiol 181:337–344

Santillana N, Arellano C, Zúñiga D (2005) PGPR capacity of *Rhizobium* on *Lycopersicon esculentum* Miller. (tomato). Ecol Aplicada 4(1, 2)

Sharma S, Kaur M (2010) Antimicrobial activities of rhizobacterial strains of *Pseudomonas* and *Bacillus* strains isolated from rhizosphere soil of carnation (*Dianthus caryophyllus* cv. Sunrise). Ind J Microbiol 50(2):229–232

Sharma SB, Sayyed RZ, Trivedi MH et al (2013) Phosphate solubilizing microbes: sustainable approach for managing phosphorus deficiency in agricultural soils. Springer Plus 2:587

Silva LR, Azevedo J, Pereira MJ, Carro L, Velazquez E, Peix A, Valentao P, Andrade PB (2014a) Inoculation of the nonlegume *Capsicum annuum* (L.) with *Rhizobium* strains. 1. Effect on bioactive compounds, antioxidant activity, and fruit ripeness. J Agric Food Chem 62(3):557–564

Silva LR, Azevedo J, Pereira MJ et al (2014b) Inoculation of the Non-legume *Capsicum annuum* L. with *Rhizobium* Strains. 2. Changes in Sterols, Triterpenes, Fatty Acids, and Volatile Compounds. J Agr Food Chem, 62(3): 565–573

Spaepen S, Vanderleyden J (2011) Auxin and plant-microbe interactions. Cold Spring Harb Perspect Biol 3:a001438

Srivastava R, Govil M (2007) Influence of biofertilizers on growth and flowering in gladiolus cv. American beauty. Acta Hortic 742:183–188

Storey EP, Boghozian R, Little JL, Lowman DW, Chakraborty R (2006) Characterization of 'Schizokinen'; a dihydroxamate-type siderophore produced by *Rhizobium leguminosarum* IARI 917. Biometals, 19(6): 637–649

Vincent JM (1970) A manual for the practical study of root nodule bacteria. Blackwell Scientific Publications, Oxford

Yanni YG, Rizk Y, El-Fattah Faiza KA et al (2001) The beneficial plant growth-promoting association of *Rhizobium leguminosarum* bv. *trifolii* with rice roots. Australian J Plant Physiol 28:845–870

Zulueta-Rodriguez R, Cordoba-Matson MV, Hernandez-Montiel LG et al (2014) Effect of *Pseudomonas putida* on growth and anthocyanin pigment in two Poinsettia (*Euphorbia pulcherrima*) cultivars. Sci World J Article ID 810192

Chapter 3
Analysis of the PGPB Potential of Bacterial Endophytes Associated with Maize

Lorena Celador-Lera, Esther Menéndez, Jose D. Flores-Félix, Pedro F. Mateos and Raúl Rivas

Abstract Bacterial endophytes live inside plants for at least part of their life cycle, because plants are very attractive as nutrient reservoirs for such bacteria. Plants need the presence of those associated bacteria for their growth and adaptation to different ecosystems. This is the case of maize, which is a well-known cereal crop for hosting a great diversity of endophytic bacteria. Thus, microbes take advantage of the plant nutrients, whereas plants receive benefits from associated bacteria, such as nitrogen and phosphorous uptake. To study bacterial diversity, we isolated the endophytic bacteria from stalk and root of maize plants (*Zea mays*) growing in a soil of Ciudad Rodrigo (NW Spain). Bacterial isolates were analyzed by RAPD fingerprinting, allowing the differentiation among strains of the same species. Our results showed the high diversity of isolates inside maize rhizosphere and endosphere, establishing many different RAPD types. However, despite the usefulness of RAPD profiles in bacterial diversity analysis, 16S rRNA gene sequencing was carried out and subsequently analyzed. A total of 25 different genera were isolated. Moreover, we determined the ability of these strains to promote plant growth, performing in vitro PGPB mechanism analysis: (i) phosphate solubilization, (ii) siderophores production, and (iii) IAA and/or precursors biosynthesis. Here, we report the infraspecific diversity of bacterial endophytes isolated from *Z. mays*, which is higher than we expected. Moreover, the analysis of in vitro PGPB mechanisms indicates that these strains can promote plant development, suggesting their possible application in biofertilization schemes.

Keywords Plant growth promotion · Endophyte · *Zea mays* · RAPD · Biodiversity

L. Celador-Lera · E. Menéndez (✉) · J.D. Flores-Félix · P.F. Mateos · R. Rivas
Department of Microbiology and Genetics, University of Salamanca, Salamanca, Spain
e-mail: esthermenendez@usal.es

P.F. Mateos · R. Rivas
Associated Unit I+D University of Salamanca—CSIC 'Plant-Microbe Interactions Group', Salamanca, Spain

© Springer International Publishing Switzerland 2016

F. González-Andrés and E. James (eds.), *Biological Nitrogen Fixation and Beneficial Plant–Microbe Interactions*, DOI 10.1007/978-3-319-32528-6_3

3.1 Introduction

Bacterial endophytes correspond to the population of bacteria that colonize the inner tissues of healthy plants and subsequently establish a harmonious relationship with the host plant (Kloepper and Beauchamp 1992). They have been isolated from a range of plant species, from woody trees to herbaceous plants, suggesting an ubiquitous existence in nearly all higher plants (Luo et al. 2011). Distinctive microbial communities colonize different plant compartments, being found in various plant organs, such as roots, stems, leaves, or flowers, as well as in fruits and seeds (Sun et al. 2010; Compant et al. 2011).

Bacteria associated with plants are important for an adequate growth of the plants and its adaptation to different ecosystems. Thus, microbes and plants gain advantages mutually; microbes obtain an enhanced availability of nutrients, whereas plants receive benefits, such as plant growth enhancement or stress reduction. Therefore, mutualistic interactions between host plants and associated microorganisms could have emerged as a result of the clear positive selection exerted on these associations (Thrall et al. 2007).

The diversity and composition of bacterial communities in the endosphere is probably governed by stochastic events, which are influenced by deterministic processes of colonization (Battin et al. 2007). The majority of plant-associated bacteria derive from the soil environment. They may migrate to the rhizosphere and subsequently, the rhizoplane of their hosts before show any beneficial effect. Some rhizoplane-colonizing bacteria can also penetrate plant roots, and some strains may move to aerial plant parts through the vascular system, with a decreasing bacterial density in comparison with rhizosphere or root-colonizing populations (Compant et al. 2011).

Among these colonizer bacteria, plant growth-promoting rhizobacteria (PGPR) are frequent soil bacteria able to benefit plants by improving their productivity and immunity. These bacteria are associated with the rhizosphere, the part of soil under the influence of plant roots and their exudates. According to their interactions with plants, PGPRs can be divided into symbiotic bacteria, which live inside plants and exchange metabolites with them directly, and free-living rhizobacteria, which live outside plant cells (Gray and Smith 2005). Most symbiotic bacteria live in the intercellular spaces of the host plant, but there are some bacteria able to form truly mutualistic interactions with their hosts and penetrate plant cells (García-Fraile et al. 2015). To get better understanding of the interactions between symbiotic bacteria and host plants and their ecological roles in ecosystems, it is crucial to delve into the knowledge on the abundance and composition of endophytic populations in plants, such as cereals. Maize (*Zea mays*) is one of the most important and worldwide-distributed crops, due to its fast growth and higher biomass production (Ruiz et al. 2009; Wuana and Okieimen 2010). It is one of the most important crops of Castille and Leon, being the first Spanish region in maize agricultural surface with 121,965 hectares cultivated and a production of 1,292,927 tons in 2013 (MAGRAMA 2013).

Also, maize is a well-known cereal crop, hosting a great diversity of endophytic bacteria. For this reason, our objectives were to isolate and characterize the bacterial communities (bacterial endophytes) from roots and stems of *Z. mays* plants growing in an agricultural soil.

Furthermore, we aim to assess the phylogenetic diversity and phenotypically characterize culturable bacterial endophytes, performing a screening of endophytes according to their PGP abilities. Our aim is to determine the endophytes diversity in *Z. mays* and their potential abilities as PGPB.

3.2 Materials and Methods

3.2.1 Isolation of Culturable Bacterial Endophytes from Maize

Maize plants were collected from an agricultural soil in Ciudad Rodrigo, NW Spain. To isolate representative bacterial endophytes, we selected different areas from stalk and roots. Plant-surface sterilization was performing according to Luo et al. (2011) with minor modifications. Briefly, plants were washed and the stalk and the roots were cut into 5-mm pieces. Plant tissues were sterilized by sequential immersion in ethanol 70 % (v/v) for 2 min and 5 % NaClO for 2 min and then rinsed several times in sterile deionized water to remove surface sterilization agents. To confirm that the surface disinfection process was successful, one of each tissue was placed on Yeast Mannitol Agar (YMA) plates (Vincent 1970) and incubated at 28 °C for 7 days. No bacterial growth was found in any of the tests.

Root and stalk pieces were crushed with a sterile glass rod and then was spread onto YMA plates and incubated for 7 days at 28 °C. Bacterial cultures were observed daily, and different bacterial colonies were isolated based on size, shape, and color. Finally, we isolated 69 bacterial strains and one yeast, which was spread on Sabouraud Chloramphenicol Agar (SbCl) medium. Cell morphology of each isolate was examined by Gram-staining.

3.2.2 DNA Extraction

Genomic DNA from each isolate was extracted, according to Rivas et al. (2001) with some modifications. Briefly, to obtain bacterial suspensions, one fresh colony was picked and mixed into 500 µl of sterile ultrapure water. Centrifugation was carried out for 2 min at 9000 rpm. The pellet was washed subsequently with 200 µl of 0.1 % (v/v) Sarkosyl and 100 µl of 0.05 M NaOH. Cell suspensions were held in a thermoblock for 4 min at 100 °C and 200 µl of sterile ultrapure water was finally added.

3.2.3 Random Amplified Polymorphic DNA (RAPD) Fingerprints

The RAPD profiles were obtained as previously described (Rivas et al. 2002), using the M13 primer (5′ CAGGGTGGCGGTTCT 3′). PCRs were performed as follows: 0.4 µl AmpliTaq Gold (5U/µL) (Perkin–Elmer Biosystems), 2.5 µl AmpliTaq Gold Buffer (10X), 10.5 µl sterile ultrapure water, 0.25 µl dNTPs (20 mM), 1 µl bovine serum albumin (0.1 %), 2.5 µl M13 primer (20 µM), and 2 µl DNA. PCR conditions were set as follows: preheating at 95 °C for 9 min; 35 cycles of denaturing at 95 °C for 1 min; annealing at 45 °C for 1 min and extension at 75 °C for 2 min, and a final extension at 72 °C for 7 min. PCR amplification products were electrophoresed in 1.5 % agarose gel in TBE buffer (100 mM Tris, 83 mM boric acid, 1 mM EDTA, pH 8.5) during 2 h at 6 V/cm and stained with ethidium bromide (0.5 µg ml^{-1}) and visualized using the Molecular Imager Chemidoc XRS System (Biorad, USA). The RAPD patterns were obtained in the same electrophoretic conditions and using the same molecular weight standard as size marker (Standard VI; Hofmann-La Roche, Switzerland).

3.2.4 16S rRNA Parcial Sequencing of Isolates

Partial sequence of the 16S ribosomal RNA (rRNA) gene was carried out according to Rivas et al. (2002) with some modifications: 2.5 µl of GeneAmp 10x buffer, 1 µl BSA 0.1 %, 6 µl of 25 mM $MgCl_2$, 0.25 µl of 20 mM of dNTPs, 0.2 µl of 5U AmpliTaq Gold DNA polymerase (Perkin–Elmer Biosystems), 2.5 µl of 2 µM of 27F (5′ GAGGGTGGCGGTTCT 3′) and 1522R (5′ AAGGAGGTGATCCANCCRCA 3′), and 3 µl of DNA template. PCR conditions were set as follows: preheating at 95 °C for 9 min; 35 cycles of denaturing at 95 °C for 1 min; annealing at 59 °C for 1 min and extension 75 °C for 2 min; and a final extension at 72 °C for 7 min. PCR products were purified using a QIAquick PCR Purification Kit (Qiagen Ltd) according to the manufacturers specifications. Sequences have been obtained at the Sequencing Service of NUCLEUS, University of Salamanca. Sequences were compared with type strains by using EzTaxon server (Chun et al. 2007).

3.2.5 ITS Region Sequencing of Isolates

To identify yeasts, we sequenced the intergenic region ITS1-5.8S-ITS4 using the primers ITS1F (5′ GAACCTGCGGAAGGATCATT 3′) and ITS4R (5′ TCCTCCGCTTATTGATATGC 3′) and the following PCR conditions: preheating at 95 °C for 9 min; 35 cycles of denaturing at 95 °C for 1 min; annealing at 54 °C for 1 min and extension 72 °C for 1 min; and a final extension at 72 °C for 7 min.

PCR products were purified using the GeneJET Gel Extraction (Thermo Fisher Scientific Inc.) according to the manufacturers specifications. Sequences have been obtained in the Sequencing Service of NUCLEUS, University of Salamanca. Sequences were compared with type strains by using NCBI (National Center for Biotechnology Information).

3.2.6 *Analysis of in Vitro PGPB Mechanisms*

Indole acetic acid production was measured in JMM media (O'Hara et al. 1989) supplemented with 0.17 g/l tryptophan, as described (Khalid et al. 2004). Siderophores production by endophytic bacteria was determined using M9-CAS agar media as described (Schwyn and Neilands 1987; Alexander and Zuberer 1991). Strains were incubated in this media at 28 °C for at least three days. Yellow–orange halos around the colonies on M9-CAS agar were indicative of siderophore secretion. The ability to solubilize phosphate by the strains was tested in YED-P media according to Peix et al. (2004). Bacterial endophytes were inoculated in YED-P media at 28 °C for 5 days. After this incubation time, the presence or absence of halo was observed.

3.3 Results and Discussion

3.3.1 *Isolation of Endophytic Bacteria and Molecular Typing*

To study bacterial diversity, we isolated colonies from roots and stem, all with different shapes, sizes, textures, and colors. Overall, 69 bacterial endophytes were isolated from maize tissues, 58 from roots, and 11 from stem. The total number of endophytic bacteria in maize roots was higher than those found in maize stems, which is probably related to the intimate contact of plant roots with soil environment during growth and development, facilitating the entry of bacteria to the root tissues (Kobayashi and Palumbo 2000; Pereira and Castro 2014).

These endophytes represent a subgroup of rhizospheric communities, which once the rhizosphere and rhizoplane are colonized, can penetrate into the root. Once in the cortex, some of them will get into the vascular system and be transported to the aerial parts of the plant. However, there are mechanisms in the infection process, which regulates the number of endophytic bacteria during plant development. These reasons may at least partly explain the high colonization in roots compared to the stem. This difference in greater diversity in the roots and stems may also be due to rhizosphere form a transition zone between the ground and the surface of the root of the plant, being able to accumulate great numbers of different microorganisms some of which may behave as endophytes as the root exudates provide suitable ecological niches for the growth of microorganisms (Hardoim et al. 2008). Similar trends were reported by Kobayashi and Palumbo (2000); total number of endophytic bacteria in maize roots was higher than those found in stalk, which is probably related to the intimate contact of plant roots with soil environment during growth and development, facilitating the entry of bacteria to the root tissues. Similar examples were reported on *Solanum nigrum* and *Thlaspi goesingense* plants growing in Ni-contaminated soils (Luo et al. 2011; Idris et al. 2004).

RAPD profiles analysis (Radom Polymorphic Amplified DNA) stands out among the techniques currently used for the analysis of bacterial genetic diversity as one of the most useful, as it allows us to study complex genomes without prior knowledge of them (Velázquez and Rodriguez-Barrueco 2007). Here, we observed that the size of the bands obtained in all cases varies between 0.3 and 2.5 kb and few strains showed identical genetic fingerprints. Using the program Bionumerics, applying the Pearson correlation coefficient and UPGMA clustering method, we obtained the dendrogram shown in Fig. 3.1. Based on this dendrogram, our results show that the strains have a high-genetic diversity.

Therefore, these findings suggest that there are several different bacterial strains isolated from roots and stems of maize. However, despite the usefulness of RAPD profiles in bacterial diversity analysis, their suitability in taxonomy may be discussed, due to the existence of different strains of the same species or subspecies. For this reason, 16S rRNA gene sequencing was carried out and subsequently analyzed for all the isolated strains.

3.3.2 Phylogenetic Analyses of the Bacterial Endophytes

According to the high heterogeneity of the isolates, we decided to sequence all isolates for increase the reliability of the results. We obtained 16S rRNA gene partial sequences, with the exception of the strain CRZM144T, which results to be yeast, so it was identified by ITS1-5.8S-ITS4. The sequences belonging to bacterial isolates were compared with type strains sequences deposited in public databases by using Ez-Taxon server (http://www.eztaxon.org/; Chun et al. 2007). In the case of CRZM144T, the database for comparisons was GeneBank (http://www.ncbi.nml.nih.gov/blast/). Finally, 69 sequences were analyzed, as showed in the Table 3.1.

We obtained 15 different genera from the roots and nine from the stems. Interestingly, four of those genera, *Pantoea, Bacillus, Arthrobacter,* and *Microbacterium*, were common in both plant parts. Moreover, 11 genera were only isolated from the roots and five were only present in the stems.

According to the literature reviewed until now, we have observed that the following genera are not described as maize endophytes, such as *Brevundimonas* and *Leifsonia,* which are endophytic bacteria of rice, (Sun et al. 2008; Li et al. 2007); *Kitasatospora*, a *Triticum aestivum* endophyte (Conn and Franco 2004); *Leucobacter,* which is an endophyte in *Polygonum cuspidatum,* an ornamental plant

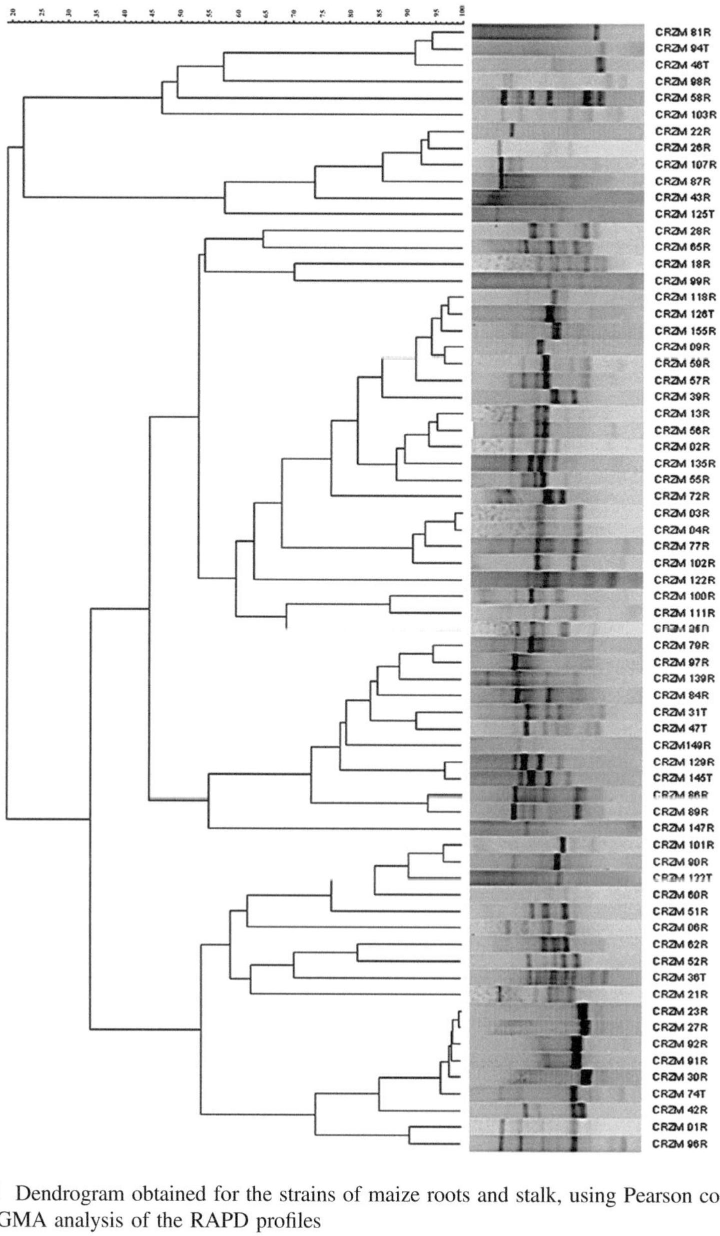

Fig. 3.1 Dendrogram obtained for the strains of maize roots and stalk, using Pearson coefficient and UPGMA analysis of the RAPD profiles

Table 3.1 Characterization of endophytes and in vitro PGPR mechanisms for IAA and siderophore production and phosphate ability

Strain	Closest relatives	Source	Gram-staining	IAA (mg l^{-1})	Siderophore production	Phosphate solubilization
CRZM 01R	*Pantoea* sp.	Root	−	260	++	−
CRZM 02R	*Bacillus* sp.	Root	+	224	−	−
CRZM 03R	*Enterobacter* sp.	Root	−	201	++	−
CRZM 04R	*Enterobacter* sp.	Root	−	169	++	−
CRZM 06R	*Serratia* sp.	Root	−	118	++	−
CRZM 09R	*Enterobacter* sp.	Root	−	208	+	−
CRZM 13R	*Bacillus* sp.	Root	+	121	−	−
CRZM 18R	*Rhizobium* sp.	Root	−	108	+	−
CRZM 21R	*Pseudomonas* sp.	Root	−	46	+	−
CRZM 22R	*Kitasatospora* sp.	Root	+	37	−	−
CRZM 23R	*Bacillus* sp.	Root	+	102	++	−
CRZM 25R	*Paenibacillus* sp.	Root	+	112	−	−
CRZM 26R	*Pseudomonas* sp.	Root	−	56	−	+
CRZM 27R	*Bacillus* sp.	Root	+	111	++	−
CRZM 28R	*Paenibacillus* sp.	Root	+	112	−	+
CRZM 30R	*Bacillus* sp.	Root	+	163	++	−
CRZM 31T	*Bacillus* sp.	Stalk	+	164	−	−
CRZM 36T	*Massilia* sp.	Stalk	−	152	−	−
CRZM 39R	*Pseudomonas* sp.	Root	−	199	+++	+
CRZM 42R	*Paenibacillus* sp.	Root	+	146	−	+
CRZM 43R	*Streptomyces* sp.	Root	+	2.0	−	−
CRZM 46T	*Arthrobacter* sp.	Stalk	+	35	+	+
CRZM 47T	*Bacillus* sp.	Stalk	+	2.0	−	−
CRZM 51R	*Pseudomonas* sp.	Root	−	43	++	+
CRZM 52R	*Rhizobium* sp.	Root	−	362	+	−
CRZM 55R	*Bacillus* sp.	Root	+	172	−	−
CRZM 56R	*Bacillus* sp.	Root	+	248	−	−
CRZM 57R	*Bacillus* sp.	Root	+	188	+	−
CRZM 58R	*Arthrobacter* sp.	Root	+	153	++	+
CRZM 59R	*Bacillus* sp.	Root	+	203	−	−
CRZM 60R	*Pseudomonas* sp.	Root	−	7.0	−	+
CRZM 62R	*Bacillus* sp.	Root	+	58	−	−
CRZM 65R	*Pseudomonas* sp.	Root	−	31	+	+
CRZM 72R	*Bacillus* sp.	Root	+	12	−	−
CRZM 74T	*Pantoea* sp.	Stalk	−	183	++	−
CRZM 77R	*Enterobacter* sp.	Root	−	462	++	−
CRZM 79R	*Paenibacillus* sp.	Root	+	335	−	−
CRZM 81R	*Brevundimonas* sp.	Root	−	42	−	−

(continued)

Table 3.1 (continued)

Strain	Closest relatives	Source	Gram-staining	IAA (mg l^{-1})	Siderophore production	Phosphate solubilization
CRZM 84R	*Paenibacillus* sp.	Root	+	21	−	+
CRZM 86R	*Burkholderia* sp.	Root	−	32	++	+
CRZM 87R	*Rhizobium* sp.	Root	−	17	−	−
CRZM 89R	*Burkholderia* sp.	Root	−	46	++	−
CRZM 90R	*Leifsonia* sp.	Root	+	17	−	−
CRZM 91R	*Rhizobium* sp.	Root	−	29	−	−
CRZM 92R	*Rhizobium* sp.	Root	−	94	+	−
CRZM 94T	*Arthrobacter* sp.	Stalk	+	19	−	+
CRZM 96R	*Pantoea* sp.	Root	−	603	++	−
CRZM 97R	*Pseudomonas* sp.	Root	−	21	+++	−
CRZM 98R	*Bacillus* sp.	Root	+	21	−	−
CRZM 99R	*Leifsonia* sp.	Root	+	21	−	−
CRZM 100R	*Paenibacillus* sp.	Root	+	21	−	−
CRZM 101R	*Paenibacillus* sp.	Root	+	52	++	−
CRZM 102R	*Enterobacter* sp.	Root	−	135	++	−
CRZM 103R	*Pantoea* sp.	Root	−	17	++	−
CRZM 107R	*Burkholderia* sp.	Root	−	289	++	+
CRZM 111R	*Arthrobacter* sp.	Root	+	165	−	−
CRZM 118R	*Microbacterium* sp.	Root	+	15	−	−
CRZM 122R	*Pantoea* sp.	Root	−	1.0	+	−
CRZM 125T	*Microbacterium* sp.	Stalk	+	14	−	−
CRZM 126T	*Sphingomonas* sp.	Stalk	−	121	+	−
CRZM 129R	*Microbacterium* sp.	Root	+	3.0	−	−
CRZM 133T	*Agrococcus* sp.	Stalk	+	2.0	−	−
CRZM 135R	*Staphylococcus* sp.	Root	+	17	−	−
CRZM 139R	*Pseudomonas* sp.	Root	−	2.0	−	+
CRZM 144T	*Aureobasidium* sp.	Stalk	−	188	++	−
CRZM 145T	*Leucobacter* sp.	Stalk	+	172	+	−
CRZM 147R	*Curtobacterium* sp.	Root	+	25	+	−
CRZM 149R	*Streptomyces* sp.	Root	+	14	+	−
CRZM 155R	*Bacillus* sp.	Root	+	11	+++	−

(−) negative, (+) positive/weak, (++) intermediate, and (+++) strong production

(Sun et al. 2013); and *Agrococcus*, *Massilia, Curtobacterium*, *Staphylococcus,* and *Sphingomonas*, which are described as endophytes in fruit trees and horticultural crops (Rosenblueth and Martínez-Romero 2006; Behrendt et al. 2008; Bulgari et al. 2011). Moreover, *Aureobasidium* was described as endophytic yeast in *Rosa hybrida* (Salgado-Salazar and Cepero 2005).

3.3.3 Assessment of PGPB Mechanisms of Bacterial Endophytes

Endophytes can promote plant growth by using more than one mechanism at different times during their life cycle, suggesting that plant growth stimulation is the net result of multiple mechanisms (Pereira and Castro 2014).

In the present study, we performed an assessment of PGPB mechanisms for all our isolates with the aim of the selection of proper PGP strains. We tested three different PGPB mechanisms: (i) IAA and/or precursors biosynthesis, (ii) siderophores production, and (iii) phosphate solubilization. Several bacterial endophytes showed ability to promote plant growth through several of those mechanisms. The results of PGPB mechanisms determination are shown in Table 3.1.

Our results show that all of the isolated strains were positive for indol-3-acetic acid (IAA) biosynthesis in the presence of L-tryptophan, being the genera *Pantoea* and *Enterobacter* the best producers (603 and 462 mg l^{-1}, respectively). Several studies reported that auxins are produced in plants from tryptophan catabolism, being this compound the usual precursor of IAA (Salisbury et al. 1984; Taiz and Zeiger 2006). Different bacteria have been also described with the ability to synthesize IAA in the presence of tryptophan, being this ability common in plant pathogenic bacteria and in beneficial plant growth promoters (Arshad et al. 1993; Vandeputte et al. 2005; Duca et al. 2014). *Pseudomonas, Rhizobium, Azospirillum, Enterobacter, Azotobacter, Klebsiella, Alcaligenes, Pantoea,* and *Streptomyces* are examples of the genera described as able to synthesize IAA (Apine and Jadhav 2011). Verma et al. (2001) reported that endophytic *Streptomyces* isolated from *Azadirachta indica* produce IAA, being the genus *Streptomyces* potential plant growth promoters. The genus *Arthrobacter* is often described in the literature as a potential plant growth promoter bacteria (Banerjee et al. 2010; Chelius and Triplett 2000; Morrisey et al. 1975). Moreover, members of the genus *Burkholderia* were described as endophytes in plants, such as *Vitis vinifera* (Compant et al. 2004), and also have the ability to promote plant growth (Mano and Morisaki 2008; Rosenblueth and Martínez-Romero 2006). Furthermore, rhizobial strains, which produce IAA, improve the growth of several crops such as *Capsicum annumm, Solanum lycopersicum, Lactuca sativa,* and *Daucus carota* (García-Fraile et al. 2012; Flores-Félix et al. 2013).

In the case of siderophores production, 52.2 % of the endophytic strains were positive, forming a yellow–orange-colored halo around the colonies. According to Idris et al. (2004), siderophore production among endophytes may be a common, since endophytes have to compete with plant cells for Fe supply, and therefore, siderophore production may be highly important for endophytic growth. In addition, siderophores form complexes with other bivalent heavy metal ions increasing the metal availability and the plant metal uptake (Carrillo-Castañeda et al. 2002; Whiting et al. 2001; Rajkumar et al. 2010). Strains CRZM 39R and CRZM 97R, belonging to the genus *Pseudomonas,* and the strain CRZM 155R, belonging to the genus *Bacillus*, showed the highest production.

Moreover, fourteen isolates showed ability to solubilize phosphate, all of them belonging to the genera *Pseudomonas, Burkholderia, Arthrobacter,* and *Paenibacillus*. Although, some studies reported that the ability to solubilize phosphate from inorganic sources is a phylogenetically extended quality and several genera of endophytic bacteria have been described as efficient phosphate solubilizers (Dastager et al. 2010; Pindi and Satyanarayana 2012), our results show that this ability is only present in 4 of the genera and in our assay conditions.

According to the results obtained in the phenotypic characterization tests, the genera *Arthrobacter* and *Burkholderia* showed a great potential as plant growth promoters.

3.4 Conclusions

The results obtained in this study show a great diversity of endophytes isolated from maize, revealing the potential of this crop as a source for endophyte isolation. Furthermore, some of the isolated bacterial strains presented proper abilities to promote plant growth.

Acknowledgements This work was supported by Junta de Castilla y Leon, Spain (JCyL SA169 U 14 project)

References

Alexander DB, Zuberer DA (1991) Use of Chrome Azurol S reagents to evaluate siderophore production by rhizosphere bacteria. Biol Fertil Soils 12:39–45

Apine OA, Jadhav JP (2011) Optimization of medium for indole-3-acetic acid production using *Pantoea agglomerans* strain PVM. J Appl Microbiol 110(5):1235–1244

Arshad M, Hussain A, Javed M et al (1993) Effect of soil applied L-methionine on growth, nodulation and chemical composition of *Albizia lebbeck* L. Plant Soil 148(1):129–135

Banerjee A, Duflo E, Glennerster R et al (2010) The miracle of microfinance? Evidence from a randomized evaluation. Bread Working Paper, 278

Battin TJ, Sloan WT, Kjelleberg S et al (2007) Microbial landscapes: new paths to biofilm research. Nat Rev Microbiol 5(1):76–81

Behrendt U, Schumann P, Ulrich A (2008) *Agrococcus versicolor* sp. nov., an actinobacterium associated with the phyllosphere of potato plants. Int J Syst Evol Micr 58:2833–2838

Bulgari D, Casati P, Crepaldi P et al (2011) Restructuring of endophytic bacterial communities in grapevine yellows-diseased and recovered *Vitis vinifera* L. plants. App Environ Microbiol 77 (14):5018–5022

Carrillo-Castañeda G, Munoz JJ, Peralta-Videa JR et al (2002) Plant growth-promoting bacteria promote copper and iron translocation from root to shoot in alfalfa seedlings. J Plant Nutr 26:1801–1814

Chelius MK, Triplett EW (2000) Diazotrophic endophytes associated with maize. In: Triplett EW (ed) Prokaryotic nitrogen fixation: a model system for the analysis of a biological process, 1st edn. Horizon Scientific Press, Wymondham, UK, pp 779–791

Chun J, Lee JH, Jung Y et al (2007) EzTaxon: a webbased tool for the identification of prokaryotes based on 16S ribosomal RNA gene sequences. Int J Syst Evol Microbiol 57:2259–2261

Compant S, Reiter B, Sessitsch A et al. (2004) Endophytic colonization of *Vitis vinifera* L. by plant growth promoting bacterium *Burkholderia* sp. strain PsJN. Appl Environ Microb 71 (4):1685–1693

Compant S, Mitter B, Colli-Mull JG et al (2011) Endophytes of grapevine flowers, berries, and seeds: identification of cultivable bacteria, comparison with other plant parts, and visualization of niches of colonization. Microb Ecol 62:188–197

Conn VM, Franco CM (2004) Analysis of the endophytic actinobacterial population in the roots of wheat (*Triticum aestivum* L.) by terminal restriction fragment length polymorphism and sequencing of 16S rRNA clones. Appl Environ Microbiol 70(3):1787–1794

Dastager SG, Deepa CK, Pandey A (2010) Isolation and characterization of novel plant growth promoting *Micrococcus* sp. NII-0909 and its interaction with cowpea. Plant Physiol Bioch 48:987–992

Duca D, Lorv J, Patten CL et al (2014) Indole-3-acetic acid in plant-microbe interactions. A Van Leeuw J Microb 106(1):85–125

Flores-Félix JD, Menéndez E, Rivera LP et al. (2013) Use of *Rhizobium leguminosarum* as a potential biofertilizer for *Lactuca sativa* and *Daucus carota* crops. J Plant Nutr Soil Sc176 (6):876–882

García-Fraile P, Carro L, Robledo M et al (2012) *Rhizobium* promotes non-legumes growth and quality in several production steps: towards a biofertilization of edible raw vegetables healthy for humans. PLoS ONE 7(5):e38122

García-Fraile P, Menéndez E, Rivas R (2015) Role of bacterial biofertilizers in agriculture and forestry. AIMS Bioeng 2(3):183–205

Gray EJ, Smith DL (2005) Intracellular and extracellular PGPR: commonalities and distinctions in the plant-bacterium signaling processes. Soil Biol Biochem 37:395–412

Hardoim PR, van Overbeek LS, van Elsas JD (2008) Properties of bacterial endophytes and their proposed role in plant growth. Trends Microbiol 16(10):463–471

Idris R, Trifonova R, Puschenreiter M et al (2004) Bacterial communities associated with flowering plants of the Ni hyperaccumulator *Thlaspi goesingense*. Appl Environ Microb 70:2667–2677

Khalid A, Arshad M, Zahir ZA (2004) Screening plant growth-promoting rhizobacteria for improving growth and yield of wheat. J Appl Microbiol 96(3):473–480

Kloepper JW, Beauchamp CJ (1992) A review of issues related to measuring colonization of plant roots by bacteria. Can J Microbiol 38:1219–1232

Kobayashi DY, Palumbo JD (2000) Bacterial endophytes and their effects on plants and uses in agriculture. In: Bacon CW, White JF (eds) Microbial endophytes, 1st edn. Marcel Dekker Inc., New York, pp 199–233

Li TY, Zeng HL, Ping Y, Lin H, Fan XL, Guo ZG, Zhang CF (2007) Construction of a stable expression vector for *Leifsonia xyli* subsp. *cynodontis* and its application in studying the effect of the bacterium as an endophytic bacterium in rice. FEMS Microbiol Lett 267(2):176–183

Luo S, Wan Y, Xiao X et al (2011) Isolation and characterization of endophytic bacterium LRE07 from cadmium hyperaccumulator *Solanum nigrum* L. and its potential for remediation. Appl Microbiol Biotechnol 89:1637–1644

MAGRAMA (2013) Ministerio de Agricultura, Alimentación y Medio Ambiente, Gobierno de España, Madrid, España

Mano H, Morisaki H (2008) Endophytic bacteria in the rice plant. Microbes Environ 23(2):109–117

Morrisey F, Dugan P, Koths JS (1975) Chitinase production by *Arthrobacter* sp. Lysing cells of *Fusarium rosetum*. Soil Biol Biochem 8:23–28

O'Hara GW, Goss TJ, Dilworth MJ et al (1989) Maintenance of intracellular pH and acid tolerance in *Rhizobium meliloti*. Appl Environ Microbiol 55(8):1870–1876
Peix A, Rivas R, Santa-Regina I et al (2004) *Pseudomonas lutea* sp. nov., a novel phosphate-solubilizing bacterium isolated from the rhizosphere of grasses. Int J Syst Evol Micr 54:847–850
Pereira SIA, Castro PML (2014) Diversity and characterization of culturable bacterial endophytes from *Zea mays* and their potential as plant growth-promoting agents in metal-degraded soils. Environ Sci Pollut R 21(24):14110–14123
Pindi PK, Satyanarayana SDV (2012) Liquid microbial consortium—a potential tool for sustainable soil health. J Biofertil Biopest 3:1–9
Rajkumar M, Prasad MNV, Freitas H (2010) Potential of siderophore producing bacteria for improving heavy metal phytoextraction. Trends Biotechnol 28:142–149
Rivas R, Velázquez E, Valverde A et al (2001) A two primers random amplified polymorphic DNA procedure to obtain polymerase chain reaction fingerprints of bacterial species. Electrophoresis 22:1086–1089
Rivas R, Velázquez E, Palomo JL et al (2002) Rapid identification of *Clavibacter michiganensis* subspecies *sepedonicus* using two primers random amplified polymorphic DNA (TP-RAPD) fingerprints. Eur J Plant pathology 108(2):179–184
Rosenblueth M, Martínez-Romero E (2006) Bacterial endophytes and their interactions with hosts. Mol Plant Microbe Interact 19(8):827–837
Ruiz E, Rodríguez L, Alonso-Azcaráte J et al (2009) Phytoextraction of metal polluted soils around a Pb–Zn mine by crop plants. Inter J Phytorem 4:360–384
Salisbury JL, Baron A, Surek B et al (1984) Striated flagellar roots: isolation and partial characterization of a calcium-modulated contractile organelle. J Cell Biol 99(3):962–970
Salgado-Salazar C, Cepero D (2005) Endophytic fungi in rose (*Rosa hybrida*) in Bogota, Colombia. Rev Iberoam Micol 22(2):99–101
Schwyn B, Neilands JB (1987) Universal chemical assay for the detection and determination of siderophores. Annal Biochem 160:47–56
Sun L, Qiu F, Zhang X et al (2008) Endophytic bacterial diversity in rice (*Oryza sativa L.*) roots estimated by 16S rDNA sequence analysis. Microb Ecol 55:415–424
Sun LN, Zhang YF, He LY et al (2010) Genetic diversity and characterization of heavy metalresistant-endophytic bacteria from two copper-tolerant plant species on copper mine wasteland. Bioresource Technol 101:501–509
Sun H, He Y, Xiao Q et al (2013) Isolation, characterization, and antimicrobial activity of endophytic bacteria from *Polygonum cuspidatum*. Afr J Microbiol Res 7(16):1496–1504
Taiz L, Zeiger E (2006) Fisiologia vegetal. Universitat Jaume I
Thrall PH, Hochberg ME, Burdon JJ et al (2007) Coevolution of symbiotic mutualists and parasites in a community context. Trends Ecol Evol 22(3):120–126
Vandeputte O, Öden S, Mol A et al (2005) Biosynthesis of auxin by the gram-positive phytopathogen *Rhodococcus fascians* is controlled by compounds specific to infected plant tissues. Appl Environ Microbiol 71:1169–1177
Velázquez E, Rodriguez-Barrueco C (eds) (2007) First international meeting on microbial phosphate solubilization, vol 102. Springer Science and Business Media
Verma SC, Ladha JK, Tripathi AK (2001) Evaluation of plant growth promoting and colonization ability of endophytic diazotrophs from deep water rice. J Biotechnol 91:127–141
Vincent JM (1970) A manual for the practical study of root nodule bacteria. Blackwell Scientific Publications, Oxford, England
Whiting NS, de Souza PM, Terry N (2001) Rhizosphere bacteriamobilize Zn for hyperaccumulation by *Thlaspi caerulescens*. Environ Sci Technol 35:3144–3150
Wuana RA, Okieimen FE (2010) Phytoremediation potential of (*Zea mays* L.). A review. Afr J Gen Agric 6:275–287

Chapter 4
Identification of Rhizobial Strains Nodulating *Pisum Sativum* in Northern Spain Soils by MALDI-TOF MS (Matrix-Assisted Laser Desorption Ionization Time-of-Flight Mass Spectrometry) Analysis

Eustoquio Martínez-Molina, Fernando Sánchez Juanes, Lorena Carro, José D. Flores-Félix, Pilar Martínez-Hidalgo, Eugenia Cerda Castillo, José Manuel González Buitrago and Encarna Velázquez

Abstract In this work, we used MALDI-TOF MS for identification of strains nodulating *Pisum sativum* L. (peas) in soils from Valladolid, Salamanca and León where the 30 % of the total Spanish production of peas is located. The results obtained showed a correct identification, with score values higher than 2.0, of strains isolated in the soil from León as *Rhizobium leguminosarum* and those isolated in soils from Valladolid and Salamanca as *Rhizobium laguerreae*. These results confirmed that MALDI-TOF MS is a reliable technique for identification at species level of fast-growing rhizobia. They also showed that this methodology is able to differentiate between sister species, not distinguishable on the basis of 16S rRNA gene analysis, as occurs in the case of *R. leguminosarum* and *R. laguerreae*.

Keywords MALDI-TOF MS · 16S rRna gene · *Rhizobium* · Identification · *Pisum sativum*

E. Martínez-Molina · L. Carro · J.D. Flores-Félix · P. Martínez-Hidalgo · E. Cerda Castillo · E. Velázquez (✉)
Departamento de Microbiología y Genética, Universidad de Salamanca, Salamanca, Spain
e-mail: evp@usal.es

F. Sánchez Juanes · J.M. González Buitrago
Instituto de Investigación Biomédica de Salamanca (IBSAL), Salamanca, Spain

F. Sánchez Juanes
Departamento de Bioquímica y Biología Molecular, Universidad de Salamanca, Salamanca, Spain

E. Martínez-Molina · E. Velázquez
Unidad Asociada Universidad de Salamanca-CSIC 'Interacción Planta-Microorganismo', Salamanca, Spain

© Springer International Publishing Switzerland 2016
F. González-Andrés and E. James (eds.), *Biological Nitrogen Fixation and Beneficial Plant–Microbe Interactions*, DOI 10.1007/978-3-319-32528-6_4

4.1 Introduction

MALDI-TOF MS (matrix-assisted laser desorption ionization time-of-flight mass spectrometry) is a reliable method for bacterial identification mainly applied to date to clinical isolates (Patel 2015). In 2011 we extended by first time the database Biotyper 2.0 with the species from family *Rhizobiaceae* (Ferreira et al. 2011) including the type strains of species *Rhizobium leguminosarum* USDA 2370^{T} and *Rhizobium pisi* DSM 30132^{T} isolated from *Pisum* (Ramírez-Bahena et al. 2008) and *Rhizobium fabae* LMG 23997^{T} isolated from *Vicia* (Tian et al. 2008). Now, we have added to this database several recently described species of genus *Rhizobium*, including those isolated from nodules of *Pisum*, *Vicia*, or *Lens*, such as *Rhizobium laguerreae* (Saïdi et al. 2014), *Rhizobium anhuiense* (Zhang et al. 2015), *Rhizobium bangladeshense, Rhizobium lentis* and *Rhizobium binae* (Rashid et al. 2015). All these mentioned legumes are worldwide used in human diet, but *Pisum sativum* (peas) is particularly appreciated by the high nutritional value of its seeds being widely cultivated in north Spain where more than 36,000 ha are cultivated in "Castilla y León" region.

P. sativum establishes nitrogen-fixing symbiosis with fast-growing rhizobia belonging to genus *Rhizobium*, but this symbiosis has been scarcely studied in most countries. Currently, the identification of the *P. sativum* endosymbionts at species level is carried out by housekeeping gene analysis because several species nodulating this legume have identical 16S rRNA genes as occurs in the case of *R. leguminosarum* (Ramírez-Bahena et al. 2008), *R. laguerreae* (Saïdi et al. 2014), and *R. anhuiense* (Zhang et al. 2015) or very closely related such as *R. leguminosarum* and *R. pisi* (Ramírez-Bahena et al. 2008).

Since in the previous work we showed that MALDI-TOF MS is able to differentiate among phylogenetically close species, the aim of this work was to analyze the potential of MALDI-TOF MS and the new extended database for the identification of strains nodulating *P. sativum* in different soils of north Spain where, despite they are important producers of this legume, there are no data about its rhizobial endosymbionts.

4.2 Materials and Methods

4.2.1 Isolation of Rhizobial Strains

A total of 46 strains (Table 4.1) able to form effective nodules on *P. sativum* used as trap plant were isolated according to Vincent (1970) on YMA medium in three soils located at Riego de la Vega (42°23′21″N 5°58′56″W) from León Province, Aldea de San Miguel (41°27′37″N 4°37′0″W) from Valladolid Province, and Pedrosillo el Ralo (41°03′46″N 5°32′59″W) from Salamanca Province. In the soil from Valladolid, *P. sativum* is commonly cultivated; in the soil from Salamanca, the

Table 4.1 Identification with MALDI-TOF MS analysis of strains isolated from *Pisum sativum* nodules

Strain	Soil	Closest type strain	Score value
AMPS01	Aldea San Miguel (Valladolid)	*Rhizobium laguerreae* FB206^T	2.384
AMPS02	Aldea San Miguel (Valladolid)	*Rhizobium laguerreae* FB206^T	2.193
AMPS04	Aldea San Miguel (Valladolid)	*Rhizobium laguerreae* FB206^T	2.444
AMPS05	Aldea San Miguel (Valladolid)	*Rhizobium laguerreae* FB206^T	2.228
AMPS08	Aldea San Miguel (Valladolid)	*Rhizobium laguerreae* FB206^T	2.151
AMPS09	Aldea San Miguel (Valladolid)	*Rhizobium laguerreae* FB206^T	2.247
AMPS10	Aldea San Miguel (Valladolid)	*Rhizobium laguerreae* FB206^T	2.105
AMPS12	Aldea San Miguel (Valladolid)	*Rhizobium laguerreae* FB206^T	2.087
AMPS13	Aldea San Miguel (Valladolid)	*Rhizobium laguerreae* FB206^T	2.079
AMPS14	Aldea San Miguel (Valladolid)	*Rhizobium laguerreae* FB206^T	2.345
AMPS15	Aldea San Miguel (Valladolid)	*Rhizobium laguerreae* FB206^T	2.212
AMPS16	Aldea San Miguel (Valladolid)	*Rhizobium laguerreae* FB206^T	2.022
AMPS17	Aldea San Miguel (Valladolid)	*Rhizobium laguerreae* FB206^T	2.308
AMPS18	Aldea San Miguel (Valladolid)	*Rhizobium laguerreae* FB206^T	2.061
AMPS20	Aldea San Miguel (Valladolid)	*Rhizobium laguerreae* FB206^T	2.311
AMPS21	Aldea San Miguel (Valladolid)	*Rhizobium laguerreae* FB206^T	2.068
AMPS22	Aldea San Miguel (Valladolid)	*Rhizobium laguerreae* FB206^T	2.077
AMPS23	Aldea San Miguel (Valladolid)	*Rhizobium laguerreae* FB206^T	2.324
AMPS26	Aldea San Miguel (Valladolid)	*Rhizobium laguerreae* FB206^T	2.389
AMPS25	Aldea San Miguel (Valladolid)	*Rhizobium laguerreae* FB206^T	2.389
AMPS27	Aldea San Miguel (Valladolid)	*Rhizobium laguerreae* FB206^T	2.175
AMPS28	Aldea San Miguel (Valladolid)	*Rhizobium laguerreae* FB206^T	2.371
AMPS30	Aldea San Miguel (Valladolid)	*Rhizobium laguerreae* FB206^T	2.218
AMPS33	Aldea San Miguel (Valladolid)	*Rhizobium laguerreae* FB206^T	2.296
AMPS34	Aldea San Miguel (Valladolid)	*Rhizobium laguerreae* FB206^T	2.032
AMPS35	Aldea San Miguel (Valladolid)	*Rhizobium laguerreae* FB206^T	2.035
RPS07	Riego de la Vega (León)	*Rhizobium leguminosarum* USDA 2370^T	2.090
RPS17	Riego de la Vega (León)	*Rhizobium leguminosarum* USDA 2370^T	2.021
RPS20	Riego de la Vega (León)	*Rhizobium leguminosarum* USDA 2370^T	2.100
RPS22	Riego de la Vega (León)	*Rhizobium leguminosarum* USDA 2370^T	2.044
RPS24	Riego de la Vega (León)	*Rhizobium leguminosarum* USDA 2370^T	2.096
RPS25	Riego de la Vega (León)	*Rhizobium leguminosarum* USDA 2370^T	2.070
RPS27	Riego de la Vega (León)	*Rhizobium leguminosarum* USDA 2370^T	2.052
RPS30	Riego de la Vega (León)	*Rhizobium leguminosarum* RVS03	2.292

(continued)

Table 4.1 (continued)

Strain	Soil	Closest type strain	Score value
RPS31	Riego de la Vega (León)	*Rhizobium leguminosarum* RVS03	2.295
RPS34	Riego de la Vega (León)	*Rhizobium leguminosarum* RVS03	2.235
RPS39	Riego de la Vega (León)	*Rhizobium leguminosarum* USDA 2370^T	2.056
RPS43	Riego de la Vega (León)	*Rhizobium leguminosarum* USDA 2370^T	2.029
PEPS01	Pedrosillo el Ralo (Salamanca)	*Rhizobium laguerreae* $FB206^T$	2.368
PEPS09	Pedrosillo el Ralo (Salamanca)	*Rhizobium laguerreae* $FB206^T$	2.364
PEPS10	Pedrosillo el Ralo (Salamanca)	*Rhizobium laguerreae* $FB206^T$	2.337
PEPS11	Pedrosillo el Ralo (Salamanca)	*Rhizobium laguerreae* $FB206^T$	2.235
PEPS12	Pedrosillo el Ralo (Salamanca)	*Rhizobium laguerreae* $FB206^T$	2.099
PEPS13	Pedrosillo el Ralo (Salamanca)	*Rhizobium laguerreae* $FB206^T$	2.292
PEPS15	Pedrosillo el Ralo (Salamanca)	*Rhizobium laguerreae* $FB206^T$	2.161

traditional crop is *Lens culinaris*; and in León soil, *P. sativum* has been never cultivated.

4.2.2 Sample Preparation for MALDI-TOF MS

Strains were cultivated on TY medium (Beringer 1974) for 24 h. As indicated by the manufacturer, we used amounts of biomass between 5 and 100 mg to be transferred from the plate to a 1.5-ml tube (Eppendorf, Germany) with a pipette tip and mixed thoroughly in 300 μl of water to resuspend the bacterial cells. Then, 900 μl of absolute ethanol was added, the mixture was centrifuged at 15,500 *g* for 2 min, and the supernatant was discarded. The pellet was air-dried at room temperature for 1 h. Subsequently, 50 μl of formic acid (70 % v/v) was added to the pellet and mixed thoroughly before the addition of 50 μl acetonitrile to the mixture. The mixture was centrifuged again at 15,500 *g* for 2 min. One microliter of the supernatant was placed onto a spot of the steel target and air-dried at room temperature. Each sample was overlaid with 1 μl of matrix consisting of a saturated solution of alpha-cyano-4-hydroxycinnamic acid in 50 % acetonitrile and 2.5 % trifluoroacetic acid, and air-dried.

4.2.3 MALDI-TOF MS Performing and Data Analysis

The sample preparation and the MALDI-TOF MS performing were carried out as previously published (Ferreira et al. 2011). The calibration masses was the Bruker

Bacterial Test Standards (BTS) which were as follows (masses as averages): RL36, 4365.3 Da; RS22, 5096.8 Da; RL34, 5381.4 Da; RL33meth, 6255.4 Da; RL29, 7274.5 Da; RS19, 10300.1 Da; RNase A, 13683.2 Da; and myoglobin, 16952.3 Da. MALDI-TOF MS identifications were classified using the score values proposed by the manufacturer: a score value between 2.3 and 3.00 indicated species identification; a score value between 2.0 and 2.299 indicated genus identification and possible species identification; a score value between 1.7 and 1.999 indicated genus identification; and a score value <1.7 indicated no identification.

4.2.4 Phylogenetic Analysis of rrs, recA, *and* atpD *Genes*

Sequences were aligned using the Custal X software (Thompson et al. 1997). The distances were calculated according to Kimura's two-parameter model (Kimura 1980). A phylogenetic tree was inferred using the neighbor-joining analysis (Saitou and Nei 1987). MEGA5 software (Tamura et al. 2011) was used for all analyses.

4.3 Results and Discussion

4.3.1 Phylogenetic Analyses

P. sativum L. (peas) is commonly nodulated by fast-growing rhizobial strains belonging to the genus *Rhizobium*, being *R. leguminosarum* the first described endosymbiont of this legume (Peix et al. 2015). Nevertheless, in the past decade, a problem was detected with the type strains of this species conserved in different culture collections, and the characterization of these strains resulted in the description of a new species named *R. pisi* (Ramírez-Bahena et al. 2008). Both species nodulate *Vicia* and *Pisum* because their type strains belong to the same symbiovar viciae (Marek-Kozaczuk et al. 2013), but they have different *rrs* gene sequences (Ramírez-Bahena et al. 2008). However, several recently described species nodulating *Vicia* and *Pisum*, such as *R. laguerreae* (Saïdi et al. 2014) and *R. anhuiense* (Zhang et al. 2015) or *R. bangladeshense* and *R. binae* (Rashid et al. 2005), have identical 16S rRNA genes, but they are distinguishable on the basis of their housekeeping gene analysis (Fig. 4.1).

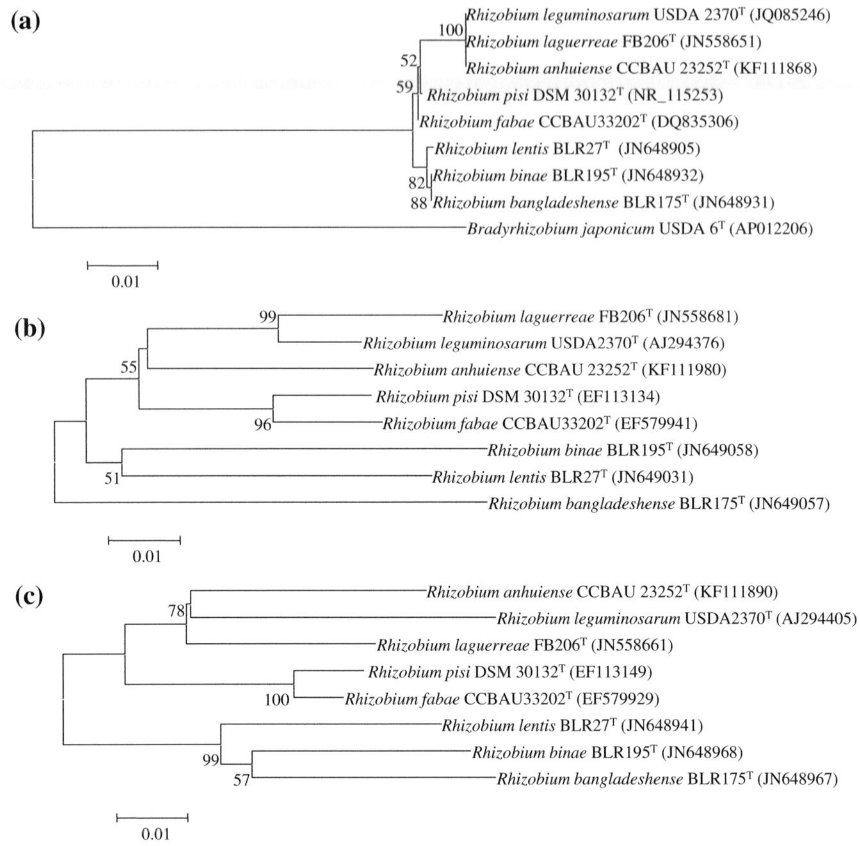

Fig. 4.1 Neighbour-joining phylogenetic rooted trees based on 16S rRNA gene sequences (**a**), *recA* gene sequences (**b**), and *atpD* gene sequences (**c**) of the *Rhizobium* species isolated from *Pisum* and *Vicia.* Bootstrap values calculated for 1000 replications are indicated. Bar, 1 nt substitution per 100 nt

4.3.2 MALDI-TOF MS Analysis

In the previous work, we showed that strains from family *Rhizobiaceae* with phylogenetically divergent housekeeping genes can be differentiated by MALDI-TOF MS (Ferreira et al. 2011). Therefore, in the present work, we used this technique to identify fast-growing rhizobial strains nodulating *P. sativum* in north Spain. They were isolated from nodules of this legume in soils from Valladolid, Salamanca, and León, three provinces of Castilla y León where the 30 % of the total Spanish production of peas is located, being Valladolid the major producer with more than 16,000 ha cultivated with this legume.

The results obtained after MALDI-TOF MS analysis showed that the strains were identified as *R. leguminosarum* or *R. laguerreae* (Table 4.1). The type strains

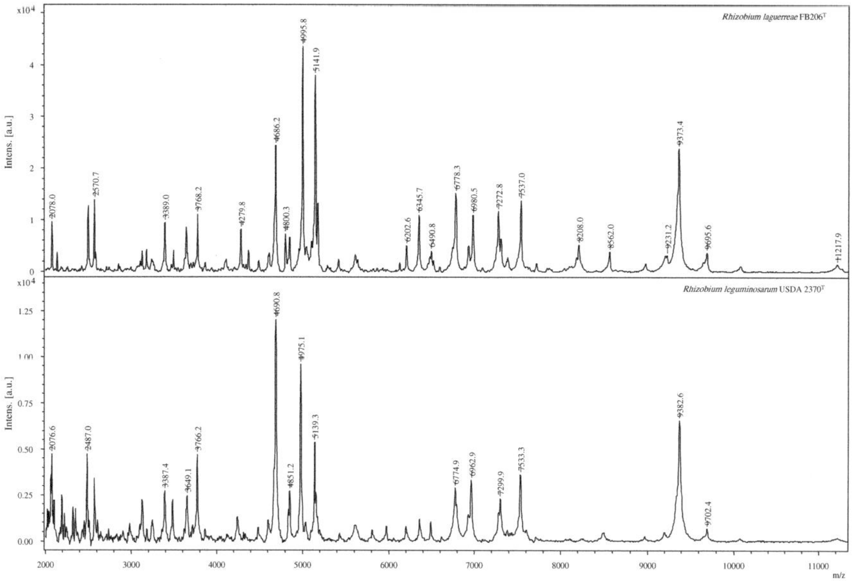

Fig. 4.2 MALDI-TOF MS spectra of whole-cell extracts obtained from the type strains of *Rhizobium leguminosarum* USDA 2370^T (**a**) and *R. laguerreae* FB206^T (**b**)

of *R. laguerreae* FB206^T and *R. leguminosarum* USDA 2370^T matched with score values lower than 2.0, and a comparison of their spectra showed that these two species have different peaks (Fig. 4.2). The strains isolated in the soils from Valladolid and Salamanca matched with the type strain of *R. laguerreae* FB206^T with score values higher than 2.0, whereas those isolated in the soil from León matched with score values higher than 2.0 with the type strain of *R. leguminosarum* USDA 2370^T, except three strains that matched with the strain *R. leguminosarum* RVS03 isolated from *Vicia sativa* and were previously identified as *R. leguminosarum* by MALDI-TOF MS (Ferreira et al. 2011).

These results confirm MALDI-TOF MS as a reliable technique for identification at species level of fast-growing rhizobia being able to differentiate between sister species that are not distinguishable on the basis of 16S rRNA gene analysis, as occurs in the case of *R. leguminosarum* and *R. laguerreae* (Saïdi et al. 2014). This is a very relevant finding since the description of sister species is currently common within genus *Rhizobium* since, in addition to those already mentioned, other species able to nodulate *Vicia* and *Pisum*, such as *R. bangladeshense* and *R. binae*, have identical 16S rRNA genes (Rashid et al. 2015). The rapidity and reliability of MALDI-TOF MS together with its ability to differentiate sister species make this technique the most suitable to identify fast-growing rhizobial strains particularly when large populations of nodule isolates should be analyzed.

Acknowledgements This work was supported by the Grant AGL2013-48098-P from the MINECO (Spanish Government).

References

Beringer JE (1974) R factors transfer in *Rhizobium leguminosarum*. J Gen Microbiol 84:188–198

Ferreira L, Sánchez-Juanes F, García-Fraile P et al (2011) MALDI-TOF mass spectrometry is a fast and reliable platform for identification and ecological studies of species from family *Rhizobiaceae*. PLoS One 6:e20223

Kimura M (1980) A simple method for estimating evolutionary rates of base substitutions through comparative studies of nucleotide sequences. J Mol Evol 16:111–120

Marek-Kozaczuk M, Leszcz A, Wielbo J et al (2013) *Rhizobium pisi* sv. trifolii K3.22 harboring nod genes of the *Rhizobium leguminosarum* sv. trifolii cluster. Syst Appl Microbiol 36:252–258

Patel R (2015) MALDI-TOF MS for the diagnosis of infectious diseases. Clin Chem 61:100–111

Peix A, Ramírez-Bahena MH, Velázquez E et al (2015) Bacterial Associations with Legumes. Crit Rev Plant Sci 34:17–42.

Ramírez-Bahena MH, García-Fraile P, Peix A et al (2008) Revision of the taxonomic status of the species *Rhizobium leguminosarum* (Frank 1879) Frank 1889AL, *Rhizobium phaseoli* Dangeard 1926AL and *Rhizobium trifolii* Dangeard 1926AL. *R. trifolii* is a later synonym of *R. leguminosarum*. Reclassification of the strain *R. leguminosarum* DSM 30132 (=NCIMB 11478) as *Rhizobium pisi* sp. nov. Int J Syst Evol Microbiol 58:2484–2490

Rashid MH, Young JP, Everall I et al (2015) Average nucleotide identity of genome sequences supports the description of *Rhizobium lentis* sp. nov., *Rhizobium bangladeshense* sp. nov. and *Rhizobium binae* sp. nov. from lentil (*Lens culinaris*) nodules. Int J Syst Evol Microbiol 65:3037–3045

Saïdi S, Ramírez-Bahena MH, Santillana N et al (2014) *Rhizobium laguerreae* sp. nov. nodulates *Vicia faba* on several continents. Int J Syst Evol Microbiol 64:242–247

Saitou N, Nei M (1987) The neighbor-joining method: a new method for reconstructing phylogenetic trees. Mol Biol Evol 4:406–425

Tamura K, Peterson D, Peterson N et al (2011) MEGA5: molecular evolutionary genetics analysis using maximum likelihood, evolutionary distance, and maximum parsimony methods. Mol Biol Evol 28:2731–2739

Tian CF, Wang ET, Wu LJ et al (2008) *Rhizobium fabae* sp. nov., a bacterium that nodulates *Vicia faba*. Int J Syst Evol Microbiol 58:2871–2875

Thompson JD, Gibson TJ, Plewniak F et al (1997) The clustal_X windows interface: flexible strategies for multiple sequence alignment aided by quality analysis tools. Nucleic Acid Res 25:4876–4882

Vincent JM (1970) The cultivation, isolation and maintenance of rhizobia. In: Vincent JM (ed) A Manual for the Practical Study of Root-Nodule. Blackwell Scientific, Oxford, pp 1–13

Zhang YJ, Zheng WT, Everall I et al (2015) *Rhizobium anhuiense* sp. nov., isolated from effective nodules of *Vicia faba* and *Pisum sativum*. Int J Syst Evol Microbiol 65:2960–2967

Chapter 5
Analysis of Cultivable Endophytic Bacteria in Roots of Maize in a Soil from León Province in Mainland Spain

Esther Menéndez, Martha Helena Ramirez-Bahena, Alvaro Peix, Carmen Tejedor, Rebeca Mulas, Fernando González-Andrés, Eustoquio Martínez-Molina and Encarna Velázquez

Abstract In the present study, we isolated 14 bacterial strains from the roots of maize plants growing in a soil located at León (Spain). All these strains presented some of the following in vitro plant growth promotion mechanisms: (i) phosphate solubilization, (ii) ACC deaminase activity, (iii) siderophore and/or (iv) IAA production. These strains displayed 11 different RAPD profiles, showing they were genetically diverse. Representative strains from each RAPD type were identified on the basis of the 16S rRNA gene sequencing analysis, showing that most of them belonged to gamma-Proteobacteria and, specifically, to the families Enterobacteriaceae and Pseudomonadaceae (genus *Pseudomonas*). Most of the isolated strains belong to species that can be pathogenic for humans, such as *Pantoea agglomerans*, *Pantoea ananatis*, *Lelliottia amnigena* (formerly *Enterobacter amnigenus*), *Kosakonia cowanii* (formerly *Enterobacter cowanii*), *Enterobacter ludwigii* and *Rahnella aquatilis,* or for plants, such as *Pseudomonas corrugata*, *Pseudomonas brassicacearum* and *Agrobacterium larrymoorei.* These results showed the need of a correct identification of bacterial endophytes before selecting strains for the design of biofertilizers, which must be pathogen-free.

E. Menéndez (✉) · C. Tejedor · E. Martínez-Molina · E. Velázquez
Department of Microbiology and Genetics, University of Salamanca,
Salamanca, Spain
e-mail: esthermenendez@usal.es

M.H. Ramirez-Bahena · A. Peix
IRNASA-CSIC, Salamanca, Spain

M.H. Ramirez-Bahena · A. Peix · E. Martínez-Molina · E. Velázquez
Plant-Microbe Interactions Group, Associated Unit I+D University of Salamanca—CSIC,
Salamanca, Spain

R. Mulas · F. González-Andrés
Institute of Environment, Natural Resources and Biodiversity,
University of León, León, Spain

© Springer International Publishing Switzerland 2016
F. González-Andrés and E. James (eds.), *Biological Nitrogen Fixation and Beneficial Plant–Microbe Interactions*, DOI 10.1007/978-3-319-32528-6_5

Keywords Endophytic bacteria · Plant growth promotion · Maize · Identification · 16S rRNA gene

5.1 Introduction

The plant microbiome includes rhizospheric and endophytic microorganisms, and its structure depends on the soil and the plant and its health and development stage (Gaiero et al. 2013). Endophytic microorganisms constitute the plant microbiome subset that most directly affects the growth and health of plants, due to the location of these microorganisms in inner tissues (Gaiero et al. 2013).

The endophytic microorganisms live in the inner tissues of plants without causing damage, and they can be strict or facultative (Reinhold-Hurek and Hurek 2011). The strict endophytic bacteria are not able to grow outside the plants and can be transferred from parental plants to their progeny through the seeds of the vegetative parts of the plants, whereas the facultative endophytes can enter the plant along its life cycle (Reinhold-Hurek and Hurek 2011). Endophytic microorganisms can present plant growth promotion mechanisms via phytostimulation through the production of IAA or the ACC deaminase activity, nutrients uptake through nitrogen fixation or phosphate solubilization and biocontrol through siderophores or antibiotic production (Gaiero et al. 2013). Only cultivable and facultative endophytes can be used as biofertilizers, and they must be non-pathogenic for humans, animals or plants (García-Fraile et al. 2012), making necessary their identification before their use in biofertilization schemes.

The identification of bacteria is currently based on the sequencing and analysis of the 16S rRNA gene (Woese et al. 2000), and therefore, this methodology is the most commonly used for identification of endophytic bacteria (Hardoim et al. 2015). They have been analysed in several plants, highlighting those present in cereals, such as maize (Chelius and Triplett 2001; Ikeda et al. 2013) but, up to the date, there are no studies about the endophytic bacteria from these plants in Spain. Therefore, the aim of this study was the isolation and identification of maize endophytes in a soil from León, a province in north-west Spain, where this cereal is commonly cultivated in rotation with common beans, potatoes and sugar beet.

5.2 Materials and Methods

5.2.1 Strains Isolation

Strains were isolated from roots of maize cultivated in a soil from Riego de la Vega (León, NW Spain). The maize plants (cv. Antalya) were harvested directly from the soil, six months after sowing. The roots were surface-disinfected with 70 % ethanol for 1 min, 2 % NaClO for 3 min and 70 % ethanol for 30 s, following rinse five times with sterile distilled water. Then, the roots were crushed into a tube containing 10 ml of phosphate-buffered saline (PBS), pH 7–7.4. Samples were incubated at room temperature and 160 rpm during 1 h. One hundred microlitre per plate of the corresponding serial dilutions were spread onto TSA (Difco) plates, which were daily evaluated for pure strains isolation after a subsequent purification process.

5.2.2 Analysis of Plant Growth Promotion Mechanisms in Vitro

Phosphate solubilization was analysed on YED-P plates containing 2 % $CaHPO_4$, which were incubated for 7 days at 28 °C (Peix et al. 2001). Siderophore production was evaluated in M9-CAS-AGAR (Schwyn and Neilands 1987), modified with the addition of a cationic solvent, HDTMA, to stabilize the Fe–CAS complex providing its characteristic colour (Alexander and Zuberer 1991). Indole acetic acid (IAA) production was evaluated in JMM medium (O'Hara et al. 1989) supplemented with 107 mg l^{-1} of tryptophan. After 7 days of incubation, the supernatants were recovered by centrifugation at 5000*g* and filtered using 0.22 μm Millipore filters (Millipore Co., Amicon, USA). Then 1 ml of Salkowsky reagent was added to 2 ml of supernatant, and the red colour formed was measured by spectrophotometry at 550 nm using an ATI Unicam 8625 Spectrometer (Mattson, USA) (Khalid et al. 2004). The ACC deaminase was determined according to Penrose and Glick (2003), measuring the released amounts of α-ketobutyrate (μmol α-ketobutyrate · mg $protein^{-1}$ h^{-1}).

5.2.3 RAPD Fingerprinting

RAPD patterns were obtained as previously described (Rivas et al. 2006), using the primer M13 (5′-GAGGGTGGCGGTTCT-3′) and the Dream-Taq DNA Green PCR Master Mix (Fisher Scientific, USA). PCR conditions were as follows: preheating at 95 °C for 9 min; 35 cycles of denaturing at 95 °C for 1 min; annealing at 45 °C for 1 min; extension at 75 °C for 2 min; and a final extension at 72 °C for 7 min. Seventeen microlitre of each PCR products were electrophoresed on 1.5 % (w/v) agarose gel in TBE buffer (100 mM Tris, 83 mM boric acid, 1 mM EDTA, pH 8.5) at 6 V/cm, stained in a solution containing 0.5 g/ml ethidium bromide, and photographed under UV light. GeneRuler 100 bp (Fisher Scientific, USA) was used as a size marker.

5.2.4 16S rRNA Gene Sequencing and Analysis

The 16S rRNA gene was amplified using the primers 27F: 5′-AGA GTTTGATCTGGCTCAG-3′ (*Escherichia coli* positions 8–27) and 1522R: 5′-AAGGAGGTGATCCANCCRCA-3′ (*E. coli* positions 1509–1522) at a final concentration of 0.2 mmol l^{-1}. PCR amplifications were performed with a REDExtract-N-Amp PCR Kit (Sigma Co., USA) following the manufacturers' instructions. Amplification and sequencing conditions were performed as previously described by Rivas et al. (2007). The obtained sequences were compared to the sequences held in EzTaxon-e server (Kim et al. 2012).

5.3 Results and Discussion

5.3.1 Genetic Diversity of Strains

RAPD fingerprinting is a technique widely used to analyse the bacterial diversity that allowed the differentiation of strains from the same and different species (Williams et al. 1990). Therefore, in this work, we used this methodology to analyse the genetic diversity of endophytic bacteria isolated from maize, in which 14 endophytic bacterial strains from maize roots were isolated. They displayed 11 different RAPD profiles (Fig. 5.1), showing that they were genetically diverse and allowing us the selection of representative strains for identification.

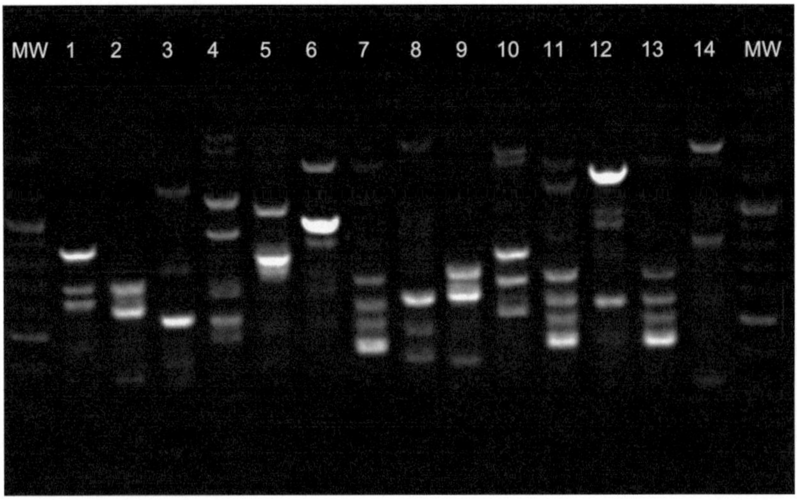

Fig. 5.1 RAPD fingerprinting analyses of the isolates. RZME02 (lane 1), RZME03 (lane 2), RZME04 (lane 3), RZME05 (lane 4), RZME06 (lane 5), RZME07 (lane 6), RZME09 (lane 7), RZME10 (lane 8), RZME13 (lane 9), RZME14 (lane 10), RZME16 (lane 11), RZME17 (lane 12), RZME18 (lane 13) and RZME20 (lane 14)

5.3.2 Plant Growth Promotion Mechanisms

The results of the analysis of in vitro plant growth promotion mechanisms presented by the strains isolated from maize roots are shown in Table 5.1. They showed that, with the exception of the strain RZME20, all strains produced IAA in concentrations ranging from 0.016 to 0.128 mg l^{-1}. The strains able to produce higher amounts of this phytohormone were RZME02 and RZME03 with 0.182 and 0.100 mg l^{-1}, respectively. Almost 80 % of the strains showed ACC deaminase activity, with the highest values being showed by the strains RZME04, RZME09

Table 5.1 Characteristics of strains isolated from maize roots

Strains	RAPD type	Closest related species	IAA (mg l^{-1})	Phosphate solubilization	Siderophore production	ACC deaminase[a]
Phylum alpha-Proteobacteria, Family *Rhizobiaceae*						
RZME10	VIII	*Agrobacterium larrymoorei*	0.075	–	–	0.000
RZME17	X	*Agrobacterium larrymoorei*	0.043	–	–	0.000
Phylum gamma-Proteobacteria, Family *Enterobacteriaceae*						
RZME02	I	*Pantoea agglomerans*	0.128	–	–	0.060
RZME03	II	*Lelliottia amnigena*	0.100	+	–	2.082
RZME04	III	*Pantoea ananatis*	0.073	–	+	3.334
RZME07	VI	*Enterobacter ludwigii*	0.043	–	+	0.867
RZME09	VII	*Rahnella aquatilis*	0.048	+	–	3.205
RZME13	II	*Lelliottia amnigena*	0.073	+	–	1.700
RZME16	VII	*Rahnella aquatilis*	0.032	+	–	0.000
RZME18	VII	*Rahnella aquatilis*	0.066	+	–	0.129
RZME20	X1	*Kosakonia cowanii*	0.000	–	+	4.157
Phylum gamma-Proteobacteria, Family *Pseudomonadaceae*						
RZME05	IV	*Pseudomonas. chlororaphis*	0.081	+	+	0.640
RZME06	V	*Pseudomonas corrugata*	0.077	w	+	1.620
RZME14	IX	*Pseudomonas brassicacearum*	0.016	w	+	0.869

[a]μmol α-ketobutyrate mg $protein^{-1}$ h^{-1}; + positive; – negative; *w* weak

and RZME20. Over 50 % of the strains were able to solubilize phosphate, being the strains RZME18 and RZME09, the most active phosphate solubilizers, showing halos of 24 and 20 mm in diameter, respectively, around the colonies. All the strains were able to grow in the M9-CAS indicator medium, and the colonies were surrounded by a yellow orange halo indicative of siderophore production in almost 50 % of strains. The strains RZME05, RZME06 and RZME07 were the most active in siderophore production with orange halos of 40 mm in diameter around the colonies. With these results, we can conclude that all the isolates from maize roots presented some in vitro plant growth promotion mechanisms and that the strains belonging to the genus *Pseudomonas* presented all the analysed mechanisms although in different degrees. Therefore, many of these strains could be suitable for their use as biofertilizers.

5.3.3 16S rRNA Gene Sequencing

The comparison of 16S rRNA gene sequences with those held in EzTaxon-e database showed that most of isolates belonged to gamma-Proteobacteria and especially to the families Enterobacteriaceae and Pseudomonadaceae. Only two strains belonged to the alpha-Proteobacteria and concretely to the family Rhizobiaceae (Table 5.1). With the exception of the strain RZME05, identified as *Pseudomonas chlororaphis*, a non-pathogenic species (Solaiman et al. 2015), and which showed positive results in every in vitro plant growth promotion mechanism tested, the remaining strains isolated from maize roots in this work were pathogenic for humans, plants or both types of organisms.

The strain RZME02 was identified as *Pantoea agglomerans*, which stands out by its ability to produce IAA; however, it has been found as opportunist pathogen in immunocompromised (Shubov et al. 2011) and neonate patients (Tiwari and Beriha 2015). The strains RZME03 and RZME13, which presented identical RAPD pattern and produced high amounts of IAA and showed high ACC deaminase activity, were identified as *E. amnigenus*, recently reclassified into a new genus with the name of *L. amnigena* (Brady et al. 2013). This species is an unusual opportunistic human pathogen (Capdevila et al. 1998) that has been found in endophthalmitis (Westerfeld et al. 2009). The strain RZME07, which produced siderophores and moderate amounts of IAA and showed moderate ACC deaminase activity, was identified as *E. ludwigii*, a species described on the basis of the characterization of several clinical isolates (Hoffmann et al. 2005). The strains RZME09 and RZME18, displaying the same RAPD pattern, were able to solubilize phosphate, and they showed a moderate ACC deaminase activity and a moderate production of IAA. These strains slightly differed in the RAPD profile from the strain RMZE16, but the differences between them in the in vitro PGP activity were great. They were identified as *R. aquatilis*, which has been isolated from immunocompromised patients (Matsukura et al. 1996; Gaitan and Bronze 2010; Kuzdan et al. 2015).

Several strains were identified with bacterial species isolated from clinic and vegetal samples, as occurs in the case of the strain RMZE20, which stands out by its ACC deaminase activity, and the production of siderophores was identified as *E. cowanii*, a species originally isolated from clinical specimens (Inoue et al. 2000) and recently reclassified into *K. cowanii* (Brady et al. 2013), which has been cited as the causing agent of the bacterial spot on the forest tree *Mabea fistulifera* (Furtado et al. 2012). The strain RZME04, which produced siderophores and IAA, and showed a high ACC deaminase activity, was identified as *P. ananatis*, a plant pathogen which causes diseases in a wide range of plants (Coutinho and Venter 2009) and whose pan-genome incorporates a large number of genes encoding proteins enabling it to colonize, persist and potentially cause symptoms in a wide range of plant and animal hosts (De Maayer et al. 2014). Although the human infection by this bacterium is rare, it has been reported a case of bacteraemia in an immunocompromised patient (De Baere et al. 2004).

Finally, several strains were identified with plant pathogenic species, such as the strain RZME06 identified as *P. corrugata*, a species causing pith necrosis in tomato (Trantas et al. 2015), and the strain RZME14 identified as *Pseudomonas brassicacearum*, a species that can also cause pith necrosis in tomato, although non-pathogenic strains are plant growth promoters (Belimov et al. 2007). These two strains have all the in vitro mechanisms analysed in this study, unlike the strains RZME10 and RZME17, closely related to *A. larrymoorei*, a species isolated from tumours in leaves of *Ficus benjamina* (Bouzar and Jones 2001), that only were able to produce moderate amounts of IAA.

5.4 Conclusions

Although it has been previously reported that bacterial endophytes can include pathogenic species (Rosenblueth and Martínez-Romero 2006), in this study, we found that surprisingly most of strains isolated from maize roots are related to bacterial species, which are pathogens of humans or plants. Only one strain (RMZE05, *P. chlororaphis*), which expressed in different degrees the four in vitro PGP activities analysed, can be used for biofertilization since it was identified as the non-pathogenic bacteria. Therefore, despite all the isolated endophytic strains could be good plant growth-promoting bacteria, only one of them was suitable for biofertilization highlighting the need of a correct identification of bacterial endophytes before the design of biofertilizers. This is especially important not only for the farmers who will manage these inputs, but also for the final consumers, especially in the case of fresh vegetables.

Acknowledgements This work was supported by the Project AGL2013-48098-P from the MINECO (Spanish Government).

References

Alexander DB, Zuberer DA (1991) Use of chrome azurol S reagents to evaluate siderophore production by rhizosphere bacteria. Biol Fertil Soils 12:39–45

Belimov AA, Dodd IC, Safronova VI et al (2007) *Pseudomonas brassicacearum* strain Am 3 containing 1-aminocyclopropane-1-carboxylate deaminase can show both pathogenic and growth-promoting properties in its interaction with tomato. J Exp Bot 58:1485–1495

Bouzar H, Jones JB (2001) *Agrobacterium larrymoorei* sp. nov., a pathogen isolated from aerial tumours of *Ficus benjamina*. Int J Syst Evol Microbiol 51:1023–1026

Brady C, Cleenwerck I, Venter S et al (2013) Taxonomic evaluation of the genus *Enterobacter* based on multilocus sequence analysis (MLSA): proposal to reclassify *E. nimipressuralis* and *E. amnigenus* into *Lelliottia* gen. nov. as *Lelliottia nimipressuralis* comb. nov. and *Lelliottia amnigena* comb. nov., respectively, *E. gergoviae* and *E. pyrinus* into *Pluralibacter* gen. nov. as *Pluralibacter gergoviae* comb. nov. and *Pluralibacter pyrinus* comb. nov., respectively, *E. cowanii*, *E. radicincitans*, *E. oryzae* and *E. arachidis* into *Kosakonia* gen. nov. as *Kosakonia cowanii* comb. nov., *Kosakonia radicincitans* comb. nov., *Kosakonia oryzae* comb. nov. and *Kosakonia arachidis* comb. nov., respectively, and *E. turicensis*, *E. helveticus* and *E. pulveris* into *Cronobacter* as *Cronobacter zurichensis* nom. nov., *Cronobacter helveticus* comb. nov. and *Cronobacter pulveris* comb. nov., respectively, and emended description of the genera *Enterobacter* and *Cronobacter*. Syst Appl Microbiol 36:309–319

Capdevila JA, Bisbe V, Gasser I et al (1998) *Enterobacter amnigenus*: an unusual human pathogen. Enferm Infecc Microbiol Clin 16:364–366

Chelius MK, Triplett EW (2001) The diversity of archaea and bacteria in association with the roots of *Zea mays* L. Microb Ecol 41:252–263

Coutinho TA, Venter SN (2009) *Pantoea ananatis*: an unconventional plant pathogen. Mol Plant Pathol 10:325–335

De Baere T, Verhelst R, Labit C et al (2004) Bacteremic infection with *Pantoea ananatis*. J Clin Microbiol 42:4393–4395

De Maayer P, Chan WY, Rubagotti E et al (2014) Analysis of the *Pantoea ananatis* pan-genome reveals factors underlying its ability to colonize and interact with plant, insect and vertebrate hosts. BMC Genom 15:404

Furtado GQ, Guimarães LMS, Lisboa DO et al (2012) First report of *enterobacter cowanii* causing bacterial spot on *Mabea fistulifera*, a native forest species in Brazil. Plant Dis 96:1576

Gaiero JR, McCall CA, Thompson KA, Day NJ, Best AS, Dunfield E (2013) Inside the root microbiome: bacterial root endophytes and plant growth promotion. Am J Bot 100:1738–1750

Gaitán JI, Bronze MS (2010) Infection caused by *Rahnella aquatilis*. Am J Med Sci 339:577–579

García-Fraile P, Carro L, Robledo M et al (2012) *Rhizobium* promotes non-legumes growth and quality in several production steps: towards a biofertilization of edible raw vegetables healthy for humans. PLoS One 7:e38122

Hardoim PR, van Overbeek LS, Berg G et al (2015) The hidden world within plants: ecological and evolutionary considerations for defining functioning of microbial endophytes. Microbiol Mol Biol Rev 79:293–320

Hoffmann H, Stindl S, Stumpf A et al (2005) Description of *Enterobacter ludwigii* sp. nov., a novel *Enterobacter* species of clinical relevance. Syst Appl Microbiol 28:206–212

Ikeda AC, Bassani LL, Adamoski D et al (2013) Morphological and genetic characterization of endophytic bacteria isolated from roots of different maize genotypes. Microb Ecol 65:154–160

Inoue K, Sugiyama K, Kosako Y et al (2000) *Enterobacter cowanii* sp. nov., a new species of the family Enterobacteriaceae. Curr Microbiol 41:417–420

Khalid A, Arshad M, Zahir ZA (2004) Screening plant growth-promoting rhizobacteria for improving growth and yield of wheat. J Appl Microbiol 96:473–480

Kim OS, Cho YJ, Lee K et al (2012) Introducing EzTaxon-e: a prokaryotic 16S rRNA gene sequence database with phylotypes that represent uncultured species. Int J Syst Evol Microbiol 62:716–721

Kuzdan C, Soysal A, Özdemir H et al (2015) *Rahnella aquatilis* sepsis in a premature newborn. Case Rep Pediatr 2015:860671

Matsukura H, Katayama K, Kitano N et al (1996) Infective endocarditis caused by an unusual gram-negative rod, *Rahnella aquatilis*. Pediatr Cardiol 17:108–111

O'Hara GW, Goss TJ, Dilworth MJ et al (1989) Maintenance of intracellular pH and acid tolerance in *Rhizobium meliloti*. Appl Environ Microbiol 55:1870–1876

Peix A, Rivas-Boyero AA, Mateos PF et al (2001) Growth promotion of chickpea and barley by a phosphate solubilizing strain of *Mesorhizobium mediterraneum* under growth chamber conditions. Soil Biol Biochem 33:103–110

Penrose DM, Glick BR (2003) Methods for isolating and characterizing ACC deaminase-containing plant growth-promoting rhizobacteria. Physiol Plant 118:10–15

Reinhold-Hurek B, Hurek T (2011) Living inside plants: bacterial endophytes. Curr Opin Plant Biol 14:435–443

Rivas R, Peix A, Mateos PF, Trujillo ME, Martínez-Molina E, Velázquez E. (2006) Biodiversity of populations of phosphate solubilizing rhizobia that nodulate chickpea in different Spanish soils. Plant Soil 287: 23–33

Rivas R, García-Fraile P, Mateos PF et al (2007) Characterization of xylanolytic bacteria present in the bract phyllosphere of the date palm *Phoenix dactylifera*. Lett Appl Microbiol 44:181–187

Rosenblueth M, Martínez-Romero E (2006) Bacterial endophytes and their interactions with hosts. Mol Plant Microbe Interact 19:827–837

Schwyn B, Neilands JB (1987) Universal chemical assay for the detection and determination of siderophores. Anal Biochem 160:47–56

Shubov A, Jagannathan P, Chin-Hong PV (2011) *Pantoea agglomerans* pneumonia in a heart-lung transplant recipient: case report and a review of an emerging pathogen in immunocompromised hosts. Transpl Infect Dis 13:536–539

Solaiman DK, Ashby RD, Gunther NW et al (2015) Dirhamnose-lipid production by recombinant nonpathogenic bacterium *Pseudomonas chlororaphis*. Appl Microbiol Biotechnol 99:4333–4342

Tiwari S, Beriha SS (2015) *Pantoea* species causing early onset neonatal sepsis: a case report. J Med Case Rep 9:188

Trantas EA, Licciardello G, Almeida NF et al (2015) Comparative genomic analysis of multiple strains of two unusual plant pathogens: *Pseudomonas corrugata* and *Pseudomonas mediterranea*. Front Microbiol 6:811

Williams JG, Kubelik AR, Livak KJ, Rafalski JA, Tingey SV (1990). DNA polymorphisms amplified by arbitrary primers are useful as genetic markers. Nucleic Acids Res 18(22):6531–6535

Westerfeld C, Papaliodis GN, Behlau I, Durand ML, Sobrin L (2009) *Enterobacter amnigenus* endophthalmitis. Retin Cases Brief Rep 3:409–411

Woese CR, Olsen GJ, Ibba M, Söll D (2000). Aminoacyl-tRNA synthetases, the genetic code, and the evolutionary process. Microbiol Mol Biol Rev 64(1):202–236

Part II
Genetics, Genomics and Proteomics of PPM and Their Associated Plants

Chapter 6
Molecular Determinants of Negative Regulation of the *Bradyrhizobium diazoefficiens* Transcription Factor $FixK_2$

Noemí Fernández, Juan J. Cabrera, Sergio Salazar, Sergio Parejo, M. Carmen Rodríguez, Andrea Lindemann, Mariette Bonnet, Hauke Hennecke, Eulogio J. Bedmar and Socorro Mesa

Abstract $FixK_2$ is a key regulator involved in the control of different lifestyles of the facultative soybean endosymbiont *Bradyrhizobium diazoefficiens*. Expression of the *$fixK_2$* gene is activated by the superimposed two-component regulatory system FixLJ in response to low oxygen (at or below 5 % oxygen in the gas phase) and repressed by its own product (directly or indirectly) by a still unknown mechanism. $FixK_2$ is an unusual member of the CRP/FNR family of bacterial transcription factors, because it is active in vitro without an additional effector molecule and is negatively regulated posttranslationally by the oxidation of its singular cysteine residue (Cys-183). $FixK_2$ is also regulated by proteolysis by the $ClpAP_1$ chaperone-protease system. Remarkably, the recently solved $FixK_2$ structure in complex with DNA revealed the molecular basis for the posttranslational control of $FixK_2$. Specifically, both Cys-183 and the last 12 amino acids of $FixK_2$ are surface-exposed, rendering the protein accessible for both oxidation and proteolysis, respectively. Here, we compile the updated knowledge about the critical elements involved in the negative control of *$fixK_2$* gene expression and $FixK_2$ protein activity.

Keywords Clp chaperone-proteases · CRP/FNR superfamily · Microoxia · Negative autoregulation · Posttranslational control · Transcription factor

N. Fernández · J.J. Cabrera · S. Salazar · S. Parejo · M.C. Rodríguez
E.J. Bedmar · S. Mesa (✉)
Department of Soil Microbiology and Symbiotic Systems,
Estación Experimental del Zaidín, Consejo Superior de Investigaciones Científicas,
ES-18080, Granada, Spain
e-mail: socorro.mesa@eez.csic.es

A. Lindemann · M. Bonnet · H. Hennecke
ETH Zurich, Institute of Microbiology, Vladimir-Prelog-Weg 4, 8093 Zurich, Switzerland

© Springer International Publishing Switzerland 2016
F. González-Andrés and E. James (eds.), *Biological Nitrogen Fixation and Beneficial Plant–Microbe Interactions*, DOI 10.1007/978-3-319-32528-6_6

Abbreviations

Clp	Caseinolytic protease
CRP/FNR superfamily	cyclic AMP receptor protein/fumarate-nitrate reductase regulator superfamily
GFP	Green fluorescent protein
H-T-H	Helix-turn-helix
IVT	In vitro transcription
OD_{600}	Optical density at 600 nm
PCR	Polymerase chain reaction
qRT-PCR	Quantitative reverse transcription PCR
ROS	Reactive oxygen species

6.1 Introduction

Bradyrhizobium diazoefficiens (formerly named *Bradyrhizobium japonicum*) (Delamuta et al. 2013) is an alpha-proteobacterium capable of living facultatively within the infected cells of soybean root nodules (endosymbiotic) and of fixing molecular nitrogen as a nitrogen source for the host plant through nitrogen fixation. *B. diazoefficiens* is also able to denitrify both under free-living and symbiotic conditions (Mesa et al. 2004; Bedmar et al. 2005).

The switch from the free-living state to the endosymbiotic state challenges rhizobia, including *B. diazoefficiens*, to adapt their physiology to the highly specialized environment in a eukaryotic cell. For instance, they need to cope with the extremely low partial pressure of free oxygen (O_2) within nodules (microoxia; Kuzma et al. 1993), and they must overcome oxidative stress generated at different steps of the symbiotic interaction (Pauly et al. 2006; Chang et al. 2009). Accordingly, rhizobia must have a set of transcription regulators to reprogram gene expression in order to cope with these environmental stimuli.

In *B. diazoefficiens*, a sophisticated regulatory network consisting of two linked regulatory cascades coordinates the expression of genes required for microoxic respiration and denitrification (the FixLJ-FixK_2 cascade) and for nitrogen fixation (the RegSR-NifA cascade; Fig. 6.1). In these two cascades, different O_2-sensing mechanisms are responsible for a stepwise activation of downstream events (Sciotti et al. 2003; Torres et al. 2014). At a concentration ≤ 5 % O_2 in the gas phase, the FixLJ two-component regulatory system activates the expression of several genes (Nellen-Anthamatten et al. 1998; Mesa et al. 2008), including the transcription factor gene *fixK*$_2$. FixK_2, in turn, activates the transcription of more than 200 genes such as the *fixNOQP* operon coding for the cbb_3-type high-affinity terminal oxidase needed for respiration under microoxia, or the *nnrR* gene, encoding a denitrification regulator (Nellen-Anthamatten et al. 1998; Mesa et al. 2003, 2008). Further, FixK_2 exerts a negative feedback, directly or indirectly, on the expression of its own gene

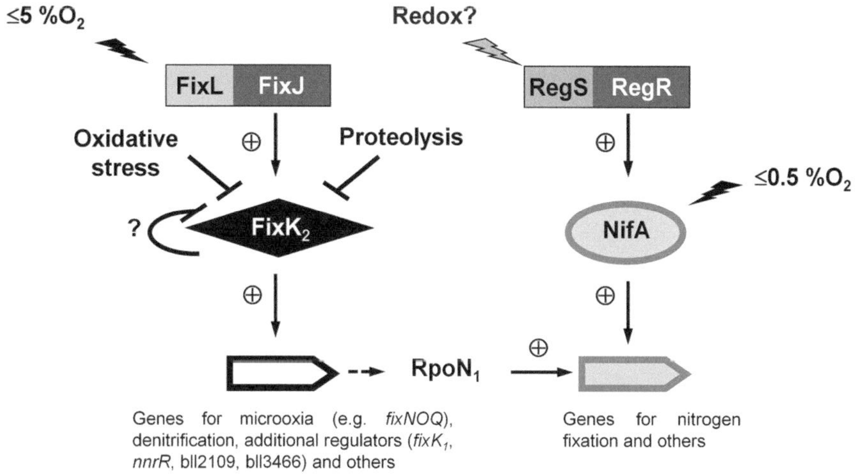

Fig. 6.1 Schematic representation of the two regulatory cascades controlling genes required for microoxic and anoxic lifestyles and for nitrogen fixation in *B. diazoefficiens*. The mechanism how *fixK*$_2$ gene is negatively autoregulated remains enigmatic. Details are explained in the text. *Plus* positive regulation; *minus* negative regulation

by a still unknown mechanism (Nellen-Anthamatten et al. 1998; Reutimann et al. 2010).

6.2 FixK$_2$ Is a Peculiar CRP/FNR Protein

FixK$_2$ belongs to the cyclic AMP receptor protein/fumarate-nitrate reductase regulator (CRP/FNR) superfamily of transcription factors that mainly act as activators in a wide range of bacteria (for review, see Körner et al. 2003). These regulators control expression of specific sets of genes implicated in perception of a broad range of metabolic and environmental signals. Despite their low sequence identity (<25 %), they all are predicted to be structurally similar to CRP with an amino-terminal sensor domain, a β-barrel which makes contact with the RNA polymerase, a long α-helix for protein dimerization, and a DNA-binding domain with a helix-turn-helix (H-T-H) motif. In all cases studied, the active form consists of homodimeric proteins which generally bind to a twofold symmetric consensus DNA sequence located at distinct distances in the promoter region of regulated genes.

The response to the respective environmental or intracellular stimuli is usually transduced through an interaction between a signaling molecule and the sensory domain, whereby a conformational change is induced that leads to the binding of the active dimer to operators near the promoters of the target genes (reviewed in Browning and Busby 2004). Three modes of signal perception have been described, as reviewed in Körner et al. (2003): (i) direct perception of a stressor, as in the *Lactobacillus casei* FLP protein; (ii) dependency on a prosthetic group such as a $[4Fe\text{-}4S]^{2+}$ cluster or heme in *Escherichia coli* FNR or *Rhodospirillum rubrum* CooA, respectively; and (iii) binding of a small effector molecule such as cyclic AMP for *E. coli* CRP or 2-oxoglutarate for cyanobacterial NtcA (Valladares et al. 2008).

Based on the DNA-binding domain and the sensor domain, the CRP/FNR members were classified into 21 major groups (Körner et al. 2003). Structural data have allowed a better understanding of the allosteric mechanism triggered by the binding of cofactors and revealed that the action mechanism seems to be different for each subgroup.

$FixK_2$ is one of the 16 CRP/FNR proteins that had been predicted to be present in the *B. diazoefficiens* genome (Mesa et al. 2006). It belongs to the FixK subgroup, together with Bll2109, Bll3466, and Bll7696, proteins very similar to the FNR clade but lack one or more of the four cysteines necessary to bind the $[4Fe\text{-}4S]^{2+}$ cluster. Based on its primary sequence, $FixK_2$ indeed lacks an obvious sensory module. More importantly, purified protein from oxically grown $FixK_2$-over-producing *E. coli* cells is active as a homodimer in an in vitro transcription (IVT) activation assay without a recognizable coactivator (Mesa et al. 2005).

6.3 Negative Autoregulation of *$FixK_2$* Gene Expression

Nellen-Anthamatten and coworkers reported that expression of a chromosomally integrated *$fixK_2$'-'lacZ* translational fusion led to a more than 10-fold higher β-Galactosidase activity in a Δ*$fixK_2$* strain as compared to the wildtype, when cells of both strains were grown under microoxic conditions (0.5 % O_2 in the gas phase) for 72 h (Nellen-Anthamatten et al. 1998). This observation was later confirmed by microarray-assisted transcriptome analyses of microoxically grown cells at early exponential phase (optical density at 600 [OD_{600}] around 0.5); again, transcription from *$fixK_2$* promoter was found to be upregulated by about twofold in the Δ*$fixK_2$* strain (Mesa et al. 2008). This suggested that $FixK_2$ negatively regulates, directly or indirectly, its own structural gene.

In other bacteria, such as *Ensifer meliloti* and *Caulobacter crescentus*, FixT-like single domain response regulators are involved in the repression of genes encoding FixK-type proteins via an interaction at the FixLJ level (Foussard et al. 1997; Garnerone et al. 1999; Crosson et al. 2005; Jenal and Galperin 2009). A *fixT*-like gene (bll2758), located between *fixLJ* and *$fixK_2$*, was also identified in *B. diazoefficiens*. Although this gene is a direct target for $FixK_2$, however, the Bll2758 protein turned out not to be involved in the negative control of the *$fixK_2$* gene (Reutimann et al. 2010). Whole-genome transcriptome profiling of microoxically grown bll2758 mutant cells showed that neither *$fixK_2$* itself nor any of the *bona fide* direct $FixK_2$ targets was significantly affected in their expression (Reutimann et al. 2010), which speaks against a negative regulatory role of Bll2758 in *B. diazoefficiens*.

Interestingly, we noticed differences in *fixK*$_2$ induction in the Δ*fixK*$_2$ strain when comparing the expression of the chromosomally integrated *fixK*$_2$'-'*lacZ* fusion and the microarray-assisted data. In order to elucidate whether the reason of this it is due to the different methodologies applied or that the cells were grown up to different phases (stationary vs. early exponential), we performed a kinetics of expression of a chromosomally integrated *fixK*$_2$'-'*lacZ* fusion in cells grown under oxic and microoxic conditions (Fig. 6.2). Since the former *fixK*$_2$'-'*lacZ* fusion (plasmid pRJ9054; Nellen-Anthamatten et al. 1998) comprises the complete bll2758 gene, we constructed a new *fixK*$_2$'-'*lacZ* fusion (plasmid pMB1109; Fig. 6.2a) by deletion of a 136-bp *Sma*I fragment in plasmid pRJ9054 within the bll2758 coding region. Plasmid pMB1109 was then integrated by homologous recombination into the chromosome of *B. diazoefficiens* 110*spc*4 (wildtype; Regensburger and Hennecke 1983), 9043 (Δ*fixK*$_2$; Nellen-Anthamatten et al. 1998), and 7414 (bll2758::*aphII*; Anthamatten and Hennecke 1991), resulting in strains 1109, 9043-1109, 7414-1109, respectively.

The expression pattern of the *fixK*$_2$'-'*lacZ* fusion in the 7414-1109 strain was similar to that found in the wild-type strain under all growth conditions and time points (Fig. 6.2b), which confirm that Bll2758 does not appear to interfere with the negative regulation of *fixK*$_2$. While expression of the *fixK*$_2$'-'*lacZ* fusion was mildly induced in strain 9043-1109 compared to the wildtype, both grown microoxically at 24 h (about twofold induction), after 48 and 72 h of incubation, expression of *fixK*$_2$'-'*lacZ* was significantly enhanced six- to sevenfold in strain 9043-1109 in comparison with the wildtype (Fig. 6.2b). Interestingly, *fixK*$_2$'-'*lacZ* expression was also higher (two- to threefold) in the 9043-1109 strain in comparison with that found in the 1109 strain, when both strains were grown in oxic conditions at 48 and 72 h of incubation (Fig. 6.2b). This is an indication that the negative feedback of *fixK*$_2$ expression in *B. diazoefficiens* cells also occurs in environmental conditions (such as oxia) where the FixK$_2$ protein is supposed not to be active. Further, *fixK*$_2$ repression seems to be growth phase-dependent, both in oxic and microoxic conditions (Fig. 6.2b).

6.3.1 *Are other Transcription Factors Involved in* FixK$_2$ *Negative Autoregulation?*

Once the role of the Bll2758 protein as putative repressor of *fixK*$_2$ was excluded, it was rationalized that other transcription factors whose genes are activated by FixK$_2$ might be involved in the negative feedback of *fixK*$_2$. A close examination of the comprehensive FixK$_2$ regulon elaborated by Mesa et al. (2008) uncovered a total of eight genes that code for transcription factors. Three of them, i.e., the alternative sigma factor RpoN$_1$, the CRP/FNR-type protein FixK$_1$, and the denitrification regulator NnrR, do not seem to be involved in downregulating *fixK*$_2$ (Hauser et al. 2007; Mesa et al. 2008; Reutimann et al. 2010). Next, we studied the potential involvement in *fixK*$_2$ autoregulation of two out the five remaining regulatory genes,

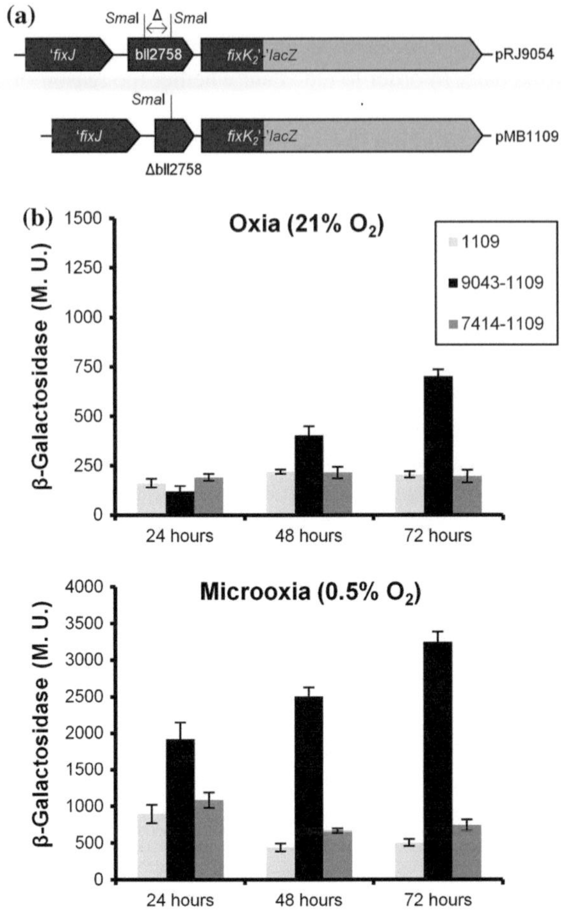

Fig. 6.2 Negative feedback of *fixK*$_2$ is phase-dependent, both in oxic and in microoxic conditions. **a** Structure of the translational *fixK*$_2$*'-'lacZ* fusion. Note that in order to delete an internal 136-bp *Sma*I fragment of bll2758, plasmid pRJ9054 harboring the former *fixK*$_2$*'-'lacZ* fusion (Nellen-Anthamatten et al. 1998) was digested and religated to yield plasmid pMB1109. This plasmid was transferred by biparental conjugation to *B. diazoefficiens* wildtype and strains 7414 (bll2758::*aphII*), and 9043 (Δ*fixK*$_2$) to yield strains 1109, 7414-1109, and 9043-1109, respectively. **b** Expression of chromosomally integrated *fixK*$_2$*'-'lacZ* in different backgrounds. Cells were grown oxically and microoxically (0.5 % O_2) and β-Galactosidase activity was determined at 24, 48, and 72 h. Values are the means with standard errors from at least two independent experiments with two cultures assayed in quadruples (Miller units: M.U.)

i.e., bll2109 and bll3466, which code for CRP/FNR-type transcription factors. Strains 9552 (bll2109::*aphII*; Lindemann et al. 2007) and 2007 (Δbll3466; in frame deletion of an internal fragment of bll3466 between genomic coordinates 3834754 and 3835271) were grown microoxically to an OD_{600} of about 0.7 (early stationary phase), and *fixK*$_2$ expression was analyzed by quantitative reverse transcription PCR (qRT-PCR) with primers fixK2-14 (5′-TGCGCAGCTACAAGCTTCTC-3′)

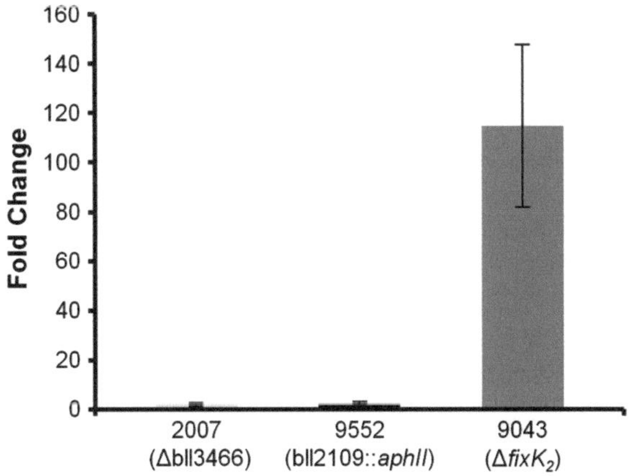

Fig. 6.3 Neither Bll3466 nor Bll2109 is involved in the negative autoregulation of *fixK*$_2$. Expression of the *fixK*$_2$ gene was analyzed by qRT-PCR in *B. diazoefficiens* wildtype, and 2007, 9552, and 9043 strains grown microoxically (0.5 % O_2) to an OD_{600} of about 0.7. The relevant genotype of host strains is indicated in *parentheses*. Cell harvest, RNA isolation, cDNA synthesis, and qRT-PCR reactions were performed as previously described (Mesa et al. 2008; Torres et al. 2014). Fold-change values were calculated according to Pfaffl (2001) and refer to differences of expression when the mutant strains were compared to the wildtype. The primary sigma factor gene *sigA* was used as reference for normalization. Values are the means of reactions with three cDNA concentrations run in triplicates

and fixK2-15 (5′-CTGGTGTCGATGATGGCTTC-3′). For comparison, the wild-type and Δ*fixK*$_2$ (9043; Nellen-Anthamatten et al. 1998) strains were included as controls. While *fixK*$_2$ gene expression was drastically increased in the Δ*fixK*$_2$ strain in comparison with the wildtype (115 ± 33, relative fold change ± standard deviation; Fig. 6.3), the corresponding changes in expression in strains 9552 and 2007 were only 1.9 ± 0.6 and 2.2 ± 0.9, respectively. Therefore, neither Bll2109 nor Bll3466 are involved in the negative feedback of *fixK*$_2$..

6.4 Posttranslational Control of FixK$_2$ by Oxidation and Proteolysis

Besides the transcriptional control of *fixK*$_2$ expression, an oxidation-mediated posttranslational control occurs at FixK$_2$ (Fig. 6.1), whereby a singular cysteine at position 183 (Cys-183) near the H-T-H motif of the DNA-binding domain is a target for oxidation, which results in the inactivation of the protein (Mesa et al. 2009). Oxidation of Cys-183 in vitro triggers both the formation of a reversible intermolecular disulfide bridge between two monomers and the conversion of the cysteine thiol into irreversible sulfinic or sulfonic acid derivatives (Mesa et al. 2009). In vivo, however, the more likely oxidation mechanism is the latter.

The significance of $FixK_2$ oxidation in *B. diazoefficiens* cells is based on indirect evidence. Particularly, microarray analyses had shown that most of the direct $FixK_2$-activated target genes showed decreased expression in hydrogen peroxide-treated *B. diazoefficiens* wild-type cells, but not in a strain having a $FixK_2$ variant with Cys-183 replaced by alanine. $FixK_2$ oxidation might be relevant for instantly shutting off its activity in response to ROS during symbiotic interaction with host plants, i.e., (i) in the early stage of root hair infection, (ii) in the process of endosymbiotic respiration, and (iii) late during nodule senescence (Pauly et al. 2006; Chang et al. 2009). In all of these circumstances, $FixK_2$ activity might not be either needed or even detrimental.

Lately, Sugawara et al. (2011) reported on a regulatory link between sulfur metabolism and $FixK_2$-mediated regulation. When comparing the transcriptional profile of cells grown in minimal medium with cysteine as sulfur source with that of cells grown in sulfate or sulfonate, 85 % of the genes differentially expressed in such comparison were also previously shown to be regulated by the FixLJ-$FixK_2$ cascade (Mesa et al. 2008). In contrast to the genes regulated by $FixK_2$ that showed increased expression on cysteine, expression of the *fixK*$_2$ gene increased during growth with sulfate and sulfonates. While growth with cysteine could help in maintaining $FixK_2$ in its active configuration, growth on sulfate or sulfonates can result in oxidative stress leading to the oxidation of Cys-183 in $FixK_2$.

Along to this oxidation control, $FixK_2$ is also controlled at posttranslational level by proteolysis (Fig. 6.1). The steady-state levels of $FixK_2$ remain constant irrespective of the oxic, microoxic, and anoxic modes of growth (Mesa et al. 2009). Interestingly, Delmotte et al. (2010) reported that seven out of the 16 putative *B. diazoefficiens* CRP/FNR-like proteins were identified at the RNA level but only two of them (Bll6060, Bll2244) were detected also as proteins. It was also observed that a truncated $FixK_2$ derivative is always copurified with full-length amino-terminal His-tagged $FixK_2$ derivatives from $FixK_2$-overproducing *E. coli* cells. Mass spectrometry analyses revealed that this secondary form is a carboxy-terminally cleaved $FixK_2$ derivative which lacks the last 12 amino acids (cleavage site between Val-220 and Leu-221), and it is inactive in the $FixK_2$-mediated IVT assay (Bonnet et al. 2013a). Moreover, FNR, a $FixK_2$ homolog in *E. coli,* was shown to be trapped by an inactive ClpXP complex in vivo (Flynn et al. 2003) and degraded by the proteolytic system ClpXP in vitro (Mettert and Kiley 2005). In this context, it is worth mentioning that the carboxy-terminal sequence of FNR plays a crucial role in the degradation of the inactive apo-form of this protein by ClpXP (Mettert and Kiley 2005). In particular, the region of amino acids 239-244 in FNR promotes a structural fold that limits the exposure of the proximal ClpXP site to the protease (Pan et al. 2012).

6.4.1 The $ClpAP_1$ Chaperone-Protease System Is Involved in the Degradation of $FixK_2$

The Clp (caseinolytic proteases) proteins are energy-dependent chaperone-proteases responsible for the breakdown of a large variety of cytoplasmic proteins and play a crucial role in regulation and quality control mechanisms (reviewed in Kress et al. 2009; Schmidt et al. 2009). They consist of two distinct parts: a hexameric chaperone ring belonging to the AAA+ ATPase family and a protease cylinder. Among the set of chaperone-proteases in *E. coli*, ClpAP and ClpXP display partially overlapping target substrates (Gottesman et al. 1998; Flynn et al. 2001), although the presence of their respective adaptor proteins changes their specificity. For instance, while SspB promotes the degradation of ssrA-tagged substrates by ClpXP (ssrA tag is added by the tmRNA system carboxy-terminally to polypeptide chains stalled on the ribosome; Moore and Sauer 2007), ClpS inhibits their degradation by ClpAP (reviewed in Kirstein et al. 2009). Further, ClpS modifies the substrate specificity of ClpAP, by enhancing degradation of N-rule substrates (degradation sequence at the amino-terminal end) and inhibiting degradation of specific substrates.

B. diazoefficiens chromosome (http://genome.microbedb.jp/rhizobase/) comprises genes with sequence identity with *E. coli* and *C. crescentus* proteolytic machineries and their corresponding adaptors: i.e., bll5153 (ClpA), bll4943 (ClpX), and blr6518 (SspB) (Bonnet et al. 2013b). Notably, there are two copies of ClpP- and ClpS-coding genes in the *B. diazoefficiens* genome: i.e., bll4944 ($ClpP_1$), blr0611 ($ClpP_2$), bll5154 ($ClpS_1$), and bll2636 ($ClpS_2$) (Bonnet et al. 2013b).

Inspired by *E. coli* FNR which is proteolytically controlled by ClpXP (Mettert and Kiley 2005), it was therefore studied, the possibility of $FixK_2$ proteolysis by the *B. diazoefficiens* Clp chaperone-proteases. The Clp proteins from *B. diazoefficiens*, specifically $ClpAP_1$ and $ClpXP_1$, were purified and tested for their ability to degrade $FixK_2$ (Bonnet et al. 2013b). Contrary to FNR, $FixK_2$ is not a substrate for $ClpXP_1$ but it was found to be a substrate for $ClpAP_1$ in vitro (Fig. 6.4a). This degradation was specific and was inhibited by the $ClpS_1$ adaptor protein (Fig. 6.4a), indicating that $FixK_2$ is a novel direct substrate for $ClpAP_1$.

Remarkably, only three substrates (ClpA, RepA, RepB) were previously shown to be directly recognized and degraded by *E. coli* ClpAP (Wickner et al. 1999; Wagner 2007; Maglica et al. 2008); however, no consensus sequence has been identified for ClpA recognition. While the *E. coli* ClpA autodegradation signal is encoded in the last nine carboxy-terminal amino acids (Maglica et al. 2008), the recognition sequence of RepA is located within the first 15 amino-terminal residues (Hoskins et al. 2000). The location of the recognition site for ClpA on RepE is still unknown. In order to identify the element on $FixK_2$ that is recognized by ClpA, the nine amino-terminal (Fn_9-GFP), the 12 carboxy-terminal (GFP-Fc_{12}) amino acids of $FixK_2$ or both (Fn_9-GFP-Fc_{12}) were fused to green fluorescent protein (GFP) (Fig. 6.4b). Although the last 12 amino acids of $FixK_2$ play the main role in the recognition by ClpA, the combination of the amino- and carboxy-termini fused to GFP was additive (Fig. 6.4b) and equaled the degradation rate of $FixK_2$ itself (Bonnet et al. 2013b).

(a)

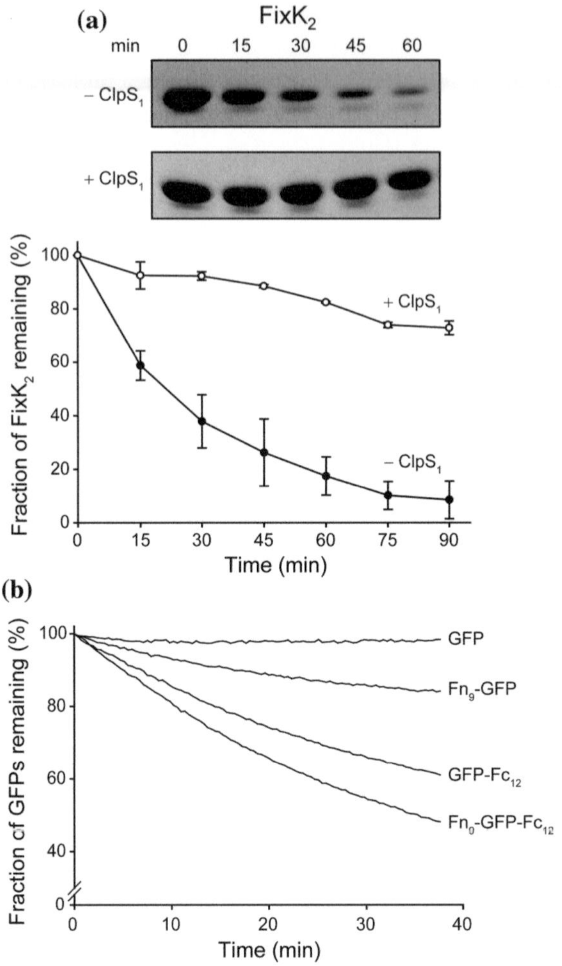

(b)

Fig. 6.4 **a** $FixK_2$ is a direct substrate for $ClpAP_1$. $FixK_2$ degradation assay by $ClpAP_1$ in the absence or presence of $ClpS_1$. The degradation assays were followed by Blue Coomassie-stained SDS–PAGE (*top panel*) and quantified with AlphaImager (*bottom panel*). Values are the means and standard deviations from three independent experiments. **b** The carboxy-terminus of $FixK_2$ is the main recognition site for $ClpAP_1$. Degradation of different GFP derivatives was followed by fluorescence. Adapted from Bonnet et al. (2013b)

6.4.2 Molecular Basis for the Posttranslational Control of $FixK_2$

To better understand the molecular mechanism underlying the posttranslational control of $FixK_2$ by oxidation and proteolysis and its interaction with target DNA, crystallization was attempted (Bonnet et al. 2013a). Although several CRP/FNR-type proteins were

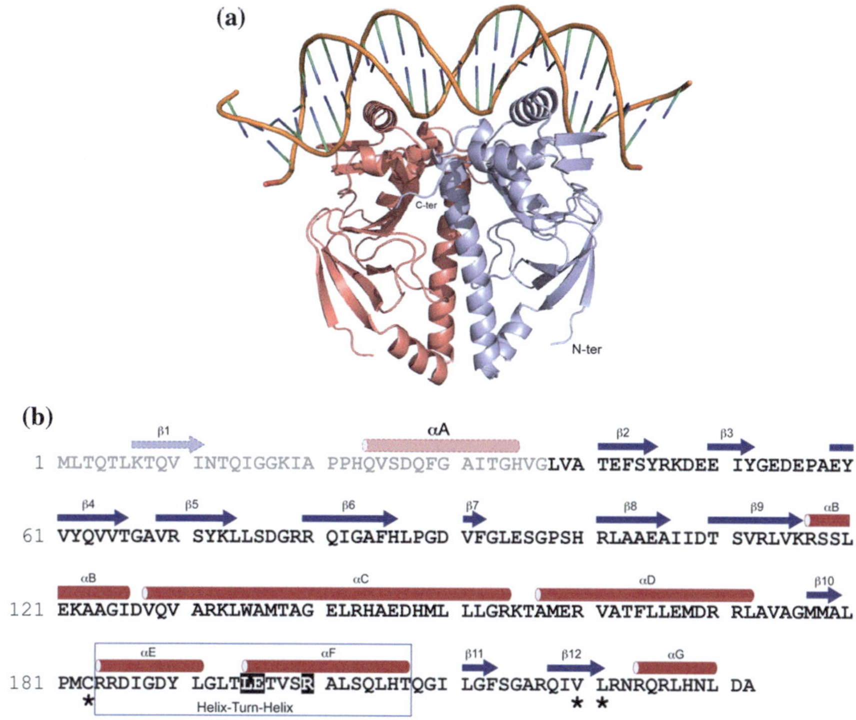

Fig. 6.5 a Ribbon representation of the C183S $FixK_2His_6$ dimer bound to the $FixK_2$ box-containing target DNA. The structure was solved by single wavelength anomalous dispersion and refined to 1.77 Å. *N-ter* amino-terminus, *C-ter* carboxy-terminus. **b** Schematic representation of $FixK_2$ secondary structure based of its crystal structure. The secondary structure elements (β-sheet, α-helix) are depicted. Cys-183 involved in oxidation and the cleavage site between Val-220 and Leu-221 are marked with *asterisks*. The H-T-H DNA-binding motif consisting of α helices E and F is *boxed*. Amino acid residues in *black boxes* are predicted to make base specific contacts. For details, see Bonnet et al. (2013a)

previously crystallized, no structure of protein sensitive to O_2 was available. Crystallization of $FixK_2$ succeeded only when (i) an oxidation- and protease-insensitive protein variant (C183S $FixK_2His_6$) was used and (ii) this protein was allowed to form a complex with a 30-mer double-stranded target DNA ($FixK_2$-binding site present at the promoter of the cytochrome oxidase cbb_3-encoding genes *fixNOQP*).

The structure of the $FixK_2$-DNA complex was solved at a resolution of 1.77 Å, at which the protein formed a homodimer with a perfect twofold symmetry (Fig. 6.5a; Bonnet et al. 2013a). Each monomer in the dimeric $FixK_2$ protein contains seven α-helices (αA–αG) and 12 β-sheets (β1–β12) (Fig. 6.5b).[1] The full-length protein comprises two major domains: (i) the effector-binding domain at

[1]Figures 6.5 and 6.6 were originally published in Bonnet et al. (2013a). © The American Society for Biochemistry and Molecular Biology.

the amino-terminal half (residues 1–127: αA–αB and β1–β9) containing the double-stranded β-roll structure (residues 41–116: β2–β9), and (ii) the carboxy-terminal DNA-binding domain (residues 155–236: αD–αG and β10–β12) with the H-T-H motif (residues 184–216: αE and αF). The two major domains are connected by the long α-helix (residues 128–154: αC), which is involved in dimerization of the two monomers.

In $FixK_2$-DNA structure, the oxidation-sensitive Cys-183 is surface-exposed and in close proximity to DNA (Bonnet et al. 2013a). Therefore, the oxo acids generated after oxidation of the cysteine thiol causes not only a steric hindrance, but also an electrostatic repulsion of the opposite negatively charged DNA phosphate, preventing $FixK_2$ binding to DNA or at least DNA bending. These explain why ROS inhibit $FixK_2$ posttranslationally. Oxidation-sensitive cysteine thiols have also been described for the *Desulfitobacterium dehalogenans* CprK protein, but these form either intra- or intersubunit disulfide bridges upon oxidation (Levy et al. 2008).

The carboxy-terminal helix (αG) of $FixK_2$ is also surface-exposed, which might explain its sensitivity to protein cleavage and $ClpAP_1$-mediated degradation (Fig. 6.6). Cleavage between Val-220 and Leu-221 disrupts the carboxy-terminal end of β-strand 12 (Fig. 6.5b). Interestingly, β12 is the intermediate β-strand of a small β-sheet formed by β10, β11, and β12 near the DNA-binding domain (Fig. 6.6). Although this β-sheet does not directly interact with the bound DNA, hydrogen bonds are formed between β12 and αG and between αG and αE of the H-T-H motif. Therefore, disrupting these hydrogen bonds by proteolytic removal of the last 12 amino acids might affect the DNA-binding domain, leading to inactivation of the protein. The carboxy-terminal helix (αG) is also present in some other members of the CRP/FNR superfamily such as *D. dehalogenans* CprK (Levy et al. 2008), *Listeria*

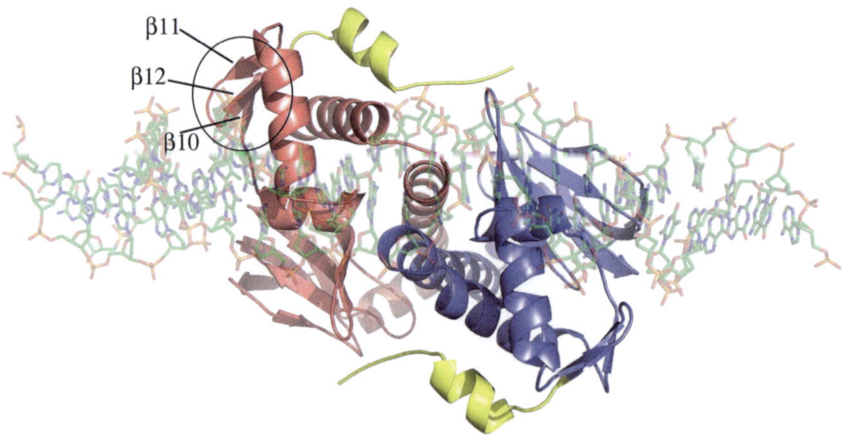

Fig. 6.6 Top view of $FixK_2$-DNA structure. The 12 carboxy-terminal amino acids of the protein are colored in *light gray*. They are surface-exposed, rendering the protein accessible to proteolysis. The three β-strands near the protease-sensitive site (between Val-220 and Leu-221) are marked on one monomer, and the entire β-sheet is *encircled*. Adapted from Bonnet et al. (2013a)

monocytogenes PrfA (Eiting et al. 2005), and *Thermus termophilus* SdrP (Agari et al. 2008) and therein plays a crucial role in DNA binding and activity. For instance, in CprK, the carboxy-terminus interacts directly with DNA in the on-state but is unstructured in the apo-form (Levy et al. 2008). In PrfA, mutations of the last helices impair DNA binding, and consequently activation of target genes (Herler et al. 2001).

6.5 Concluding Remarks

Here, we underscore how transcription factor $FixK_2$ a key regulator of genes for microoxic, anoxic, and symbiotic growth in *B. diazoefficiens* is exquisitely regulated by different means. Expression of *fixK*$_2$ gene is a balance between FixLJ-mediated induction and negative autoregulation. The mechanism underlying the latter remains enigmatic, despite the fact that the regulatory modules (FixL, FixJ, FixT, FixK) are conserved. However, we unexpectedly discovered that the negative feedback of *fixK*$_2$ gene expression occurs also in oxically grown *B. diazoefficiens* cells. Further, *fixK*$_2$ repression seems to be phase-dependent, i.e., more pronounced in stationary than exponential phase, in both oxic and microoxic conditions. The physiological role of these two findings is still not uncovered, as well as the regulatory mechanism of such control.

The posttranslational regulation of $FixK_2$ by oxidation and proteolysis provides a second important means to counterbalance the cellular pool of active $FixK_2$ and, therefore, to control $FixK_2$ activity, in addition to the regulation at transcriptional level. ROS are produced at different stages of the symbiotic interaction with host plats. Hence, oxidation of $FixK_2$ at Cys-183 is a manner of shutting off $FixK_2$ activity in circumstances when it might not be either required or even not desirable. The $ClpAP_1$ chaperone-protease system specifically degrades $FixK_2$ in vitro and it might also be involved in the turnover of $FixK_2$ in vivo. The degradation of $FixK_2$ might therefore be an additional way of regulating the protein activity, either by continuously degrading the protein amount in the cells or under a specific set of environmental conditions.

Finally, the structure of $FixK_2$ in complex with its cognate DNA provides not only the molecular basis for the posttranslational control of $FixK_2$ by oxidation and proteolysis, but also valuable information for the better understanding of the molecular mechanism of $FixK_2$ (e.g., hallmark of $FixK_2$-DNA interaction, dimerization interface). In particular, investigation of the oxidation-mediated control of $FixK_2$ is the first of a member of the CRP/FNR superfamily for which a structural explanation has been found this kind of modification. Thus, the functional characterization of $FixK_2$ based on its structure might serve as model to understand the molecular mechanism of other CRP/FNR proteins and consequently, further expand the spectrum of possibilities how activity of these transcription factors can be controlled.

Acknowledgements This work was supported by grants from ETH Zurich, the Swiss National Science Foundation, and the European Regional Development Fund (ERDF)-cofinanced grants AGL2011-23383 from Ministerio de Economía y Competitividad of Spain and P12-AGR-1968 from Consejería de Innovación, Ciencia y Empresa of Junta de Andalucía (Spain). Support from Junta de Andalucía to group BIO275 is also acknowledged.

References

Agari Y, Kashihara A, Yokoyama S, Kuramitsu S, Shinkai A (2008) Global gene expression mediated by *Thermus thermophilus* SdrP, a CRP/FNR family transcriptional regulator. Mol Microbiol 70:60–75

Anthamatten D, Hennecke H (1991) The regulatory status of the *fixL*- and *fixJ*-like genes in *Bradyrhizobium japonicum* may be different from that in *Rhizobium meliloti*. Mol Gen Genet 225:38–48

Bedmar EJ, Robles EF, Delgado MJ (2005) The complete denitrification pathway of the symbiotic, nitrogen-fixing bacterium *Bradyrhizobium japonicum*. Biochem Soc Trans 33:141–144

Bonnet M, Kurz M, Mesa S, Briand C, Hennecke H, Grutter MG (2013a) The structure of *Bradyrhizobium japonicum* transcription factor $FixK_2$ unveils sites of DNA binding and oxidation. J Biol Chem 288:14238–14246

Bonnet M, Stegmann M, Maglica Z, Stiegeler E, Weber-Ban E, Hennecke H, Mesa S (2013b) $FixK_2$, a key regulator in *Bradyrhizobium japonicum*, is a substrate for the protease ClpAP in vitro. FEBS Lett 587:88–93

Browning DF, Busby SJ (2004) The regulation of bacterial transcription initiation. Nat Rev Microbiol 2:57–65

Chang C, Damiani I, Puppo A, Frendo P (2009) Redox changes during the legume-rhizobium symbiosis. Mol Plant 2:370–377

Crosson S, McGrath PT, Stephens C, McAdams HH, Shapiro L (2005) Conserved modular design of an oxygen sensory/signaling network with species-specific output. Proc Natl Acad Sci USA 102:8018–8123

Delamuta JR, Ribeiro RA, Ormeno-Orrillo E, Melo IS, Martinez-Romero E, Hungria M (2013) Polyphasic evidence supporting the reclassification of *Bradyrhizobium japonicum* group Ia strains as *Bradyrhizobium diazoefficiens* sp. nov. Int J Syst Evol Microbiol 63:3342–3351

Delmotte N, Ahrens CH, Knief C, Qeli E, Koch M, Fischer HM, Vorholt JA, Hennecke H, Pessi G (2010) An integrated proteomics and transcriptomics reference data set provides new insights into the *Bradyrhizobium japonicum* bacteroid metabolism in soybean root nodules. Proteomics 10:1391–1400

Eiting M, Hageluken G, Schubert WD, Heinz DW (2005) The mutation G145S in PrfA, a key virulence regulator of *Listeria monocytogenes*, increases DNA-binding affinity by stabilizing the HTH motif. Mol Microbiol 56:433–446

Flynn JM, Levchenko I, Seidel M, Wickner SH, Sauer RT, Baker TA (2001) Overlapping recognition determinants within the ssrA degradation tag allow modulation of proteolysis. Proc Natl Acad Sci USA 98:10584–10589

Flynn JM, Neher SB, Kim YI, Sauer RT, Baker TA (2003) Proteomic discovery of cellular substrates of the ClpXP protease reveals five classes of ClpX-recognition signals. Mol Cell 11:671–683

Foussard M, Garnerone AM, Ni F, Soupene E, Boistard P, Batut J (1997) Negative autoregulation of the *Rhizobium meliloti* fixK gene is indirect and requires a newly identified regulator, FixT. Mol Microbiol 25:27–37

Garnerone AM, Cabanes D, Foussard M, Boistard P, Batut J (1999) Inhibition of the FixL sensor kinase by the FixT protein in *Sinorhizobium meliloti*. J Biol Chem 274:32500–32506

Gottesman S, Roche E, Zhou Y, Sauer RT (1998) The ClpXP and ClpAP proteases degrade proteins with carboxy-terminal peptide tails added by the SsrA-tagging system. Genes Dev 12:1338–1347

Hauser F, Pessi G, Friberg M, Weber C, Rusca N, Lindemann A, Fischer HM, Hennecke H (2007) Dissection of the *Bradyrhizobium japonicum* NifA+ σ^{54} regulon, and identification of a ferredoxin gene (*fdxN*) for symbiotic nitrogen fixation. Mol Genet Genomics 278:255–271

Herler M, Bubert A, Goetz M, Vega Y, Vazquez-Boland JA, Goebel W (2001) Positive selection of mutations leading to loss or reduction of transcriptional activity of PrfA, the central regulator of *Listeria monocytogenes* virulence. J Bacteriol 183:5562–5570

Hoskins JR, Kim SY, Wickner S (2000) Substrate recognition by the ClpA chaperone component of ClpAP protease. J Biol Chem 275:35361–35367

Jenal U, Galperin MY (2009) Single domain response regulators: molecular switches with emerging roles in cell organization and dynamics. Curr Opin Microbiol 12:152–160

Kirstein J, Moliere N, Dougan DA, Turgay K (2009) Adapting the machine: adaptor proteins for Hsp100/Clp and AAA+ proteases. Nat Rev Microbiol 7:589–599

Körner H, Sofia HJ, Zumft WG (2003) Phylogeny of the bacterial superfamily of Crp-Fnr transcription regulators: exploiting the metabolic spectrum by controlling alternative gene programs. FEMS Microbiol Rev 27:559–592

Kress W, Maglica Z, Weber-Ban E (2009) Clp chaperone-proteases: structure and function. Res Microbiol 160:618–628

Kuzma MM, Hunt S, Layzell DB (1993) Role of oxygen in the limitation and inhibition of nitrogenase activity and respiration rate in individual soybean nodules. Plant Physiol 101:161–169

Levy C, Pike K, Heyes DJ, Joyce MG, Gabor K, Smidt H, van der Oost J, Leys D (2008) Molecular basis of halorespiration control by CprK, a CRP-FNR type transcriptional regulator. Mol Microbiol 70:151–167

Lindemann A, Moser A, Pessi G, Hauser F, Friberg M, Hennecke H, Fischer HM (2007) New target genes controlled by the *Bradyrhizobium japonicum* two-component regulatory system RegSR. J Bacteriol 189:8928–8943

Maglica Z, Striebel F, Weber-Ban E (2008) An intrinsic degradation tag on the ClpA C-terminus regulates the balance of ClpAP complexes with different substrate specificity. J Mol Biol 384:503–511

Mesa S, Bedmar EJ, Chanfon A, Hennecke H, Fischer HM (2003) *Bradyrhizobium japonicum* NnrR, a denitrification regulator, expands the FixLJ-FixK_2 regulatory cascade. J Bacteriol 185:3978–3982

Mesa S, Alché JD, Bedmar EJ, Delgado MJ (2004) Expression of *nir*, *nor* and *nos* denitrification genes from *Bradyrhizobium japonicum* in soybean root nodules. Physiol Plant 120:205–211

Mesa S, Ucurum Z, Hennecke H, Fischer HM (2005) Transcription activation in vitro by the *Bradyrhizobium japonicum* regulatory protein FixK_2. J Bacteriol 187:3329–3338

Mesa S, Hennecke H, Fischer HM (2006) A multitude of CRP/FNR-like transcription proteins in *Bradyrhizobium japonicum*. Biochem Soc Trans 34:156–159

Mesa S, Hauser F, Friberg M, Malaguti E, Fischer HM, Hennecke H (2008) Comprehensive assessment of the regulons controlled by the FixLJ-FixK_2-FixK_1 cascade in *Bradyrhizobium japonicum*. J Bacteriol 190:6568–6579

Mesa S, Reutimann L, Fischer HM, Hennecke H (2009) Posttranslational control of transcription factor FixK_2, a key regulator for the *Bradyrhizobium japonicum*-soybean symbiosis. Proc Natl Acad Sci USA 106:21860–21865

Mettert EL, Kiley PJ (2005) ClpXP-dependent proteolysis of FNR upon loss of its O_2-sensing [4Fe-4S] cluster. J Mol Biol 354:220–232

Moore SD, Sauer RT (2007) The tmRNA system for translational surveillance and ribosome rescue. Annu Rev Biochem 76:101–124

Nellen-Anthamatten D, Rossi P, Preisig O, Kullik I, Babst M, Fischer HM, Hennecke H (1998) *Bradyrhizobium japonicum* FixK_2, a crucial distributor in the FixLJ-dependent regulatory cascade for control of genes inducible by low oxygen levels. J Bacteriol 180:5251–5255

Pan Q, Shan Y, Yan A (2012) A region at the C-terminus of the *Escherichia coli* global transcription factor FNR negatively mediates its degradation by the ClpXP protease. Biochemistry 51:5061–5071

Pauly N, Pucciariello C, Mandon K, Innocenti G, Jamet A, Baudouin E, Herouart D, Frendo P, Puppo A (2006) Reactive oxygen and nitrogen species and glutathione: key players in the legume-*Rhizobium* symbiosis. J Exp Bot 57:1769–1776

Pfaffl MW (2001) A new mathematical model for relative quantification in real-time RT-PCR. Nucleic Acids Res 29:e45

Regensburger B, Hennecke H (1983) RNA polymerase from *Rhizobium japonicum*. Arch Microbiol 135:103–109

Reutimann L, Mesa S, Hennecke H (2010) Autoregulation of *fixK*$_2$ gene expression in *Bradyrhizobium japonicum*. Mol Genet Genomics 284:25–32

Schmidt R, Bukau B, Mogk A (2009) Principles of general and regulatory proteolysis by AAA+ proteases in *Escherichia coli*. Res Microbiol 160:629–636

Sciotti MA, Chanfon A, Hennecke H, Fischer HM (2003) Disparate oxygen responsiveness of two regulatory cascades that control expression of symbiotic genes in *Bradyrhizobium japonicum*. J Bacteriol 185:5639–5642

Sugawara M, Shah GR, Sadowsky MJ, Paliy O, Speck J, Vail AW, Gyaneshwar P (2011) Expression and functional roles of *Bradyrhizobium japonicum* genes involved in the utilization of inorganic and organic sulfur compounds in free-living and symbiotic conditions. Mol Plant-Microbe Interact 24:451–457

Torres MJ, Argandoña M, Vargas C, Bedmar EJ, Fischer HM, Mesa S, Delgado MJ (2014) The global response regulator RegR controls expression of denitrification genes in *Bradyrhizobium japonicum*. PLoS ONE 9:e99011

Valladares A, Flores E, Herrero A (2008) Transcription activation by NtcA and 2-oxoglutarate of three genes involved in heterocyst differentiation in the cyanobacterium *Anabaena* sp. strain PCC 7120. J Bacteriol 190:6126–6133

Wagner JN (2007) The HSP100 protein ClpA: significance of the N-terminal domain in substrate recognition and binding and characterization of the novel substrate RepE. Albert-Ludwigs University Freiburg, Germany

Wickner S, Maurizi MR, Gottesman S (1999) Posttranslational quality control: folding, refolding, and degrading proteins. Science 286:1888–1893

Chapter 7
Matrix-Assisted Laser Desorption Ionization Time-of-Flight Mass Spectrometry (MALDI-TOF MS) Analysis of Rhizobia Nodulating *Phaseolus Vulgaris* in Different Soils

Lucía Isabel López Díez, Fernando Sánchez Juanes, Martha Helena Ramírez Bahena, José Manuel González Buitrago, Alvaro Peix and Encarna Velázquez

Abstract In this work, we analyzed by using MALDI-TOF MS (Matrix-Assisted Laser Desorption Ionization Time-of-Flight Mass Spectrometry) *Phaseolus vulgaris* nodulating strains isolated two different soils (soils 1 and 2) in León Province where this legume is commonly cultivated and it is protected by the GPI "La Bañeza". These strains were compared with those isolated from *P. vulgaris* nodules in a third soil in the same zone identified as *R. leguminosarum* but belonging to different *atpD* and *recA* lineages that were added to the *Rhizobiaceae* database. The MALDI-TOF MS analysis showed that most of the strains isolated in soil 1 matched with the strain RPVR24 from the group II of *R. leguminosarum*, a strain isolated in a rainfed soil cultivated with *P. vulgaris* for 20 years in the same location (soil 3). However, almost all strains isolated from soil 2 matched with the strain RPVR32, also isolated in the soil 3, and belonged to a third phylogenetic group within *R. leguminosarum*. These results showed that MALDI-TOF MS is not only a reliable technique for identification at species level of fast-growing rhizobia, but also for differentiation of subspecific groups, which is very useful for ecological studies.

L.I. López Díez · E. Velázquez (✉)
Departamento de Microbiología y Genética, Universidad de Salamanca, Salamanca, Spain
e-mail: evp@usal.es

F. Sánchez Juanes · J.M. González Buitrago
Instituto de Investigación Biomédica de Salamanca (IBSAL), Salamanca, Spain

M.H. Ramírez Bahena · A. Peix
IRNASA-CSIC, Salamanca, Spain

A. Peix · E. Velázquez
Unidad Asociada Universidad de Salamanca-CSIC 'Interacción Planta-Microorganismo', Salamanca, Spain

© Springer International Publishing Switzerland 2016
F. González-Andrés and E. James (eds.), *Biological Nitrogen Fixation and Beneficial Plant–Microbe Interactions*, DOI 10.1007/978-3-319-32528-6_7

Keywords MALDI-TOF MS · Identification · *Rhizobium* · *Phaseolus vulgaris* · Phylogeny

7.1 Introduction

Phaseolus vulgaris L. is a legume native to America that establishes nitrogen-fixing symbiosis with rhizobial strains from different genera and species (Peix et al. 2015). The *Rhizobium* species nodulating this legume are differentiated by Matrix-Assisted Laser Desorption Ionization-Time-of-Flight Mass Spectrometry (MALDI-TOF MS) analysis and then, this technique can be used in rhizobial identification schemes (Ferreira et al. 2011). Moreover, some strains from the species *R. leguminosarum* belonging to different *recA* and *atpD* lineages were differentiated by MALDI-TOF MS analysis (Ferreira et al. 2011).

Some of these strains were isolated from *P. vulgaris* nodules and belong to five *recA* and *atpD* phylogenetic lineages (García-Fraile et al. 2010). One of them (RPVF18) was coincident with that formed by the type strain of *R. leguminosarum* which constitutes the group I and other (RPVR24) was coincident with the group II of this species (Mulas et al. 2011). The remaining three lineages were represented by the strains RPVR32, RPVN04, and RPVN06 (García-Fraile et al. 2010). Therefore, strains nodulating *P. vulgaris* are good models to check the suitability of MALDI-TOF MS for infraspecific identification which was the aim of this study.

To achieve this aim, we added to the database of family *Rhizobiaceae* a representative strain from each lineage found by García-Fraile et al. (2010) in order to compare by using MALDI-TOF MS these strains with those nodulating this legume in other two soils from the same zone, one of them converted to irrigated soil onc year before sampling and cultivated with *P. vulgaris* (soil 1), and the other converted five years before sampling and cultivated with this legume in rotation with sugar beet and maize (soil 2).

7.2 Materials and Methods

7.2.1 Isolation of Rhizobial Strains

A total of 43 strains (Table 7.1) able to form effective nodules on *P. vulgaris* were isolated according to Vincent (1970) on YMA medium in two soils located at Riego de la Vega (42°23′ N 5°58′W) belonging to León Province. They were isolated in two rainfed soils converted to irrigated in order to cultivate *P. vulgaris*. The soil 1 was cultivated with this legume the year of sampling and in the soil 2, *P. vulgaris* was cultivated twice during the last five years before the sampling year 2012. The soil 3 was also located at Riego de la Vega and had a long tradition of *P. vulgaris*

Table 7.1 Results of thc identification of strains isolated n soils 1 and 2 by MALDI-TOF MS

Strain	Soil	Organism (best match)	Score value
RPV1Y01	1	*Rhizobium leguminosarum* RPVF18	2.2
RPV1Y02	1	*Rhizobium* sp. RPVR24	2.418
RPV1Y03	1	*Rhizobium* sp. RPVR24	2.381
RPV1Y04	1	*Rhizobium* sp. RPVR24	2.394
RPV1Y05	1	*Rhizobium leguminosarum* RPVF18	2.173
RPV1Y06	1	*Rhizobium* sp. RPVR24	2.415
RPV1Y08	1	*Rhizobium* sp. RPVR24	2.455
RPV1Y09	1	*Rhizobium* sp. RPVR24	2.347
RPV1Y11	1	*Rhizobium* sp. RPVR24	2.401
RPV1Y12	1	*Rhizobium* sp. RPVR24	2.259
RPV1Y15	1	*Rhizobium* sp. RPVR24	2.311
RPV1Y16	1	*Rhizobium* sp. RPVR24	2.35
RPV1Y18	1	*Rhizobium* sp. RPVR24	2.385
RPV1Y19	1	*Rhizobium* sp. RPVR24	2.287
RPV1Y20	1	*Rhizobium* sp. RPVR24	2.419
RPV1Y21	1	*Rhizobium* sp. RPVR24	2.303
RPV1Y22	1	*Rhizobium* sp. RPVR24	2.369
RPV1Y23	1	*Rhizobium leguminosarum* RPVF18	2.243
RPV1Y24	1	*Rhizobium leguminosarum* RPVF18	2.229
RPV1Y25	1	*Rhizobium* sp. RPVR24	2.376
RPV1Y26	1	*Rhizobium* sp. RPVR24	2.342
RPV5Y01	2	*Rhizobium* sp. RPVR32	2.227
RPV5Y03	2	*Rhizobium* sp. RPVR32	2.247
RPV5Y04	2	*Rhizobium* sp. RPVR32	2.17
RPV5Y05	2	*Rhizobium* sp. RPVR32	2.178
RPV5Y06	2	*Rhizobium* sp. RPVR32	2.151
RPV5Y07	2	*Rhizobium* sp. RPVR24	2.328
RPV5Y08	2	*Rhizobium* sp. RPVR32	2.258
RPV5Y09	2	*Rhizobium* sp. RPVR32	2.358
RPV5Y10	2	*Rhizobium* sp. RPVR32	2.208
RPV5Y11	2	*Rhizobium* sp. RPVR32	2.249
RPV5Y12	2	*Rhizobium* sp. RPVR32	2.256
RPV5Y13	2	*Rhizobium* sp. RPVR32	2.449
RPV5Y15	2	*Rhizobium* sp. RPVR32	2.325
RPV5Y16	2	*Rhizobium* sp. RPVR32	2.144
RPV5Y17	2	*Rhizobium* sp. RPVR32	2.325
RPV5Y18	2	*Rhizobium* sp. RPVR24	2.373
RPV5Y19	2	*Rhizobium* sp. RPVR32	2.352
RPV5Y20	2	*Rhizobium* sp. RPVR32	2.345
RPV5Y21	2	*Rhizobium* sp. RPVR32	2.265

(continued)

Table 7.1 (continued)

Strain	Soil	Organism (best match)	Score value
RPV5Y22	2	*Rhizobium* sp. RPVR32	2.228
RPV5Y23	2	*Rhizobium* sp. RPVR32	2.235
RPV5Y25	2	*Rhizobium* sp. RPVR32	2.329

cultivation, during the last 20 years, being this legume the predominant crop. The strains analyzed by García-Fraile et al. (2010) were isolated from this soil.

7.2.2 Sample Preparation for MALDI-TOF MS

Strains were cultivated on TY medium (Beringer 1974) for 24 h. As indicated by the manufacturer, we used amounts of biomass between 5 and 100 mg to be transferred from the plate to a 1.5-ml tube (Eppendorf, Germany) with a pipette tip and mixed thoroughly in 300 μl of water to resuspend the bacterial cells. Then, 900 μl of absolute ethanol was added and the mixture was centrifuged at 15,500*g* for 2 min, and the supernatant was discarded. The pellet was air-dried at room temperature for 1 h. Subsequently, 50 μl of formic acid (70 % v/v) was added to the pellet and mixed thoroughly before the addition of 50 μl acetonitrile to the mixture. The mixture was centrifuged again at 15,500*g* for 2 min. One microliter of the supernatant was placed onto a spot of the steel target and air-dried at room temperature. Each sample was overlaid with 1 μl of matrix consisting of a saturated solution of alpha-cyano-4-hydroxycinnamic acid in 50 % acetonitrile and 2.5 % trifluoroacetic acid, and then air-dried.

7.2.3 MALDI-TOF MS Performing and Data Analysis

The sample preparation and the MALDI-TOF MS performing were carried out as previously published (Ferreira et al. 2011). The calibration masses were the Bruker Bacterial Test Standards (BTS) which were as follows (masses as averages): RL36, 4365.3 Da; RS22, 5096.8 Da; RL34, 5381.4 Da; RL33meth, 6255.4 Da; RL29, 7274.5 Da; RS19, 10300.1 Da; RNase A, 13683.2 Da and myoglobin, 16952.3 Da. MALDI-TOF MS identifications were classified using the score values proposed by the manufacturer: A score value between 2.300 and 3.000 indicated species identification; a score value between 2.000 and 2.299 indicated genus identification and possible species identification; a score value between 1.700 and 1.999 indicated genus identification; and a score value <1.7 indicated no identification.

To add the strains from soil 3 to the reference library, 36 independent spectra were recorded for each of the strains (three independent measurements at twelve different spots each). Manual/visual estimation of the mass spectra was performed using Flex Analysis 3.0 (Bruker Daltonics GmbH, Germany) performing smoothing and baseline substraction. Checking existence of flatlines, outliers, or single spectra with remarkable peaks differing from the other spectra was done, taking into account that mass deviation within the spectra set shall not be more than 500 ppm. Finally, 20 spectra were selected, removing questionable spectra from the collection. To create peak lists of the spectra, the BioTyper software (Bruker Daltonics GmbH, Germany) was used as described above. The 20 independent peak lists of a strain were used for automated "main spectrum" generation with default settings of the BioTyper software. Thereby, for each library entry, a reference peak list (main spectrum) which contains information about averaged masses, averaged intensities, and relative abundances in the 20 measurements for all characteristic peaks of a given strain was created, so a main spectrum displayed the most reproducible peaks typical for a certain bacterial strain.

Cluster analysis was performed based on comparison of strain-specific main spectra created as described above. The dendrogram was constructed by the statistical toolbox of Matlab 7.1 (MathWorks Inc., USA) integrated into the MALDI Biotyper 2.0 software. The parameter settings were: "Distance Measure = Euclidian" and "Linkage = complete." The linkage function is normalized according to the distance between 0 (perfect match) and 1000 (no match).

7.2.4 Phylogenetic Analysis of recA and atpD Genes

The sequences of these genes for the strains isolated in soil 3 were obtained in a previous work (García-Fraile et al. 2010), but as several new species have been recently described in the phylogenetic group of *R. leguminosarum* we performed again the phylogenetic analysis of the *recA* and *atpD* genes of these strains. Sequences were aligned using the Custal X software (Thompson et al. 1997). The distances were calculated according to Kimura's two-parameter model (Kimura 1980). A phylogenetic tree was inferred using the neighbor-joining analysis (Saitou and Nei 1987). MEGA5 software (Tamura et al. 2011) was used for all analyses.

7.3 Results and Discussion

7.3.1 Phylogenetic Analysis of recA and atpD Genes

The strains isolated in a previous work in the soil 3 (García-Fraile et al. 2010) formed *recA* and *atpD* independent lineages (Fig. 7.1). The strains RPVR04, RPVR32, RPVN06, and RPVR24 do not coincide with none described species from genus *Rhizobium*. The strains RPVF18 and RPVR24 belong, respectively, to the groups I and II reported by Mulas et al. (2011) within *R. leguminosarum*. The group I also contains the type strain of this species USDA 2370^{T} and that of *R.*

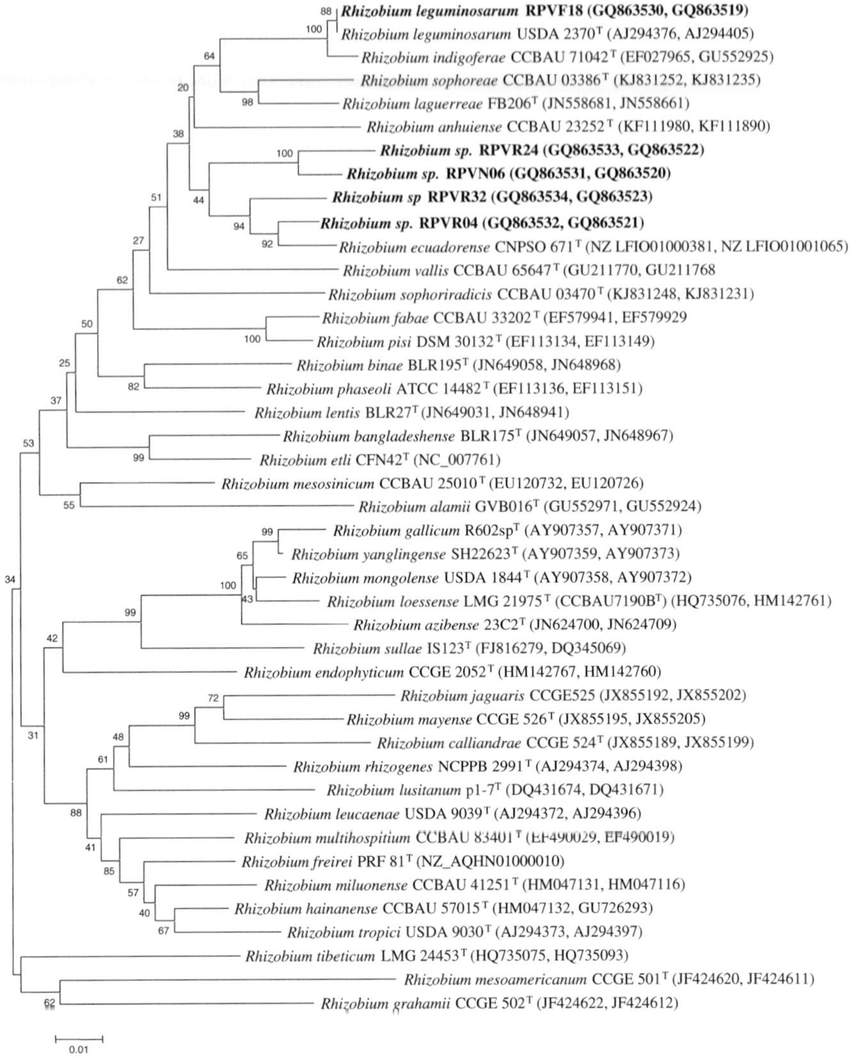

Fig. 7.1 Neighbor-joining phylogenetic-rooted tree based on concatenated *recA* and *atpD* gene sequences (about 1475 nt) showing the taxonomic location of the strains isolated from *P. vulgaris* nodules in the soil 3 that have been included in the MALDI-TOF MS database of family *Rhizobiaceae*. Bootstrap values calculated for 1000 replications are indicated. Bar, 1 nt substitution per 100 nt. Accession numbers from Genbank are given in *brackets*

indigoferae CCBAU 71042^{T} which are synonyms. The recently described species *R. laguerreae* (Saïdi et al. 2014), *R. sophorae* (Jiao et al. 2015), and *R. anhuiense* (Zhang et al. 2015), whose type strains have identical *rrs* genes, belong to divergent phylogenetic lineages in the *recA* and *atpD* genes phylogenetic tree (Fig. 7.1). Considering the phylogenetic distances found in the case of strains RPVR32 and

RPVR24 with respect to the currently described species within the phylogenetic group of *R. leguminosarum*, these strains probably constitute two undescribed species within genus *Rhizobium*.

7.3.2 MALDI-TOF MS Analysis

The MALDI-TOF MS spectra from strains RPVF18, RPVR04, RPVR32, RPVN06, and RPVR24 were different (Fig. 7.2) and, except in the case of the strain RPVF18, they did not match with any species from the *Rhizobiaceae* database with score values higher than 2.0. Two of these strains, RPVF18 and RPVR32, were previously included in this database (Ferreira et al. 2011) and the remaining ones have now been added to it, in order to compare the strains isolated in this work with these strains isolated in the soil 3 located in the same geographical area and cultivated for 20 years with *P. vulgaris*.

The results obtained after MALDI-TOF MS analysis (Table 7.1) showed that the strains isolated from soil 1 matched with score values higher than 2.0 with the strains RPFV18 and RPVR24, belonging to the groups I and II of the species *R. leguminosarum*, respectively (Mulas et al. 2011). In this soil were more abundant (85 %) the strains matching with the strain RPVR24 (group II) than those matching with the strain RPFV18 (group I).

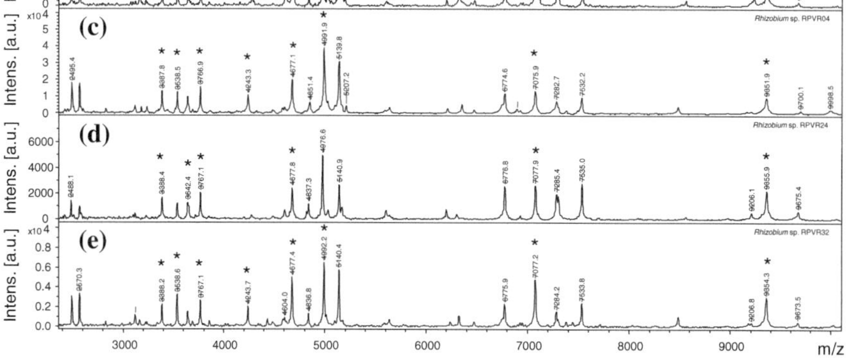

Fig. 7.2 MALDI-TOF MS spectra of whole-cell extracts obtained from the strains isolated in the soil 3: **a** RPVF18, **b** RPVN06, **c** RPVR04, **d** RPVR24, **e** RPVR32

In the soil 2, only 10 % of strains matched with the strain RPVR24 (group II), whereas most of strains (90 %) isolated from this soil matched with score values higher than 2.0 with the strain RPVR32, also isolated in the soil 3, that belonged to a different phylogenetic group related to *R. leguminosarum* (García-Fraile et al. 2010) and was not found in the soils analyzed by Mulas et al. (2011).

The mathematical analysis of spectra showed three clusters, A, B and C, each one containing the strains RPVF18, RPVR24, and RPVR32 that also belong to different *recA* and *atpD* gene phylogenetic groups (Fig. 7.3). The strains isolated in

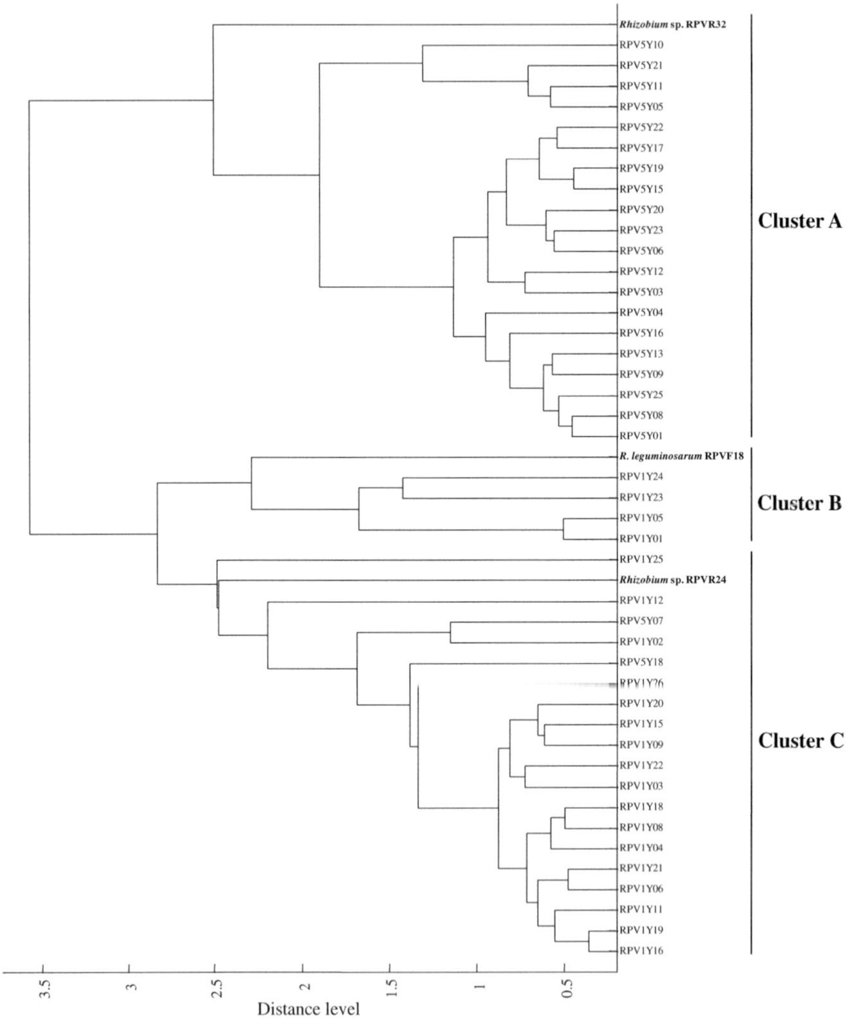

Fig. 7.3 Cluster analysis of MALDI-TOF MS spectra of the strains isolated in soils 1 and 2 compared with their closely related from soil 3. Distance is displayed in relative units

this work were distributed in these three clusters in agreement with the results of the MALDI-TOF MS analysis, nevertheless the internal similarity values found in these three clusters indicated some differences at strain level.

These differences are seen in the spectra of the strains belonging to each one of the three clusters as shown in Figs. 7.4, 7.5, and 7.6 for the strains from clusters A, B, and C, respectively.

Therefore from the results of this work, we can conclude that MALDI-TOF MS is not only a reliable technique for identification at species level of fast-growing rhizobia, but also for differentiation of subspecific groups. This implies that this technique can be used for the analysis of large populations ensuring the identification at genus, species, and strain levels, which is very useful to analyze the distribution of rhizobial strains in different ecosystems.

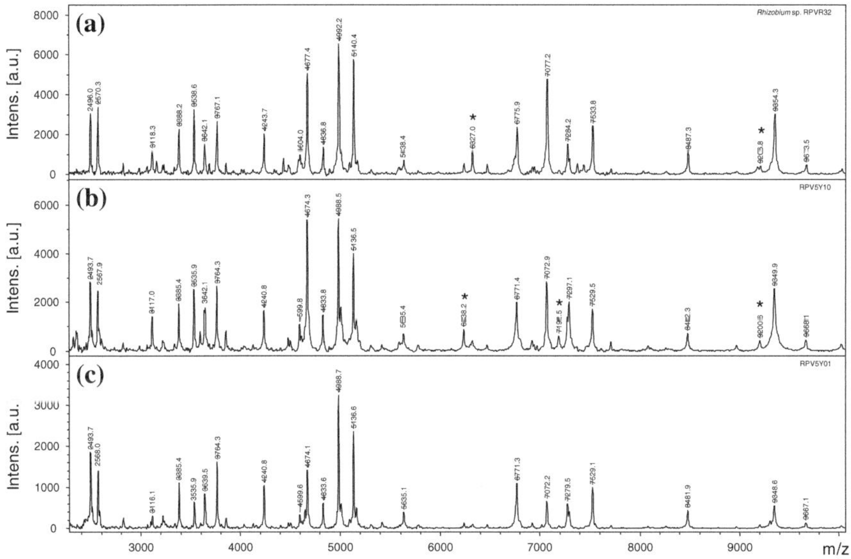

Fig. 7.4 MALDI-TOF MS spectra of whole-cell extracts obtained from representative strains of MALDI-TOF MS cluster A: **a** RPVR32, **b** RPV5Y10, **c** RPV5Y01. Peaks marked with *asterisks* are not found in all strains

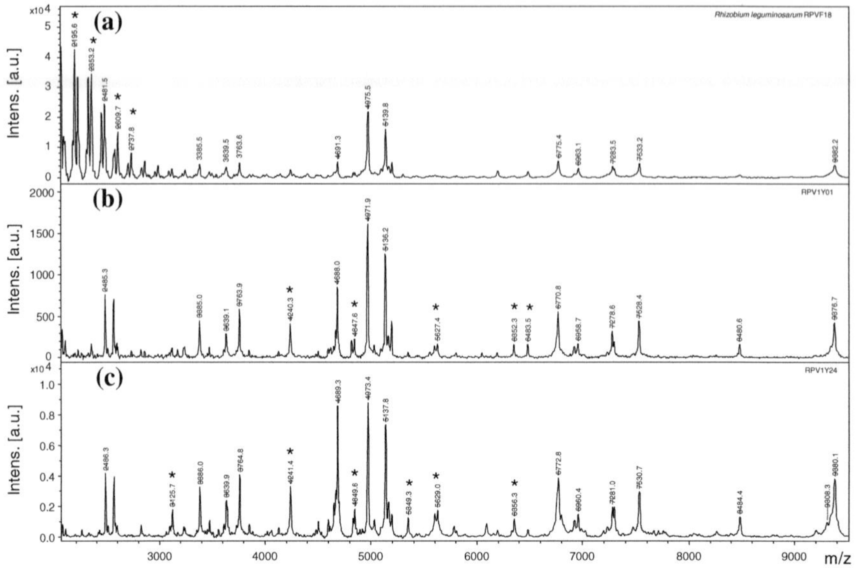

Fig. 7.5 MALDI-TOF MS spectra of whole-cell extracts obtained from representative strains of MALDI-TOF MS cluster B: **a** RPVF18, **b** RPV1Y01, **c** RPV1Y24. Peaks marked with *asterisks* are not found in all strains

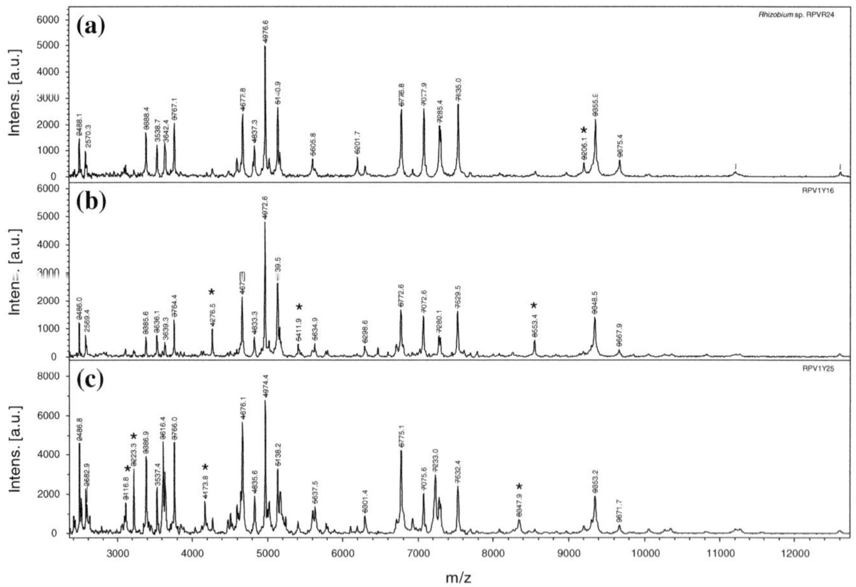

Fig. 7.6 MALDI-TOF MS spectra of whole-cell extracts obtained from representative strains of MALDI-TOF MS cluster C: **a** RPVR24, **b** RPV1Y16, **c** RPV1Y25. Peaks marked with *asterisks* are not found in all strains

References

Beringer JE (1974) R factors transfer in *Rhizobium leguminosarum*. J Gen Microbiol 84:188–198

Ferreira L, Sánchez-Juanes F, García-Fraile P et al (2011) MALDI-TOF mass spectrometry is a fast and reliable platform for identification and ecological studies of species from family *Rhizobiaceae*. PLoS ONE 6:e20223

García-Fraile P, Mulas-García D, Peix A et al (2010) *Phaseolus vulgaris* is nodulated in northern Spain by *Rhizobium leguminosarum* strains harboring two *nodC* alleles present in American *Rhizobium etli* strains: biogeographical and evolutionary implications. Can J Microbiol 56:657–666

Jiao YS, Yan H, Ji ZJ et al (2015) *Rhizobium sophorae* sp. nov. and *Rhizobium sophoriradicis* sp. nov., nitrogen-fixing rhizobial symbionts of the medicinal legume *Sophora flavescens*. Int J Syst Evol Microbiol 65:497–503

Kimura M (1980) A simple method for estimating evolutionary rates of base substitutions through comparative studies of nucleotide sequences. J Mol Evol 16:111–120

Mulas D, García-Fraile P, Carro L et al (2011) Distribution and efficiency of *Rhizobium leguminosarum* strains nodulating *Phaseolus vulgaris* in Northern Spanish soils: selection of native strains that replace conventional N fertilization. Soil Biol Biochem 43:2283–2293

Peix A, Ramírez-Bahena MH, Velázquez E et al (2015) Bacterial associations with legumes. Crit Rev Plant Sci 34:17–42

Saïdi S, Ramírez-Bahena MH, Santillana N et al (2014) *Rhizobium laguerreae* sp. nov. nodulates *Vicia faba* on several continents. Int J Syst Evol Microbiol 64:242–247

Saitou N, Nei M (1987) The neighbor-joining method: a new method for reconstructing phylogenetic trees. Mol Biol Evol 4:406–425

Tamura K, Peterson D, Peterson N et al (2011) MEGA5: molecular evolutionary genetics analysis using maximum likelihood, evolutionary distance, and maximum parsimony methods. Mol Biol Evol 28:2731–2739

Thompson JD, Gibson TJ, Plewniak F et al (1997) The clustal_X windows interface: flexible strategies for multiple sequence alignment aided by quality analysis tools. Nucleic Acid Res 25:4876–4882

Vincent JM (1970) The cultivation, isolation and maintenance of rhizobia. In: Vincent JM (ed) A manual for the practical study of root-nodule. Blackwell Scientific Publications, Oxford, pp 1–13

Zhang YJ, Zheng WT et al (2015) *Rhizobium anhuiense* sp. nov., isolated from effective nodules of *Vicia faba* and *Pisum sativum*. Int J Syst Evol Microbiol 65:2960–2967

Part III
Plant-Microorganisms Interactions

Chapter 8
Optimization of the Immunodetection of Tagged-LYK3 Nod Factor Receptor in *Medicago truncatula* Root Extracts

Maria Magdalini Chroni and Estíbaliz Larrainzar

Abstract One of the mechanisms mediating the legume–rhizobium symbiosis is the molecular recognition at the plasma membrane of signaling molecules called Nod factors. In the model legume *Medicago truncatula*, two Nod factor receptor kinases have been described: NFP and LYK3. These plasma membrane-localized proteins are present at very low amounts in root tissues, and even using antibodies specific against them, detection remains challenging. In the current work, we tested several methods for the extraction, enrichment, and immunodetection of LYK3 in root extracts using *M. truncatula* lines expressing a tagged version of LYK3, LYK3-HAST. Upon differential centrifugation, a significant fraction of LYK3 was found to precipitate at speed at which plasma membrane markers remain soluble. The sedimentation pattern of LYK3 was similar to that of markers for the endoplasmic reticulum (ER) and vacuole, suggesting that LYK3 may be also present in larger endomembrane structures. Affinity columns were shown to be effective for the enrichment of the protein, leading to an increase in the signal detected. Interestingly, LYK3 (or at least the tagged N-terminal fraction of the protein) was reproducibly identified at a range of molecular sizes, which are likely to correspond to the monomer, dimer, and a truncated form. This observation opens up the possibility of the existence of post-translational regulation mechanisms (e.g., cleavage) with likely implications in signaling.

Keywords Nod factor receptor · LYK3 · Immunodetection

M.M. Chroni · E. Larrainzar (✉)
Departamento de Ciencias del Medio Natural,
Universidad Pública de Navarra, Campus de Arrosadía,
31006 Pamplona, Spain
e-mail: estibaliz.larrainzar@unavarra.es

© Springer International Publishing Switzerland 2016

F. González-Andrés and E. James (eds.), *Biological Nitrogen Fixation and Beneficial Plant–Microbe Interactions*, DOI 10.1007/978-3-319-32528-6_8

8.1 Introduction

Legume plants are able to establish nitrogen-fixing endosymbiosis with rhizobium soil bacteria. The cultivation of legumes is associated with improvements of soil quality, and it allows the reduction of the use of mineral nitrogen fertilizers, being a key element in sustainable agricultural systems worldwide. The mutualistic legume–rhizobium symbiosis is based on a finely tuned molecular dialog between the host plant and the microsymbiont; plant root secretes flavonoid compounds that are perceived by the bacteria, inducing the biosynthesis of lipo-chitooligosaccharides called Nod factors. These Nod factors are, in turn, perceived via lysine motif (LysM) receptor kinases at the plasma membrane of roots, activating the Nod factor-signaling pathway (Madsen et al. 2010; Gough and Cullimore 2011; Venkateshwaran et al. 2013).

By mutant screening, some genes encoding putative Nod factor receptors and downstream signaling genes have been identified in the model legumes *Medicago truncatula* and *Lotus japonicus*. Among them, LYK3 and NFP have been identified in *M. truncatula* (Limpens et al. 2003; Arrighi et al. 2006), and NFR1 and NFR5 in *L. japonicus* (Madsen et al. 2003; Radutoiu et al. 2003). These genes encode transmembrane receptor-like proteins containing an extracellular region with two or three Lys motifs, which have been implicated in the molecular binding of Nod factors (Radutoiu et al. 2007). In *L. japonicus*, both NFR1 and NFR5 are essential for the initiation of responses upon Nod factor recognition and a heterodimer of these proteins is involved in Nod factor perception (Geurts et al. 2005; Broghammer et al. 2012; Moling et al. 2014). In contrast, in *M. truncatula*, only NFP is essential for all known responses to Nod factors (Amor et al. 2003), while LYK3 (and to a minor extent LYK4) has been implicated in the induction of curling of root hairs and initiation and polar growth of infection threads in a Nod factor structure-dependent fashion (Smit et al. 2007). More recently, related receptor-like kinases have been also identified in these early symbiotic stages. This is the case for *M. truncatula* LYR3, which is able to bind Nod factors (Fliegmann et al. 2013) and *L. japonicus* EPS, which controls bacterial passage through the root epidermis based on the specific binding of bacterial exopolysaccharides (Kawaharada et al. 2015).

One of the main limitations for the biochemical characterization of these LysM receptor-like kinases is that they are plasma membrane-localized proteins expressed at very low levels in legumes. Localization studies done under the regulation of native promoters have failed to detect the expressed proteins in plant tissues (Lefebvre et al. 2012), requiring high-resolution microscopy methods (e.g., spinning disk confocal microscopy) for in vivo detection (Haney et al. 2011; Moling et al. 2014). To overcome this limitation, researchers have expressed these receptors in heterologous systems (Lefebvre et al. 2010; Broghammer et al. 2012), thus avoiding the tight regulation of expression they are subjected to in legumes. However, for the identification of protein interactions and post-translational modifications, the proteins need to be analyzed in their native systems.

The development of transgenic plants expressing these receptor-like kinases under their native promoters represents a major step toward their biochemical characterization. This is the case for *M. truncatula* LYK3-HAST, a line that expresses a genomic copy of LYK3 fused to a dual-affinity tag containing three copies of the hemagglutinin epitope (HA) and a single StrepII tag complementing the LYK3 background mutation (Haney et al. 2011). This line allows the immunodetection of LYK3 in root extracts using commercial anti-HA antibodies. In the current work, we tested different extraction methods and affinity enrichment conditions to optimize immunodetection of LYK3 in *M. truncatula* in root extracts.

8.2 Materials and Methods

8.2.1 *Plant Growth*

M. truncatula plants were grown under controlled-environment conditions (14-h day/10-h night; 400 μmol m^{-2} s^{-1} light intensity; 22/16 °C day/night temperature; 70–60 % relative humidity) using a nutrient solution (Evans and Moore 1981) supplemented with 5 mM ammonium nitrate. Eight weeks after planting, root tissue was collected, weighed, flash-frozen, and stored at −80 °C for protein extraction.

8.2.2 *Protein Extraction and Affinity Purification*

Root tissue samples (approx. 10 g fresh weight) were precut and subsequently homogenized in 40 ml extraction buffer using an Ultra-Turrax T-25 homogenizer (Ika; two to three 30-s cycles on ice). Three extraction buffers were tested to maximize protein yield: buffer 1 contained 50 mM HEPES (pH 7.9), 300 mM sucrose, 150 mM NaCl, 10 mM potassium acetate, 10 mM dithiothreitol (DTT), 10 mM 2-mercaptoethanol, and 5 mM EDTA (Roppolo et al. 2011); buffer 2 contained 50 mM HEPES (pH 7.5), 150 mM NaCl, 10 mM EDTA, 300 mM sucrose, and 2 mM DTT (Antolín-Llovera et al. 2014); buffer 3 contained 50 mM MOPS (pH 7), 10 mM DTT, 10 mM 2-mercaptoethanol, 1 mM EDTA, 20 mM KCl, and 5 mM $MgCl_2$ (González et al. 1995). Buffers were supplemented with 2 % (w/v) polyvinylpolypyrrolidone, 1 % (v/v) protease inhibitors cocktail (Roche), 1 mM PMSF, and 50 mM NaF.

To remove insoluble materials, homogenates were filtered through one layer of Miracloth tissue and then centrifuged at speed ranging from 1000 to 13,000 *g* at 4 °C. Then, homogenates were ultracentrifuged for 30 min at 100,000 *g* to obtain the microsomal root fraction. Microsomal pellets were washed three times in extraction buffer, resuspended in 1 ml low-salt buffer (50 mM Tris-HCl, pH 8.0) supplemented with 1 % (v/v) Triton X-100, and solubilized for 30 min at 4 °C on a rotating wheel. To remove insoluble material and avoid column-clogging issues, the solubilized microsomal fractions were again centrifuged at 2000 *g* 4 °C for 10 min. Extracts were then incubated with 20 μl anti-HA antibodies (rat monoclonal, Roche) for 5 min at 4 °C. Subsequently, 50 μl of the μMACS anti-protein G microbead solution (Miltenyi Biotec) was applied to magnetically label the LYK3-HAST protein complexes, and fractions were incubated for further 30 min on ice. Affinity column separation was carried out following the manufacturer's instructions.

To monitor LYK3-HAST protein loss during the extraction process, 20 μl aliquots of the different fractions were collected and immunoanalyzed as described in the next section.

8.2.3 SDS-PAGE and Immunoblot Analysis

Protein samples were separated by SDS-PAGE on 10 % (w/v) polyacrylamide gels (120 V for 1 h). Gels were transferred onto nitrocellulose membranes (100 V for 75 min) and blocked with 5 % (w/v) non-fat milk powder in Tris-buffered saline (TBS). After three washes with 0.1 % (v/v) Tween-20–TBS, membranes were incubated with the following antibodies: anti-HA conjugated with horseradish peroxidase (HRP; Roche, 1:1000 dilution), anti-PIP2-7 as a marker for the plasma membrane fraction (Agrisera, 1:2000), anti-BiP (lumenal binding protein) as a marker for the ER (Agrisera, 1:2000), and anti-V-ATPase (epsilon subunit of tonoplast H+ ATPase) as a vacuolar marker (Agrisera, 1:2000). Both colorimetric and chemiluminescence-based methods were tested for optimized immunodetection of LYK3-HAST. Colorimetric detection was based on alkaline phosphatase (AP) reactions and was carried out by incubating membranes with secondary antibodies conjugated with AP (Sigma, 1:20,000). Cross-reacting bands were visualized using 5-bromo-4-chloro-3-indolyl phosphate and nitroblue tetrazolium (NBT/BCIP, Sigma). Chemiluminescence detection was performed by incubating membranes with secondary antibodies conjugated with HRP (1:100,000) using ECL plus (Amersham) as substrate.

8.3 Results and Discussion

8.3.1 LYK3-HAST Sediments at Low Centrifugal Forces

Using tagged LYK3-GFP lines, Haney et al. (2011) showed that the majority of the LYK3 signal was associated with the plasma membrane (PM). However, we had observed that a significant fraction of LYK3-HAST pelleted down when root

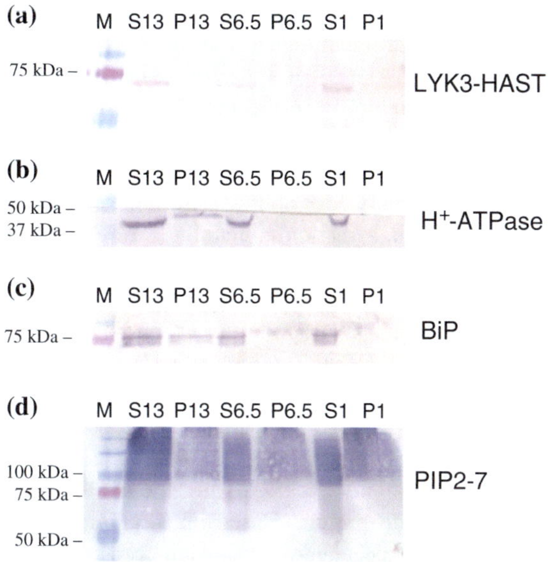

Fig. 8.1 Immunodetection of LYK3-HAST and subcellular markers at the soluble (*S*) and pellet (*P*) fractions during differential centrifugation. *M*, molecular weight markers. *1* represents a centrifugal force of 1000 *g*; *6.5* represents 6500 *g*; *13* represents 13,000 *g*

extracts were subjected to centrifugal forces larger than 2000 *g* (Larrainzar E et al. (unpublished)). Were LYK3-HAST a PM protein, it would be expected to remain in the soluble fraction unless ultracentrifugal forces were applied. To formally test this, we compared the distribution of LYK3-HAST in the soluble and pellet fractions when root extracts were subjected to a 10-min centrifugation at 1000, 6500, or 13,000 *g*. We used antibodies against the ER, vacuole, and plasma membrane/microsomal fraction to compare sedimentation properties of proteins in different subcellular fractions (Fig. 8.1). A band corresponding to LYK3-HAST was detected in the three supernatant fractions. However, the signal in the soluble fraction was most intense at 1000 *g*. At 13,000 *g*, a faint band was observed in the pellet, suggesting that a fraction of LYK3-HAST protein pool was lost (Fig. 8.1a). The vacuole marker presented a similar pattern of sedimentation, with a clear signal in the pellet corresponding to 13,000 *g* (Fig. 8.1b). The ER marker started to sediment at 6500 *g* (Fig. 8.1c), while the marker for the microsomal fraction did not show differences when comparing the signal obtained at the three centrifugal forces (Fig. 8.1d). Thus, LYK3-HAST presents a sedimentation pattern more closely related to vacuolar or ER proteins than to proteins merely located at the PM.

8.3.2 LYK3 Is Detected at Three Molecular Weights in Root Extracts

To optimize the protein extraction and solubilization method, three homogenization buffers were tested (see Materials and Methods). Buffers 1 and 2, based on Roppolo et al. (2011) and Antolín-Llovera et al. (2014), respectively, relied on sucrose and salt to stabilize membranes and maintain ionic strength. Buffer 3, based on González et al. (1995), is typically used for soluble protein extraction and lacks sucrose. Interestingly, we did not observe significant differences in LYK3-HAST protein yield when comparing root extracts obtained in the three buffers (data not shown). Thus, buffer 2 was used for subsequent experiments.

LYK3 is a very low abundant protein in the total root proteome. In order to enrich the protein fraction containing LYK3, we took advantage of the HA tag to concentrate the protein using magnetic beads and affinity columns. By using low centrifugal forces in the first centrifugation and a subsequent ultracentrifugation step, the microsomal fraction corresponding to a total of 10 *g* of roots was concentrated into a 50 μl sample, which was loaded on an SDS-PAGE gel well. Immunodetection using anti-HA antibodies showed bands at three different molecular weights: 130–140 kDa, 70 kDa, and 30–40 kDa, approx (Fig. 8.2). LYK3 has an estimated molecular weight of 68 kDa, and the HAST tag represents an addition of approx. 4 kDa. Thus, the intermediate signal is likely to correspond to the full-length protein. Based on the molecular size observed, we hypothesize that the high molecular weight signal corresponds to LYK3-HAST in the form of dimers. Experimental evidence supporting this observation has been recently found by Moling et al. (2014), who demonstrated that around 14 % of the total LYK3 pool forms dimers (or higher-order complexes) in *M. truncatula* root nodules. Interestingly, a low molecular band was consistently observed at around 30–40 kDa. Although this could correspond to random degradation products, the lower band may be associated with specific cleavage of the protein. Recently, Antolín-Llovera et al. (2014) have shown that the cleavage of a related receptor-like kinase, SYMRK, regulates the interaction of the protein with other receptor kinases, with implications in the regulation of infection thread formation in *L. japonicus*.

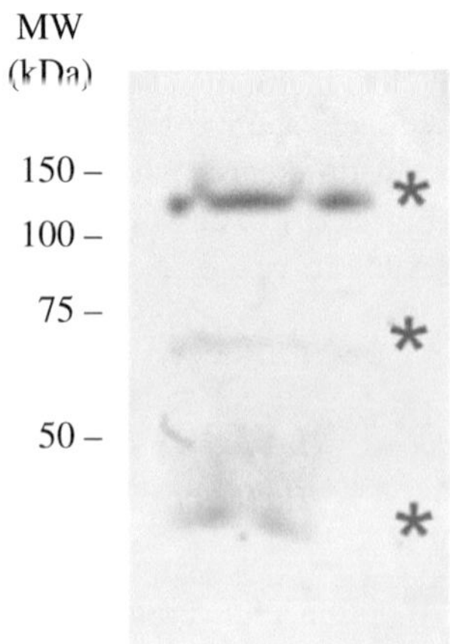

Fig. 8.2 Immunodetection of LYK3-HAST after affinity column enrichment. *Asterisk* mark the three bands cross-reacting with the anti-HA antibody. *MW*, represents molecular weight in kDa

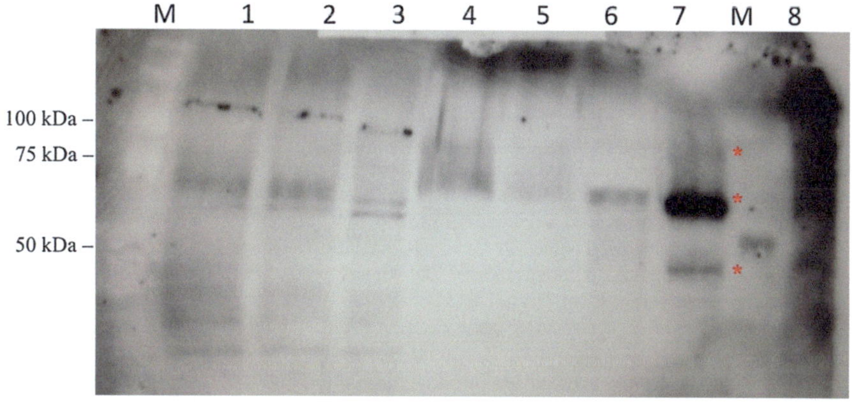

Fig. 8.3 Immunodetection of LYK3-HAST during protein purification from root microsomal fractions. *Lane 1* crude extract; *2* extract after a 2000 *g* centrifugation step; *3* supernatant of the microsomal fraction; *4* and *5* microsomal fraction before and after resuspension in buffer containing 1 % Triton X-100, respectively; *6* unbound fraction; *7* bound fraction to the affinity column; *8* insoluble fraction of the microsomal extract; *M* molecular weight markers in kDa

8.3.3 LYK3 Largely Remains in the Insoluble Fraction When Using 1 % Triton X-100

To keep proteins in their native configuration, microsomal protein solubilization was carried out using mild non-ionic detergents such as Triton X-100. Loading aliquots of the different steps of the immunoenrichment protocol, we identified two main points of protein loss. One is the residual LYK3-HAST fraction that is not bound to the affinity column (Fig. 8.3, lane 6), but the most critical step is microsomal protein solubilization. After incubation in the presence of 1 % Triton X-100, microsomal extracts were subjected to a centrifugation step to remove insoluble material and prevent column clogging. Although three clear bands were observed in the protein extract eluted from the affinity column (Fig. 8.3, lane 7), a large fraction of the LYK3-HAST signal remained in the insoluble fraction and it was only partially solubilized by strong detergents such as SDS (lane 8, Fig. 8.3). Interestingly, the HA signal in this insoluble fraction appeared to cover a range of high molecular weights. This suggests that LYK3 is present in quite a detergent-resistant fraction of the cell (e.g., lipid drafts; Lefebvre et al. 2010) and that it is likely to form high molecular weight complexes with a number of interacting proteins. Further research will be required to determine the nature of these interactions and their implications in Nod factor signaling and symbiotic regulation.

Acknowledgments M.M.C. received funding from the Public University of Navarra, and E.L. is the recipient of a Juan de la Cierva fellowship. This work has been financed by the Marie Curie IOF "SymBioSignal" project, the MINECO AGL2011-30386-C02-01/AGR project, and the Fundación Caja Navarra project CAN2015-70497.

References

Amor BB, Shaw SL, Oldroyd GE, Maillet F, Penmetsa RV, Cook D, Long SR, Dénarié J, Gough C (2003) The NFP locus of *Medicago truncatula* controls an early step of Nod factor signal transduction upstream of a rapid calcium flux and root hair deformation. Plant J 34:495–506

Antolín-Llovera M, Ried MK, Parniske M (2014) Cleavage of the symbiosis receptor-like kinase ectodomain promotes complex formation with Nod factor receptor 5. Curr Biol 24:422–427

Arrighi JF, Barre A, Amor BB, Bersoult A, Soriano LC, Mirabella R, de Carvalho-Niebel F, Journet EP, Ghérardi M, Huguet T, Geurts R, Dénarié J, Rougé P, Gough C (2006) The *Medicago truncatula* lysin motif-receptor-like kinase gene family includes NFP and new nodule-expressed genes. Plant Physiol 142:265–279

Broghammer A, Krusell L, Blaise M, Sauer J, Sullivan JT, Maolanon N, Vinther M, Lorentzen A, Madsen EB, Jensen KJ, Roepstorff P, Thirup S, Ronson CW, Thygesen MB, Stougaard J (2012) Legume receptors perceive the rhizobial lipochitin oligosaccharide signal molecules by direct binding. Proc Natl Acad Sci USA 109:13859–13864

Evans HJ, Moore TC (1981) Symbiotic nitrogen fixation in legume nodules. In: Research experiences in plant physiology. Springer, New York, p 294

Fliegmann J, Canova S, Lachaud C, Uhlenbroich S, Gasciolli V, Pichereaux C, Rossignol M, Rosenberg C, Cumener M, Pitorre D, Lefebvre B, Gough C, Samain E, Fort S, Driguez H, Vauzeilles B, Beau JM, Nurisso A, Imberty A, Cullimore J, Bono JJ (2013) Lipo-chitooligosaccharidic symbiotic signals are recognized by LysM receptor-like kinase LYR3 in the legume *Medicago truncatula*. ACS Chem Biol 8:1900–1906

Geurts R, Fedorova E, Bisseling T (2005) Nod factor signaling genes and their function in the early stages of *Rhizobium* infection. Curr Opin Plant Biol 8:346–352

González EM, Gordon AJ, James CL, Arrese-lgor C (1995) The role of sucrose synthase in the response of soybean nodules to drought. J Exp Bot 46:1515–1523

Gough C, Cullimore J (2011) Lipo-chitooligosaccharide signaling in endosymbiotic plant-microbe interactions. Mol Plant Microbe Interact 24:867–878

Haney CH, Riely BK, Tricoli DM, Cook DR, Ehrhardt DW, Long SR (2011) Symbiotic rhizobia bacteria trigger a change in localization and dynamics of the *Medicago truncatula* receptor kinase LYK3. Plant Cell 23:2774–2787

Kawaharada Y, Kelly S, Nielsen MW, Hjuler CT, Gysel K, Muszyński A, Carlson RW, Thygesen MB, Sandal N, Asmussen MH, Vinther M, Andersen SU, Krusell L, Thirup S, Jensen KJ, Ronson CW, Blaise M, Radutoiu S, Stougaard J (2015) Receptor-mediated exopolysaccharide perception controls bacterial infection. Nature 523:308–312

Lefebvre B, Timmers T, Mbengue M, Moreau S, Hervé C, Tóth K, Bittencourt-Silvestre J, Klaus D, Deslandes L, Godiard L, Murray JD, Udvardi MK, Raffaele S, Mongrand S, Cullimore J, Gamas P, Niebel A, Ott T (2010) A remorin protein interacts with symbiotic receptors and regulates bacterial infection. Proc Natl Acad Sci USA 107:2343–2348

Lefebvre B, Klaus-Heisen D, Pietraszewska-Bogiel A, Hervé C, Camut S, Auriac MC, Gasciolli V, Nurisso A, Gadella TW, Cullimore J (2012) Role of N-glycosylation sites and CXC motifs in trafficking of *Medicago truncatula* Nod factor perception protein to plasma membrane. J Biol Chem 287:10812–10823

Limpens E, Franken C, Smit P, Willemse J, Bisseling T, Geurts R (2003) LysM domain receptor kinases regulating rhizobial Nod factor-induced infection. Science 302:630–633

Madsen EB, Madsen LH, Radutoiu S, Olbryt M, Rakwalska M, Szczyglowski K, Sato S, Kaneko T, Tabata S, Sandal N, Stougaard J (2003) A receptor kinase gene of the LysM type is involved in legume perception of rhizobial signals. Nature 425:637–640

Madsen LH, Tirichine L, Jurkiewicz A, Sullivan JT, Heckmann AB, Bek AS, Ronson CW, James EK, Stougaard J (2010) The molecular network governing nodule organogenesis and infection in the model legume *Lotus japonicus*. Nat Commun 1:10

Moling S, Pietraszewska-Bogiel A, Postma M, Fedorova E, Hink MA, Limpens E, Gadella TW, Bisseling T (2014) Nod factor receptors form heteromeric complexes and are essential for intracellular infection in Medicago nodules. Plant Cell 26:4188–4199

Radutoiu S, Madsen LH, Madsen EB, Felle HH, Umehara Y, Gronlund M, Sato S, Nakamura Y, Tabata S, Sandal N, Stougaard J (2003) Plant recognition of symbiotic bacteria requires two LysM receptor-like kinases. Nature 425:585–592

Radutoiu S, Madsen LH, Madsen EB, Jurkiewicz A, Fukai E, Quistgaard EM, Albrektsen AS, James EK, Thirup S, Stougaard J (2007) LysM domains mediate lipochitin-oligosaccharide recognition and NFR genes extend the symbiotic host range. EMBO J 26:3923–3935

Roppolo D, De Rybel B, Tendon VD, Pfister A, Alassimone J, Vermeer JE, Yamazaki M, Stierhof YD, Beeckman T, Geldner N (2011) A novel protein family mediates Casparian strip formation in the endodermis. Nature 473:380–383

Smit P, Limpens E, Geurts R, Fedorova E, Dolgikh E, Gough C, Bisseling T (2007) Medicago LYK3, an entry receptor in rhizobial nodulation factor signaling. Plant Physiol 145:183–191

Venkateshwaran M, Volkening JD, Sussman MR, Ané JM (2013) Symbiosis and the social network of higher plants. Curr Opin Plant Biol 16:118–127

Chapter 9
Functional Characterization of a Chitinase Class III (CgCHI3) and a Glutathione S-Transferase (CgGST) Involved in *Casuarina glauca–Frankia* Symbiosis

Inês Graça, Márcia Guilherme, Pedro Tavares, Alice S. Pereira and Ana I. Ribeiro-Barros

Abstract Actinorhizal plants are key elements of natural ecosystems. Besides their high adaptability to extreme harsh environments, most actinorhizal plants have the capacity to obtain high levels of nitrogen due to their ability to establish symbiosis at the root level with nitrogen-fixing bacteria of the genus *Frankia*. Symbiosis is an ontogenic process that requires a highly coordinated sequence of events. One of such mechanisms is the induction of defense-related genes, whose main role during the symbiotic interaction remains to be elucidated. In this paper, we discuss the putative function of two defense-related proteins involved in the actinorhizal symbiosis established between *Casuarina glauca* and *Frankia*, namely a class III chitinase (CgCHI3) and a glutathione S-transferase (CgGST).

Keywords *Casuarina glauca* · *Frankia* · Root nodule symbiosis · Chitinase class III · Tau glutathione S-transferase

I. Graça · A.I. Ribeiro-Barros
LEAF - Linking Landscape, Environment, Agriculture and Food, PlantStress&Biodiversity Group, University of Lisbon, 1349-017 Lisbon, Portugal

A.I. Ribeiro-Barros
Instituto de Tecnologia Química e Biológica, Universidade Nova de Lisboa, 2780-157 Oeiras, Portugal

I. Graça (✉) · M. Guilherme · P. Tavares · A.S. Pereira
UCIBIO, REQUIMTE, Departamento de Química, Faculdade de Ciências e Tecnologia, Universidade Nova de Lisboa, 2829-516 Caparica, Portugal
e-mail: inesgraca@hotmail.com

© Springer International Publishing Switzerland 2016

F. González-Andrés and E. James (eds.), *Biological Nitrogen Fixation and Beneficial Plant–Microbe Interactions*, DOI 10.1007/978-3-319-32528-6_9

9.1 Introduction

Actinorhizal plants are a group of perennial dicotyledonous angiosperms distributed over eight different families and 25 genera. These plants exhibit a great resilience to extreme environments, being able to thrive under a variety of environmental stresses such as salinity, pH, or heavy metals (Dawson 1990). As such, they are of utmost environmental importance and widely used in agroforestry, coastal restoration, or prevention of desertification (Diem and Dommergues 1990; Dawson 2008). The environmental importance of actinorhizal plants is enhanced by their ability to establish nitrogen-fixing symbiosis with bacteria of the genus *Frankia*. Nodulation is a highly complex and coordinated process that involves the induction and suppression of the expression of several sets of genes and their respective products. Interestingly, there are aspects of N_2-fixing root nodule symbiosis that resemble pathogenic interactions as well as general stress responses. Indeed, it has been proposed that nodulation has evolved from a pathogenic interaction (Sprent 2007). Events such as localized cell death, production of antimicrobial compounds, and the induction of host stress and defense-related genes/proteins have been reported by several groups (Spaink 1995; Samac and Graham 2007; Tavares et al. 2007; Santos et al. 2010; Ribeiro et al. 2011). Additionally, analyses of the nodule transcriptome indicate that the percentage of genes belonging to the cell defense/resistance/rescue category is ca. 5–15 % (Ramírez et al. 2005; Lohar et al. 2006; Hocher et al. 2011). These observations suggest a significant role in defense signaling in the events leading to nodulation. Moreover, stress-like events may be related to the control of microsymbiont infection, nodule number, and host cells that accommodate the microsymbiont, as well as to nodule protection against external pathogens (Salzer et al. 2000; Kim and An 2002; Fortunato et al. 2007; Samac and Graham 2007; Tavares et al. 2007; Soto et al. 2009; Santos et al. 2008, 2010).

Transcriptome analysis in the actinorhizal plant *Casuarina glauca* indicates that around 2000 unigenes ($\sim$14 %) are upregulated or specifically induced during nodule development. This analysis also revealed that the majority of these genes codes for proteins involved in transport, metabolism, protein synthesis machinery, cell wall, defense, and response to stress (Hocher et al. 2011). Laplaze et al. (1999) identified a key gene, *CgChs* (encoding a chalcone synthase), whose expression was upregulated in *C. glauca* nodules. This gene is most likely part of the host defense response against *Frankia* through the restriction of endosymbiont invasion. Another important group of genes is that encoding two enzymes involved in reactive oxygen species (ROS) scavenging, *CgMt1* (class I type 1 metallothionein-like protein; Laplaze et al. 2002) and *CgGst* (glutathione S-transferase; Santos et al. 2010). The activation of these genes has been attributed to ROS detoxification (Obertello et al. 2007; Santos et al. 2010). A class I chitinase gene (*CgChi1*) has been implied in nodule defense against external pathogens (Kim et al. 2005; Santos et al. 2010). Svistoonoff et al. (2003) characterized a subtilisin-like protease gene

Table 9.1 *Casuarina glauca* defense genes and their putative function in actinorhizal symbioses with *Frankia*

Gene name	Protein	Putative function
CgChs	Chalcone synthetase	Protection
CgMt1	Metallothionein-like protein	Protection
CgGst	Glutathione S-transferase	Protection
CgChi1	Chitinase class I	Protection
Cg12	Subtilisin-like protease	Infection
CgPox4	Class IV peroxidase	Development
CgChi3	Class III chitinase	Development

Adapted from Santos et al. (2010)

(*Cg12*) whose putative function seems to be related to *Frankia* infection. A peroxidase-encoding gene (*CgPox4*) has been postulated to participate in the cell wall modifications that take place during nodule development (Santos et al. 2010). Finally, the nodule-specific gene, *CgChi3,* has been implied in nodule development (Fortunato et al. 2007). Table 9.1 presents a list of genes and their putative functions during *C. glauca* symbioses. In this review, we discuss the putative functions of the proteins encoded by *CgChi3* and *CgGst* in the nodules of *C. glauca*.

9.2 CgCHI3

The superfamily of glycosyl hydrolases is composed of 81 enzyme families that hydrolyze a wide range of substrates (Henrissat 1991). Among them, chitinases (EC 3.2.1.14) are ubiquitous enzymes of bacteria, fungi, animals, and plants that catalyze the hydrolytic cleavage of the β-1,4 glycosidic bond of chitin, a homopolymer of *N*-acetyl-D-glucosamine (Bhattacharya et al. 2007). Aside from being the major component of the exoskeleton of arthropods and nematodes, chitin is also the major component of the cell wall of bacteria, algae, and higher fungi such as *Basidiomycetes*, *Ascomycetes,* and *Deuteromycetes* (Boller 1987; Salzer et al. 2000; Kasprzewska 2003).

Chitinases can be divided into two categories: exochitinases, which show activity only for the non-reducing end of the chitin chain, and endochitinases, which hydrolyze internal β-1,4-glycosidic bonds (Kasprzewska 2003). They are often designated glycoside hydrolases or pathogenesis-related (PR) proteins (Yokoyama and Nishitani 2004), being classified in seven classes (classes I-VII) distributed over two families: 18 and 19 (Henrissat and Bairoch 1993; Neuhaus et al. 1996). Plant chitinases are present in all organs and tissues, both in the vacuole and in the apoplast. They can be distinguished by the primary structure of the protein, substrate specificity, catalytic mechanism, and sensitivity to inhibitors (Kasprzewska 2003). The involvement of chitinases in mechanisms of active or passive defense against pathogens has been reported (Graham and Sticklen 1994;

Carstens et al. 2003). In general, rapid induction and accumulation of chitinases correlate with resistance to pathogens (Samac and Shah 1991; Meier et al. 1993; Gerhardt et al. 1997). However, chitinases may also be involved in plant developmental processes such as cell elongation (Zhong et al. 2002), embryogenesis (Fukamizo 2000), cold tolerance (Yeh et al. 2000; De los Reyes et al. 2001; Yu and Griffith 2001), and storage (Peumans et al. 2002). In the case of symbioses with nitrogen-fixing bacteria, it is thought that chitinases are involved in the autoregulation mechanism controlling the number and formation of nodules (Vasse et al. 1993; Staehelin et al. 1995; Goormachtig et al. 1998; Kim and An 2002; Fortunato et al. 2007). Moreover, they were identified as part of the nodule protection machinery against pathogens (Salzer et al. 2000; Staehelin et al. 2000; Kim et al. 2005; Santos et al. 2010; Ribeiro et al. 2011).

Fortunato et al. (2007) have isolated a *CgChi3* gene, encoding a class III chitinase from *C. glauca* root nodules induced by nitrogen-fixing bacteria *Frankia*. CgCHI3 was found to be encoded by a single gene that was specifically activated in nodules as compared with roots and leaves. The expression of this gene was upregulated by salicylic acid and repressed by wounding. In situ hybridization analysis revealed that *CgChi3* is an early nodulin gene, expressed in the meristem and in the uninfected cortical cells of young nodules. The transcriptional activity of this gene suggests that it is specifically associated with the symbiotic process, particularly with bacterial nodule infection. The fact that *CgChi3* is expressed specifically in nodules makes it a promising symbiotic marker candidate for the elucidation of the role of defense-related genes in symbiosis as well as for the understanding of the mechanisms of evolution of symbiosis as a reminiscent of pathogenesis.

Amino acid sequence analysis revealed that CgCHI3 belongs to the family 18 (Fig. 9.1) showing 65–83 % homology with proteins from other plant species (Andrade 2014). This protein has the conserved motifs characteristic of the family, in particular the catalytic motif D-X-X-D-X-D-X-E. CgCHI3 also belongs to the group of PR proteins (family PR-8) which includes hydrolytic β-1,3-glucanases (PR-2) that are part of the response mechanism to the invasion by plant pathogens such as fungi and bacteria. Mauch and Staehelin (1989) proposed a model for the defense mechanism of plants that involves the synergistic action of β-1,3-glucanases and chitinases, in which the first would act in an early stage of infection, releasing β-1,3-glucans from fungal hyphae which then activate the expression of defense genes, in particular those encoding chitinases, that would degrade the cell walls of pathogens. Andrade (2014) reports that CgCHI3 is a bifunctional enzyme exhibiting lysozyme and chitinolytic activities, typical from plant and microbial endochitinases belonging to family 18 (Jekbl et al. 1991; Bokma et al. 2002; Ye and Ng 2005). CgCHI3 also shows ß-1,3 glucanase and ß-1,4 glucanase activities which have never been reported before. The fact that CgCHI3 presents chitinase, glucanase, and cellulase activities is unique and may correspond to a more effective defense against pathogens or to an optimized system involved in the symbiotic process. Bringing together the results of

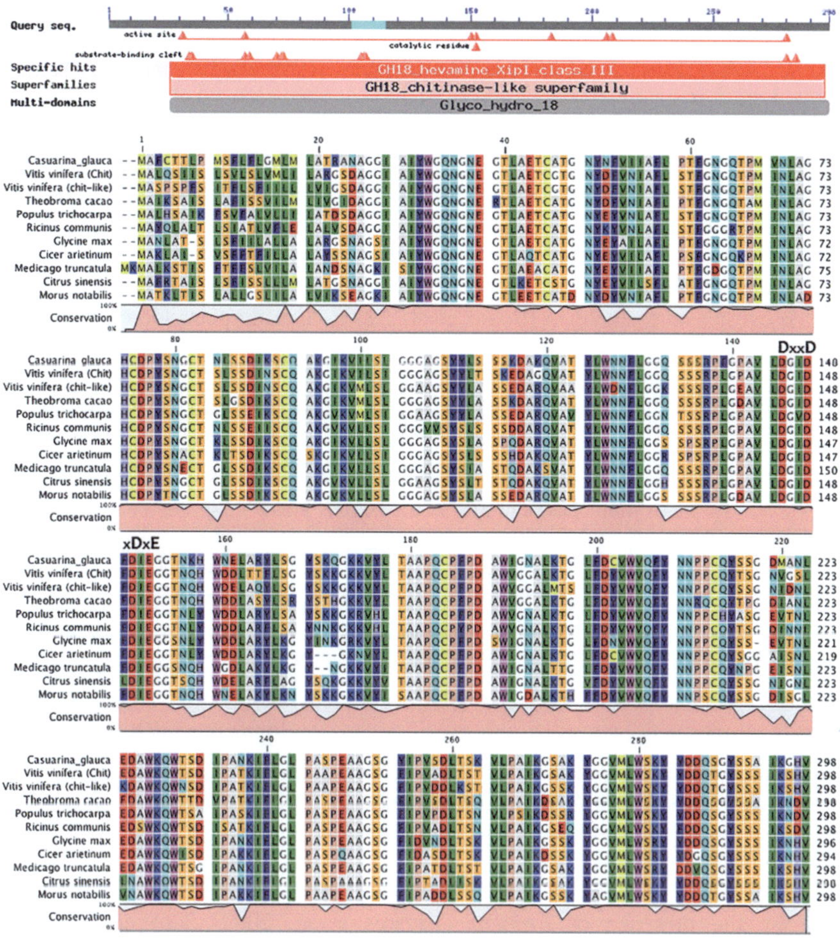

Fig. 9.1 Comparative analysis of the primary structure of CgCHI3 with other proteins identified as class III of the chitinase family 18 of glycosyl hydrolases carried out with the CLC Sequence Viewer software version 7.0.2 (www.clcbio.com). Access numbers of proteins available in the NCBI database: acidic endochitinase from *Vitis vinifera* (XP_002279147.1), acidic endochitinase-like from *Vitis vinifera* (XP_002279205.1), precursor of hevamine-A from *Theobroma cacao* (XP_007021467.1), chitinase from *Populus trichocarpa* (XP_002301383.1), putative precursor of hevamine-A from *Ricinus communis* (XP_002511934.1), glycine max hevamine-A (XP_003516317.1), acidic endochitinase-like from *Cicer arietinum* (XP_004490140.1), chitinase class III from *Medicago truncatula* (AAQ21404.1), hypothetical hevamine-A from *Citrus sinensis* (XP_006444927.1), and acidic endochitinase from *Morus notabilis* (EXB52705.1). The catalytic motif is highlighted D_{145}-X-X-D_{148}-X-D_{150}-X-E152 (in the CgCHI3 numbering)

Fortunato et al. (2007) and Andrade (2014), it is likely that CgCHI3 is involved in *Frankia* infection process, specifically in the lysis of the cell wall during intracellular penetration, in the buildup of the infection thread or in any other modification that occurs to accommodate the symbiotic bacteria.

9.3 CgGST

Glutathione S-transferases (GST, EC 2.5.1.18) are characterized as a heterogeneous group of cells detoxifying enzymes which catalyze the conjugation of the tripeptide glutathione (GSH) on electrophilic sites in a wide range of phytotoxic substrates. This group of enzymes comprises a diverse family of soluble proteins encoded from multigene families (Wilce and Parker 1994; Edwards et al. 2000; Dixon et al. 2002). GSTs are known for their response to several endogenous stresses and xenobiotics such as herbicide detoxification (Cho and Kong 2007) and are mostly soluble cytoplasmic enzymes (Board et al. 2000). Soluble plant GSTs were classified according to the sequence proximity, cross immunologic reactions, kinetic properties, and genome organization, into eight classes: phi (F), tau (U), zeta (Z), theta (T), lambda (L), dehydroascorbate reductase (DHAR), EF1Bγ, and dehalogenase tetrachlorohydroquinone (TCHQD) (Oakley 2005; Lan et al. 2009; Dixon and Edwards 2010; Jain et al. 2010). Among them, classes phi, tau, lambda, and DHAR are the main plant GSTs (Jiang et al. 2010). These enzymes can be induced by a wide variety of phytohormones including ethylene, auxin, methyl jasmonate, salicylic acid, and abscisic acid (Wagner et al. 2002; Moons 2003; Smith et al. 2003). Furthermore, plant GSTs also play a role in the endogenous metabolism including the function as glutathione peroxidases in oxidative stress (Cummins et al. 1999), flavonoid-binding protein (Mueller et al. 2000), stress signaling proteins (Loyall et al. 2000), and apoptosis regulators (Kampranis et al. 2000). All these hormones regulate many aspects of plant development which implies that GSTs may play a vital role in the growth and development of plants (Jiang et al. 2010). The genes that encode GSTs can also be induced by several exogenous factors such as attack by pathogens, heavy metals, heat shock, or wounding (Singla et al. 2007; Jiang et al. 2010).

In actinorhizal symbioses, there is evidence for the induction of an oxidative burst by the presence of *Frankia* (Tavares et al. 2007; Soto et al. 2009). Additionally, the induction of genes encoding enzymes involved in ROS scavenging has been reported (An et al. 2005; Obertello et al. 2007; Soto et al. 2009; Santos et al. 2010). The presence of GST during the symbiosis between leguminous/actinorhizal plants and N_2-fixing bacteria has been associated with the operating system and nodule protection (Minchin et al. 2008; Becana et al. 2010; Santos et al. 2010; Ribeiro et al. 2011). Santos et al. (2010) reported that the transcriptional activity *CgGST* was positively regulated in symbiotic nodules. Besides that, the expression of this gene was positively regulated by abscisic acid and wounding, suggesting a specific role in nodule protection.

Amino acid sequence analysis reveals that CgGST belongs to the tau class of specific plant glutathione S-transferase (Edwards et al. 2000) and to the cytosolic protein superfamily (Fig. 9.2) showing 64–70 % homology with proteins of other plant species (Andrade 2014). Additionally, tau CgGST shows the characteristic conserved domains of this family, in particular the functional domains at the N- and C-termini (Chi et al. 2011). Andrade (2014) reports that CgGST has the Ser13,

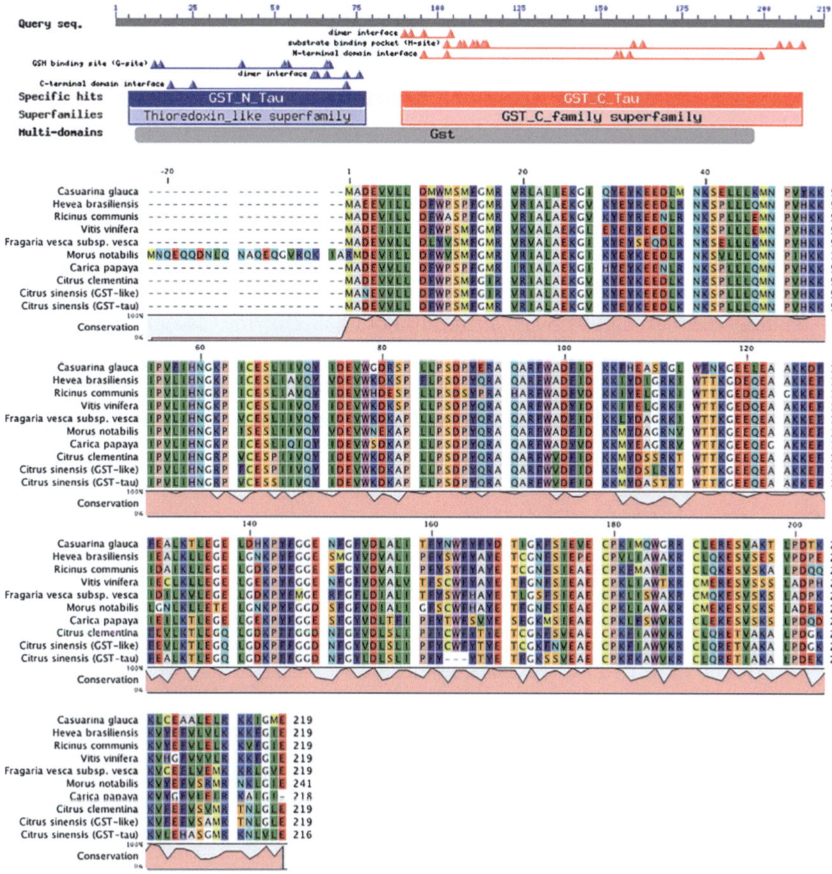

Fig. 9.2 Comparative analysis of the primary structure of CgGST with other proteins identified as tau class, glutathione S-transferases family and cytosolic protein superfamily, carried out with the CLC Sequence Viewer software version 7.0.2 (www.clcbio.com). Access numbers of proteins available in the NCBI database: tau 1 GST from *Hevea brasiliensis* (XP_004294220.1), GST from *Ricinus communis* (XP_006442220.1), GST from *Vitis vinifera* (XP_006478015.1), homolog GST from *Fragaria vesca* subsp. *vesca* (XP_002264437.1), GST from *Morus notabilis* (CAA04391.1), glutathione transferase from *Carica papaya* (ACI42271.1) hypothetical protein from *Citrus clementina* (AEF79856.1), homolog protein to GST from *Citrus sinensis* (XP_002532823.1), and tau2 GST from *Citrus sinensis* (EXB96020.1)

Lis40, Ile54, Glu66, and Ser67 residues, identified as GSH-binding active sites in other tau proteins, such PtGSTU1 from pine, TaGSTU4 from wheat, and OsGSTU1 from rice (Zeng et al. 2005; Zeng and Wang 2005). According to Zeng and Wang (2005), mutations at these sites induce significant changes in enzyme activity, substrate specificity, and thermal stability. Furthermore, Ile54 is involved in the folding process of proteins. Similar to what has been reported in other plants, CgGST protein showed high activity for 1-chloro-2,4-dinitrobenzene (CDNB) substrate (Andrade 2014). Altogether, gene expression (Santos et al. 2010), bioinformatic, and biochemical data (Andrade 2014) suggest that CgGST is an important antioxidant protein.

9.4 Conclusion

The study of events related to defense in symbiosis has increased remarkably in recent years. These genes may be involved in nodule protection or development. The nodule-specific CgCHI3 is a multifunctional enzyme associated with the process of *Frankia* infection. The presence of CgGST is probably related to the control of the nodule redox status. In the future, it would be important to exploit the involvement of these and other stress-responsive genes in nodulation.

References

An CS, Kim HB, Lee SH et al (2005) Gene expression in root nodules of *Elaeagnus umbellata*. In: Wang YP, Lin M, Tian ZX, Newton WE (eds) Biological nitrogen fixation, sustainable agriculture and the environment, vol 41. Springer, Dordrecht, pp 207–208

Andrade IMCGA (2014) Functional characterization of a chitinase class III (CgCHI3) and a glutathione S-transferase (CgGST) involved in *Casuarina glauca-Frankia* symbiosis. PhD thesis, Universidade Nova de Lisboa

Becana M, Matamoros MA, Udvardi M et al (2010) Recent insights into antioxidant defenses of legume root nodules. New Phytol 188:960–976

Bhattacharya D, Nagpure A, Gupta RK (2007) Bacterial chitinases: properties and potential. Crit Rev Biotechnol 27:21–28

Board PG, Coggan M, Chelvanayagam G et al (2000) Identification, characterization and crystal structure of the omega class of glutathione transferases. J Biol Chem 275:24798–24806

Bokma E, Rozeboom HJ, Sibbald M et al (2002) Expression and characterization of active site mutants of hevamine, a chitinase from the rubber tree *Hevea brasiliensis*. Eur J Biochem 269:893–901

Boller T (1987) Hydrolytic enzymes in plant disease resistance. In: Kosuge T, Nester EW (eds) Plant-microbe interactions, vol 2. Macmillan, New York, pp 385–413

Carstens M, Vivier MA, Pretorius IS (2003) The *Saccharomyces cerevisiae* chitinase, encoded by the CTS1-2 gene, confers antifungal activity against *Botrytis cinerea* to transgenic tobacco. Transgenic Res 12(4):497–508

Chi Y, Cheng Y, Vanitha J et al (2011) Expansion mechanisms and functional divergence of the glutathione S-transferase family in sorghum and other higher plants. DNA Res 18:1–16

Cho HY, Kong KH (2007) Study on the biochemical characterization of herbicide detoxification enzyme, glutathione S-transferase. BioFactors 30:281–287

Cummins I, Cole DJ, Edwards R (1999) A role for glutathione transferases functioning as glutathione peroxidases in resistance to multiple herbicides in black-grass. Plant J 18:285–292

Dawson JO (1990) Interactions among actinorhizal and associated plant species. In: Schwintzer CR, Tjepkema JD (eds) The biology of *Frankia* and actinorhizal plants. Academic Press, New York, pp 299–316

Dawson JO (2008) Ecology of actinorhizal plants In: Pawlowski K, Newton WE (eds) Nitrogen-fixing actinorhizal symbioses. Nitrogen fixation research: origins and progress. Springer, New York, pp 199–234

De los Reyes BG, Taliaferro CM, Anderson MP et al (2001) Induced expression of the class II chitinase gene during cold acclimation and dehydration of bermudagrass (*Cynodon* sp.). Theor Appl Genet 103:297–306

Diem HG, Dommergues YR (1990) Current and potential uses and management of Casuarinaceae in the tropics. In: Schwintzer CR, Tjepkema JD (eds) The biology of *Frankia* and actinorhizal plants. Academic Press, New York, pp 317–342

Dixon DP, Edwards R (2010) Glutathione transferases. Arabidopsis Book 8:e0131

Dixon DP, Lapthorn A, Edwards R (2002) Plant glutathione transferases. Genome Biol (rev 3004.1-3004.10)

Edwards R, Dixon DP, Walbot V (2000) Plant glutathione S-transferases: enzymes with multiple functions in sickness and in health. Trends Plant Sci 5:193–198

Fortunato A, Santos P, Graça I et al (2007) Isolation and characterization of *cgchi3*, a nodule-specific gene from *Casuarina glauca* encoding a class III chitinase. Physiol Plant 130:418–426

Fukamizo T (2000) Chitinolytic enzymes: catalysis, substrate binding, and their application. Curr Protein Pept Sci 1:105–124

Gerhardt LBD, Sachetto-Martins G, Contarini MG et al (1997) *Arabidopsis thaliana* class IV chitinase is early induced during the interaction with *Xanthomonas campestris*. FEBS Lett 419:69–75

Goormachtig S, Lievens S, van de Velde W et al (1998) *Srchi13*, a novel early nodulin from *Sesbania rostrata*, is related to acidic class III chitinases. Plant Cell 10:905–915

Graham LS, Sticklen MB (1994) Plant chitinases. Can J Bot 72:1057–1083

Henrissat B (1991) Classification of glycosyl hydrolases based on amino acid sequence similarities. Biochem J 280:309–316

Henrissat B, Bairoch A (1993) New families in the classification of glycosyl hydrolases based on amino acid sequence similarities. Biochem J 293:781–788

Hocher V, Alloisio N, Auguy F et al (2011) Transcriptomics of actinorhizal symbioses reveals homologs of the whole common symbiotic signaling cascade. Plant Physiol 156:700–711

Jain M, Ghanashyam C, Bhattacharjee A (2010) Comprehensive expression analysis suggests overlapping and specific roles of rice glutathione S-transferase genes during development and stress responses. BMC Genomics 11:73

Jekbl PA, Hartmann JBH, Beintema JJ (1991) The primary structure of hevamine, an enzyme with lysozyme/chitinase activity from *Hevea brasiliensis* latex. Eur J Biochem 200:123–130

Jiang HW, Liu MJ, Chen IC et al (2010) A glutathione S-transferase regulated by light and hormones participates in the modulation of Arabidopsis seedling development. Plant Physiol 154:1646–1658

Kampranis SC, Damianova R, Atallah M et al (2000) A novel plant glutathione S-transferase/peroxidase suppresses Bax lethality in yeast. J Biol Chem 275:29207–29216

Kasprzewska A (2003) Plant chitinases-regulation and functions. Cell Mol Biol Lett 8:809–824

Kim HB, An CS (2002) Differential expression patterns of an acidic chitinase and a basic chitinase in the root nodule of *Elaeagnus umbellata*. Mol Plant Microbe Interact 15:209–215

Kim YJ, Kim HB, Baek HE et al (2005) Constitutive expression of two endochitinases from root nodules of *Elaeagnus umbellata* confers resistance on transgenic Arabidopsis plants against the fungal pathogen *Botrytis cinerea*. J Plant Biol 48:39–46

Lan T, Yang ZL, Yang X et al (2009) Extensive functional diversification of the Populus glutathione S-transferase supergene family. Plant Cell 21:3749–3766

Laplaze L, Gherbi H, Frutz T et al (1999) Flavan-containing cells delimit *Frankia* infected compartments in *Casuarina glauca* nodules. Plant Physiol 121:113–122. doi:10.1104/pp.121.1.113

Laplaze L, Gherbi H, Duhoux E et al (2002) Symbiotic and non-symbiotic expression of *cgMT1*, a metallothionein-like gene from the actinorhizal tree *Casuarina glauca*. Plant Mol Biol 49: 81–92. doi:10.1023/A:1014415003714

Lohar DP, Sharapova N, Endre G et al (2006) Transcript analysis of early nodulation events in *Medicago truncatula*. Plant Physiol 140:221–234

Loyall L, Uchida K, Braun S et al (2000) Glutathione and a UV light-induced glutathione S-transferase are involved in signalling to chalcone synthase in cell cultures. Plant J 25: 237–245
Mauch F, Staehelin LA (1989) Functional implications of the subcellular localization of ethylene-induced chitinase and [beta]-1,3-glucanase in bean leaves. Plant Cell 1(4):447–457
Meier BM, Shaw N, Slusarenko AJ (1993) Spatial and temporal accumulation of defense gene transcripts in bean (*Phaseolus vulgaris*) leaves in relation to bacteria-induced hypersensitive cell death. Mol Plant Microbe Interact 6:453–466
Minchin FR, James EK, Becana M (2008) Oxygen diffusion, production of reactive oxygen and nitrogen species, and antioxidants in legume nodules. In: Dilworth MJ, James EK, Sprent JI, Newton WE (eds) Nitrogen-fixing symbioses. Springer, New York, pp 321–362
Moons A (2003) *Osgstu3* and *osgstu4*, encoding tau class glutathione S-transferases, are heavy metal and hypoxic stress-induced and differentially salt stress-responsive in rice roots. FEBS Lett 553:427–432
Mueller LA, Goodman CD, Silady RA et al (2000) AN9, a petunia glutathione S-transferase required for anthocyanin sequestration, is a flavonoid-binding protein. Plant Physiol 123: 1561–1570
Neuhaus JM, Fritig B, Linthorst HJM et al (1996) A revised nomenclature for chitinase genes. Plant Mol Biol Rep 14:102–104
Oakley AJ (2005) Glutathione transferases: new functions. Curr Opin Struct Biol 15:716–723
Obertello M, Wall L, Laplaze L et al (2007) Functional analysis of the metallothionein gene *CgMT1* isolated from the actinorhizal tree *Casuarina glauca*. Mol Plant Microbe Interact 20:1231–1240. doi:10.1094/MPMI-20-10-123
Peumans WJ, Proost P, Swennen RL et al (2002) The abundant class III chitinase homolog in young developing banana fruits behaves as a transient vegetative storage protein and most probably serves as an important supply of amino acids for the synthesis of ripening-associated proteins. Plant Physiol 130:1063–1072
Ramírez M, Graham MA, Lopez LB et al (2005) Sequencing and analysis of common bean ESTs. Building a foundation for functional genomics. Plant Physiol 137:1211–1227
Ribeiro A, Graça I, Pawlowski K et al (2011) Actinorhizal plant defence-related genes in response to symbiotic *Frankia*. Funct Plant Biol 38:639–644
Salzer P, Bonanomi A, Beyer K et al (2000) Differential expression of eight chitinase genes in *Medicago truncatula* roots during mycorrhiza formation, nodulation and pathogen infection. Mol Plant Microbe Interact 13:763–777
Samac DA, Graham MA (2007) Recent advances in legume-microbe interactions: recognition, defense response, and symbiosis from a genomic perspective. Plant Physiol 144:582–587
Samac DA, Shah DM (1991) Developmental and pathogen-induced activation of the *Arabidopsis* acidic chitinase promoter. Plant Cell 3:1063–1072
Santos P, Fortunato A, Ribeiro A et al (2008) Chitinases in root nodules. Plant Biotechnol 25: 299–307
Santos P, Fortunato A, Graca I et al (2010) Characterization of four defense-related genes up-regulated in root nodules of *Casuarina glauca*. Symbiosis 50:27–35
Singla B, Tyagi AK, Khurana JP et al (2007) Analysis of expression profile of selected genes expressed during auxin-induced somatic embryogenesis in leaf base system of wheat (*Triticum aestivum*) and their possible interactions. Plant Mol Biol 65:677–692
Smith AP, Nourizadeh SD, Peer WA et al (2003) *Arabidopsis AtGSTF2* is regulated by ethylene and auxin, and encodes a glutathione S-transferase that interacts with flavonoids. Plant J 36:433–442
Soto MJ, Dominguez-Ferreras A, Pérez-Mendoza D et al (2009) Mutualism versus pathogenesis: the give-and-take in plant-bacteria interactions. Cell Microbiol 11:381–388
Spaink HP (1995) The molecular basis of infection and nodulation by rhizobia: the ins and outs of sympathogenesis. Annu Rev Phytopathol 33:345–368
Sprent JI (2007) Evolving ideas of legume evolution and diversity: a taxonomic perspective of the occurrence of nodulation. New Phytol 174:11–25. doi:10.1111/j.1469-8137.2007.02015.x

Staehelin C, Schultze M, Kondorosi E et al (1995) Lipochitooligosaccharide nodulation signals from *Rhizobium meliloti* induce their rapid degradation by the host plant alfalfa. Plant Physiol 108:1607–1614

Staehelin C, Schultze M, Tokuyasu K et al (2000) N-deacetylation of *Sinorhizobium meliloti* Nod factors increases their stability in the *Medicago sativa* rhizosphere and decreases their biological activity. Mol Plant Microbe Interact 13:72–79

Svistoonoff S, Laplaze L, Auguy F et al (2003) *cg12* expression is specifically linked to infection of root hairs and cortical cells during *Casuarina glauca* and *Allocasuarina verticillata* actinorhizal nodule development. Mol Plant Microbe Interact 16:600–607. doi:10.1094/MPMI.2003.16.7.600

Tavares F, Santos CL, Sellstedt A (2007) Reactive oxygen species in legume and actinorhizal nitrogen-fixing symbioses: the microsymbiont's responses to an unfriendly reception. Physiol Plant 130:344–356

Vasse J, de Billy F, Truchet G (1993) Abortion of infection during the *Rhizobium meliloti* alfalfa symbiotic interactions is accompanied by a hypersensitive reaction. Plant J 4:555–566

Wagner U, Edwards R, Dixon DP et al (2002) Probing the diversity of the *Arabidopsis* glutathione S-transferase gene family. Plant Mol Biol 49:515–532

Wilce MC, Parker MW (1994) Structure and function of glutathione S-transferases. Biochim Biophys Acta 1205:1–18

Ye X, Ng TB (2005) A chitinase with antifungal activity from the mung bean. Protein Expr Purif 40:230–236

Yeh S, Moffatt BA, Griffith M et al (2000) Chitinase genes responsive to cold encode antifreeze proteins in winter cereals. Plant Physiol 124:1251–1263

Yokoyama R, Nishitani K (2004) Genomic basis for cell-wall diversity in plants. A comparative approach to gene families in rice and Arabidopsis. Plant Cell Physiol 45:1111–1121

Yu XM, Griffith M (2001) Winter rye antifreeze activity increases in response to cold and drought, but not abscisic acid. Physiol Plant 112:78–86

Zeng Q, Wang X (2005) Catalytic properties of glutathione-binding residues in a tau class glutathione transferase (PtGSTU1) from *Pinus tabulaeformis*. FEBS Lett 579:2657–2662

Zeng QY, Lu H, Wang XR (2005) Molecular characterization of a glutathione transferase from *Pinus tabulaeformis* (Pinaceae). Biochimie 87:445–455

Zhong R, Kays SJ, Schroeder BP et al (2002) Mutation of a chitinase-like gene causes ectopic deposition of lignin, aberrant cell shapes, and overproduction of ethylene. Plant Cell 14: 165–179

Chapter 10
Effective Colonization of Spinach Root Surface by *Rhizobium*

Alejandro Jiménez-Gómez, Esther Menéndez, José D. Flores-Félix, Paula García-Fraile, Pedro F. Mateos and Raúl Rivas

Abstract Plant growth-promoting rhizobacteria are a group of bacteria able to promote plant growth and increase crop productivity. For a successful and effective bacteria–plant association, microorganism colonization and its arrangement and abundance on plant root surfaces are very important. Here, we analysed the ability of the strain PEPV12 identified as *Rhizobium* sp., to colonize spinach (*Spinacia oleracea*) root surfaces, and evaluated three parameters of in vitro plant growth promotion: (i) siderophore production, (ii) phosphate solubilization and (iii) indole acetic acid production. Moreover, we tested in vitro cellulose production and biofilm formation. Our results showed that this strain colonizes effectively and adheres to abiotic surfaces. Root colonization and adherence events were observed under fluorescence microscopy by using GFP-tagged PEPV12 strain, which was inoculated on spinach seedlings. Interestingly, we observed morphologic changes in root hairs, such as deformations and redirections at the tip in early stages of plant development of PEPV12 inoculated plants. Furthermore, we proceeded to evaluate in vitro plant development after inoculation of spinach seedlings with the strain of this study, showing significant differences in shoot length with respect to uninoculated plants. Our results showed that *Rhizobium* sp. PEPV12 actively and successfully colonizes spinach root surfaces, producing changes in root hairs and an increase in plant growth, suggesting its potentiality as a biofertilizer for *Spinacia oleracea*.

Keywords Root colonization · *Rhizobium* · *Spinacia oleracea* · Fluorescence microscopy · Plant growth promotion

A. Jiménez-Gómez · E. Menéndez (✉) · J.D. Flores-Félix · P.F. Mateos · R. Rivas
Department of Microbiology and Genetics, University of Salamanca, Salamanca, Spain
e-mail: esthermenendez@usal.es

P. García-Fraile
Institute of Microbiology ASCR, Academy of Sciences of the Czech Republic, Videnska 1083, 142 20 Prague 4, Czech Republic

P.F. Mateos · R. Rivas
Associated Unit I+D University of Salamanca—CSIC Plant-Microbe Interactions Group, Salamanca, Spain

© Springer International Publishing Switzerland 2016

F. González-Andrés and E. James (eds.), *Biological Nitrogen Fixation and Beneficial Plant–Microbe Interactions*, DOI 10.1007/978-3-319-32528-6_10

10.1 Introduction

Conventional agriculture systems and its practices involve serious environmental problems, caused mainly by the massive application of chemical fertilizers, which are industrially manipulated, and their exploitation causes serious problems such as air, ground and water pollution. In the past decades, many efforts have been done to avoid these problems, expecting to lead a change on this system towards the production of more quantity and high-quality food in a sustainable way, involving the application of microbial fertilizers in agriculture (Bhardwaj et al. 2014). Currently, the demand is increased worldwide due to the number of laws protecting the environment and the expanding demand to produce ecological food "biosafe" (García-Fraile et al. 2015). To fulfil these society demands, the study of the rhizosphere and their associated microorganisms is becoming the key for a sustainable development of new mindfulness agriculture.

10.1.1 The Rhizosphere and Its Effects

Since the mention of the rhizosphere concept by Lorenz Hiltner in 1904, there are many soil plant grown studies analysing microbial diversity. The rhizosphere zone is relatively rich in nutrients, which are released by the roots, and its microbial communities differ from those outside of root influence. Rhizosphere is the zone where microbes can develop and interact (Nihorimbere et al. 2011).

In general, the rhizosphere is compound by a variety of microbial community, including endophytes, pathogens and beneficial microbes. Those beneficial microbes are generally classified into two broad groups based on their effects: (i) microorganisms with direct effects on plant growth promotion (plant growth-promoting microbes, PGPM) and (ii) biological control agents (BCA) that indirectly assist with plant productivity through the control of plant pathogens (Avis et al. 2008).

Plant growth-promoting rhizobacteria (PGPR) is the term used to describe a group of beneficial bacteria capable of colonizing the rhizosphere and able to benefit plants by improving their productivity and immunity. PGPR can promote plant growth with a broad variety of mechanisms, which can be grouped according to their mode of action: (i) the synthesis of substances that can be assimilated directly by plants, (ii) the mobilization of nutrients, (iii) the induction of plant stress resistance, and (iv) the prevention of plant diseases (García-Fraile et al. 2015).

10.1.2 Rhizosphere and Root Colonization

The successful root colonization is an important and initial step in the interaction of beneficial bacteria with plants. In order to acquire the beneficial effects from the microbial communities, it is crucial an effective colonization of the plant root area (Yuan et al. 2015).

For many years, a number of different methods have analysed the colonization by beneficial bacteria in the rhizosphere. Several markers were developed to study localization and quantification of PGPR: (i) antibiotic resistance (Gamalero et al. 2003; Bonaldi et al. 2015) and (ii) fluorescent markers for colonization studies (Cao et al. 2011; Krzyzanowska et al. 2012). These studies focus on a basic knowledge between PGPR and the host plant in axenic conditions. Keep in mind the potential microbial diversity in in vivo conditions, competition between beneficial strains and other microorganisms present in the same ecological niche may reduce the colonization in many cases (Bonaldi et al. 2015).

Root colonization does not follow a defined pattern, explained by the action of different factors, such as root exudates, bacterial quorum-sensing effects and amino acid synthesis, among others (Compant et al. 2010). Yuan et al. (2015) described how organic acids from banana roots exudates could attract a beneficial strain of *Bacillus amyloliquefaciens* and promote its biofilm formation. Veliz-Vallejos et al. (2014) observed an increase in nodule number in *Medicago truncatula* under the impact of 3-oxo-C14-HSL, an N-acyl-homoserine lactone (the major group of quorum-sensing signal molecules in gram-negative bacteria) produced by *Ensifer meliloti*. Xie et al. (2014) described the effects of different concentrations of spermidine and implicated a promotion of tobacco plant growth with an optimum concentration of exogenous spermidine.

In summary, successful bacterial root colonization is a mix of physical, biological and chemical characters, as well as properties of the microorganism itself. It required a clear knowledge of each part (Dutta and Podile 2010). Among these bacterial colonizers, several genera of rhizobia are catalogued as effective colonizers, such as members of the genus *Rhizobium* (Avis et al. 2008).

10.1.3 Rhizobium as Effective Colonizer

Rhizobia are soilborne bacteria with a key role in the biological nitrogen fixation process as microbial symbiotic partners of legumes. *Rhizobium* spp. populations reside in soils and are constant evolution, being able to adapt to different habitats. The colonization success of new environments depends on their attachment abilities to biotic and abiotic substrates (Wielvo et al. 2015).

Wang and Chen (2004) described how rhizobia colonize new environments, facing two different problems. In the first place, rhizobia need to establish a population in the local community among groups of indigenous bacteria, and there exists a competition between space and nutrients. Another factor is the adaptation of rhizobia to environmental factors, such as humidity, salt content, pH and/or temperature, among others.

Since some rhizobial strains were described as PGPR, there are a huge number of studies showing their abilities to increase legume and non-legume crops production (Antoun et al. 1998; Alami et al. 2000; Mishra et al. 2006; García-Fraile et al. 2012; Flores-Félix et al. 2013). These studies analysed the ability and effect to establish communities and produce an important growth and effective colonization from rhizobial strains. García-Fraile et al. (2012) observed that two species of rhizobia firmly attached to tomato and pepper root surfaces, colonizing intercellular spaces. Moreover, Flores-Félix et al. (2013) showed that a *Rhizobium leguminosarum* strain colonize root epidermis and intercellular spaces in *Lactuca sativa* and *Daucus carota* plants.

In this chapter, we aim to determine whether a rhizobial strain is able to colonize effectively spinach roots and analyse the effects on the growth. This study will generate further knowledge of plant–microorganisms interaction and colonization process in the context of microbial fertilizers.

10.2 Materials and Methods

10.2.1 Bacterial Isolation and Plant Material

Strain PEPV12 was isolated from an effective nodule of *Phaseolus vulgaris,* growing in a soil from Salamanca (NW, Spain) as described in Vincent (1970) on YMA plates (yeast–mannitol agar) at 28 °C. A commercial variety of spinach, *Spinacia oleracea* L. var "viking", was selected for this study.

10.2.2 Determination of In Vitro PGPR Mechanisms

10.2.2.1 Inorganic Phosphate Solubilization

The ability to solubilize inorganic phosphate was tested by growing the strain on YED-P medium, according to Peix et al. (2004). The formation of transparent zones around the bacterial colonies was indicative of inorganic phosphate solubilization on YED-P.

10.2.2.2 Siderophore Production

Siderophore production was analysed using M9-CAS-AGAR medium, as described in Schwyn and Neilands (1987) and modified by Alexander and Zuberer (1991). After growing the bacteria at 28 °C for 3 days, colonies were observed for the formation of yellowish orange halos, indicating siderophore secretion.

10.2.2.3 Indole Acetic Acid (IAA) Biosynthesis

Indole acetic acid production was measured by growing the bacteria in JMM medium supplemented with L-tryptophan (0.17 g l^{-1}) (O'Hara et al. 1989; Khalid et al. 2004).

10.2.3 16S rRNA Parcial Sequencing of Strain PEPV12

Partial sequence of the 16S ribosomal RNA gene was carried out as follows: 8.5 μl of miliQ sterile water, 10 μl REDExtract Ready Mix, 0.5 μl of 27 F (5′ GAGGGTGGCGGTTCT 3′) and 0.5 μl 1522R (5′AAGGAGGTGATCCANCC RCA 3′) and 0.5 μl of DNA template. PCR conditions were as follows: preheating at 95 °C for 9 min; 35 cycles of denaturing at 94 °C for 1 min; annealing at 55 °C for 1′ 30 min; extension at 72 °C for 2 min; and a final extension at 72 °C for 7 min.

PCR product was purified using QIAquick PCR purification kit (Qiagen) according to the manufacturer's specifications. The sequence reaction was performed on an ABI PRISM sequencer (Applied Biosystem®, EEUU) using a "Dye Terminator Cycle Sequencing Ready Reaction Kit" (Sequencing Service, University of Salamanca). Sequence was compared with type strain by using EzBiocloud Server (Chun et al. 2007).

10.2.4 Qualitative Determination of Cellulose Production

Cellulose production was determined by growing the strain PEPV12 in YMA medium supplemented with 25 mg l^{-1} of Congo red and incubated at 28 °C for 3 days. Qualitative analysis was performed in comparison with the "Royal Horticultural Society Color Chart" as described (Robledo et al. 2012).

10.2.5 Determination of Biofilm Formation

Biofilm formation was analysed as described in O'Toole (1999) and modified by Fujishige et al. (2006). Briefly, a pre-inoculum of strain PEPV12 was grown in YMB medium for 2 days until an OD_{600} of 2.0. Bacterial attachment to PVC plates was assayed by pipetting 100 μl of a normalized suspension of an OD_{600} of 0.2. Plates were incubated statically at 28 °C and revealed with 0.3 % crystal violet at defined times (24 and 48 h). Plate wells were washed, and the remaining dye was solubilized with 80 % ethanol–20 % acetone. Absorbance at 570 nm was measured in an ASYS UVM340 Microtiter Plate Reader (Biochrom, UK) and subsequently analysed.

10.2.6 GFP Labelling of Strain PEPV12

GFP-tagged strain was obtained by biparental mating with *E. coli* S17.1, carrying pHC60 plasmid (Cheng and Walker 1998) as described in García-Fraile et al. (2012). Tetracycline was added to the media when required (10 μg/mL).

10.2.7 Plant Assays

Spinach (*Spinacia oleracea* L.) seeds were surfaced-sterilized with 70 % ethanol for 30 s and 5 % NaClO for 5 min. Then, seeds were washed with sterilized fivefold and spread on 1 % agar plates. Seedlings were maintained in dark for during 3 days and, then, transferred to 1 % agar square plates and inoculated with 250 μl plant^{-1} of a bacterial suspension with a turbidity of 6 in McFarland Standards (1.5×10^6 CFU ml^{-1}). Plants were incubated in a growth chamber and observed at 3, 6 and 9 days post-inoculation. Uninoculated treatment was included as negative control.

For the observation of colonization process and morphologic changes in root hairs, plants inoculated with 250 μl plant^{-1} of a suspension of the GFP-tagged PEPV12 as indicated above. Colonization events and root hair deformations were observed at 5 and 7 days post-inoculation under NIKON Eclipse 80i epifluorescence microscope.

10.2.8 Statistical Analysis

All data were analysed with one-way ANOVA using Statview 5.0 software (SAS Institute Inc.) with a post hoc test using Fisher's protected least significant difference ($p \geq 0.05$).

10.3 Results and Discussion

10.3.1 Identification of Strain PEPV12

On the basis of the partial sequenciation of the 16rDNA gene, the strain PEPV12 was classified into the genus *Rhizobium,* showing 100 % of identity with two *Rhizobium leguminosarum/laguerreae* type strains. This result is in agreement with Mulas et al. (2011), who described that several strains from the same species (*Rhizobium leguminosarum/laguerreae*) isolated in north Spain are able to nodulate common bean crops.

10.3.2 Rhizobium Exhibits In Vitro Plant Growth Promotion Mechanisms

We have observed that the PEPV12 strain forms colonies in M9-CAS-AGAR medium surrounded by a yellow halo (0.5 cm radium around colonies). This mechanism has been already reported in rhizobial strains, such as some strains of *R. leguminosarum* (Chabot et al. 1996; García-Fraile et al. 2012). In YED-P medium, phosphate solubilization was not detected. Previous studies showed members of the genus *Rhizobium* and, specifically *R. leguminosarum,* are able to solubilize tricalcium phosphate under in vitro conditions (Peix et al. 2001). However, the conditions may vary, and further analyses should be performed to elucidate if this strain can solubilize phosphate, both in vitro and *in plant* conditions.

Moreover, we observed IAA production from L-tryptophan in the strain PEPV12 (0.092 mg l^{-1}). Several authors have been demonstrated that this phytohormone has a role as efficient tool enhancing plant development (Kamilova et al. 2006). The concentration of indoleacetic acid from bacteria should be low to prevent undesirable effects, such as the effects produced under the action of pathogenic bacteria in plants. The effect of low-level indoleacetic acid contributes to root and shoot elongation, due to the induction of plant cell proliferation (Vacheron et al. 2013).

Despite of the absence of phosphate solubilization, the results of the in vitro plant growth promotion analysis showed that strain PEPV12 is able to secrete siderophores, as well as to produce proper levels of IAA, suggesting that this strain has a high potential as PGPR, alone and/or in combination with other bacterial strains in biofertilization schemes.

10.3.3 Cellulose Production

Bacterial cellulose biosynthesis is a complex process, involving the action of several enzymes (Menéndez et al. 2015). Due to its interests in several industries, cellulose is obtaining a relevant attention in scientific community.

Initially, the role of cellulose in bacteria was poorly understood due to its complexity. However, the function has been already described in *Rhizobium* strains. Mateos et al. (1995) showed that *Rhizobium leguminosarum* bv. trifolii polymerizes β-1-4 glucose residues into cellulose microfibrils that are important in the attachment to the roots of its host plant, *Trifolium repens*.

Congo red stain binds to β-1-4 glucosidic bonds, such as the ones present in cellulose (Teather and Wood 1982). However, there are other polymers containing β-1-4 glucosidic bonds, which bind Congo red, such as amyloid fibres (Khurana et al. 2001). The significant reduction in Congo red binding to *Rhizobium* strains is related to their impairment in cellulose production, suggesting that these bacteria do not produce other substances that bind Congo red and contribute to their red colony pigmentation. Our results showed that PEPV12 strain could be classified as middling cellulose producer comparing the red intensity with the established in the "Royal Horticultural Society Color Cart" (Robledo et al. 2012).

10.3.4 Biofilm Formation

The bacterial attachment to a substrate is the initial step in biofilm formation, which is followed by the establishment of colonies; the structure is finished when is covered by exopolymer and other matrix materials (Robledo et al. 2012). Three-dimensional structure formation, described as biofilms, protects microorganisms from adverse agent action, involving an increase of nutrient availability. Their extracellular matrix protects cells from external attacks acting as a barrier against low-weight molecules diffusion (O'Toole 1999; Ramey et al. 2004). For that reason, biofilm formation is presented as an outstanding strategy of microorganisms with clear advantages in colonization process. In this study, strain PEPV12 biofilm formation was determined and compared with a positive producer, *Rhizobium cellulosilyticum* ALA10B2T, which is a natural cellulose producer. This strain is used due to the similarity in growth rate with the strain of study. The results showed that PEPV12 established mature biofilms gradually, showing an increase at 48 h post-inoculation.

The ability to form biofilms in different biotic or abiotic substrates by members of the genus *Rhizobium*, as we showed in this study, has been already described (Williams et al. 2008; Janczarek et al. 2010). Janczarek et al. (2010) reported a considerably decreased in root hairs attachment by bacterial strains, described as a relevant effect in the first stage of the colonization process. Moreover, Williams et al. (2008) analysed the importance of polysaccharides production in relation to biofilm formation and its effect in root attachment by *R. leguminosarum* strain. Those studies are in agreement with our results.

10.3.5 Plant Growth Assays

The interaction between *Spinacia oleracea* seedlings and PEPV12 strain was tested in in vitro conditions. To observe the potential plant growth-promoting effects of the strain PEPV12, we performed an assay in plates. The results of seedling inoculation assays showed that PEPV12 strain promotes growth of *Spinacia oleracea* under microcosmos conditions, evaluating effects on edible parts and root length.

Plants were measured at 3, 6 and 9 days post-inoculation (Table 10.1). The results showed that, in general, inoculated plants presented an increase in root and shoot length in each time point analysed. However, values were not significantly different between inoculated and uninoculated plants, excepting the values obtained for shoot length at 6 dpi, which were significatively higher than uninoculated plants

Table 10.1 Effect of PEPV12 strain on roots length and shoot part in inoculated and uninoculated treatments. Each data are the mean (±S.E.*) of at least 15 plants of three different assays. Values followed by the same letter in each treatment are not significantly different from each other at $P \geq 0.05$ according to Fisher's protected LSD (least significant differences)

Treatment	3 dpi		6 dpi		9 dpi	
	Root length (cm)	Shoot length (cm)	Root length (cm)	Shoot length (cm)	Root length (cm)	Shoot length (cm)
Control	5.57 ± 0.77 a	0.55 ± 0.07 a	8.2 ± 0.12 a	2.1 ± 0.35 a	9.87 ± 0.19 a	2.47 ± 0.09 a
PEPV12	6.9 ± 0.28 a	1.1 ± 0.13 a	8.75 ± 0.25 a	3.23 ± 0.09 b	10.08 ± 0.09 a	3.43 ± 0.47 a

**S.E.* Standard error

at the same time point. However, it is important to emphasize that all the values in the case of inoculated plant were higher than uninoculated treatment. Nevertheless, further studies should be done in glasshouse conditions to confirm that *Rhizobium* sp. PEPV12 is enhancing plant growth in a significant manner.

To observe colonization and the effects of this strain in root surfaces and root hairs, GFP-tagged PEPV12 was inoculated in similar conditions, and then, spinach roots were observed under fluorescence microscopy (Fig. 10.1). Rhizobial attachment to root surfaces was gradually increased during observations. This strain was able to colonize the root surfaces, occupying intercellular spaces in cortical cells. PEPV12 was distributed on the root, also occupying intercellular spaces (Fig. 10.1a–c). Moreover, typical microcolonies of biofilm formation can be observed (Fig. 10.1d–f). Our results revealed that PEPV12 rhizobial strain inoculation presented the same distribution, which has been reported in many studies (Chabot et al. 1996; García-Fraile et al. 2012; Flores-Félix et al. 2013). Furthermore, Perrine-Walker et al. (2007) showed that *Rhizobium* strains form similar patterns in *Oryza sativa* roots and lateral roots than in spinach roots (Fig. 10.1a–c).

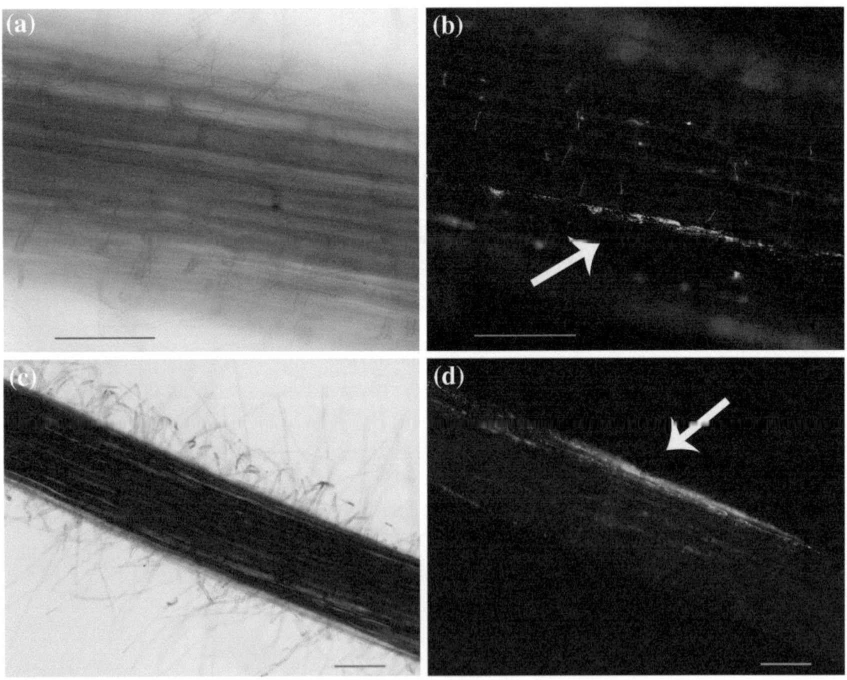

Fig. 10.1 Scanning micrographs of spinach seedling roots 7 days after inoculation with GFP-tagged cells of PEPV12 strain. Images obtained in **a** and **c** by transmitted light in bright field mode, **b** and **d** in epifluorescence. **a**–**b**: Micrographs show uniform colonization along root cell depressions. Bar: 100 μm. **c**–**d**: micrographs show typical microcolonies of biofilm initiation. Bar: 100 μm

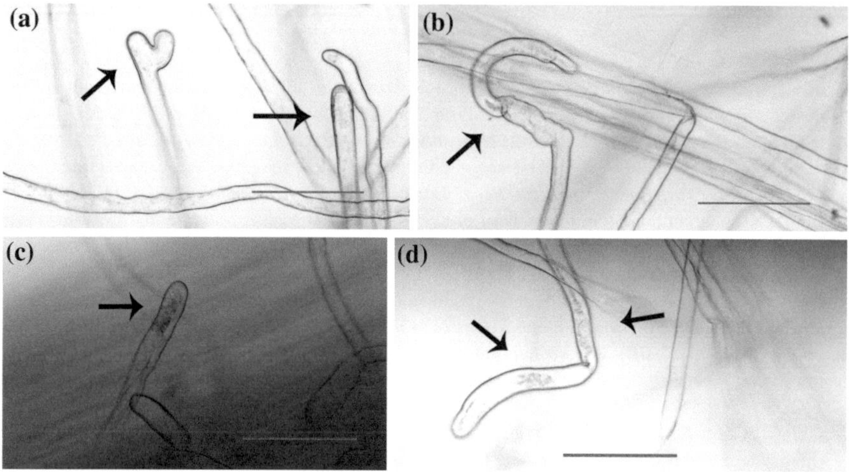

Fig. 10.2 Scanning micrographs of spinach seedling roots 7 days after inoculation with PEPV12 strain. Images obtained by transmitted light in bright field mode. **a–b**: micrographs show redirections at the root hairs. **a–c–d**: micrographs show bacterial aggregates. Bar: 100 μm

Root hair colonization and adherence events were also monitored (Fig. 10.2). Two different events were observed in inoculated seedlings: (i) deformations and redirections at the root hair tips (Fig. 10.2a, b) and (ii) bacterial aggregates in roots hairs (Fig. 10.2a–c–d).

The previous studies reported that root hairs deformations might occur also in non-legume crops, such as rice, after inoculation with *Rhizobium* sp. strains (Perrine-Walker et al. 2007, 2009). These studies described that rhizobial strains produced curled root hairs as result of the infection and colonization processes. Our results suggest that the colonization by PEPV12 strain has similar effects in spinach.

In some cases, microbial cells are attached to the tip of root hairs (Fig. 10.2), but this kind of colonization appears and could be correlated with the dying off of infected rice seedlings (Perrine-Walker et al. 2007).

10.4 Conclusion

The results of this present study show that *Rhizobium* sp. PEPV12, a strain that colonizes the root surfaces of *Spinacia oleracea*, increase plant growth initial development, being a potential biofertilizer for this non-legume crop.

References

Alami Y, Achouak W, Marol C et al (2000) Rhizosphere soil aggregation and plant growth promotion of sunflowers by an exopolysaccharide-producing *Rhizobium* sp. strain isolated from sunflower roots. Appl Environ Microb 66:3393–3398

Alexander DB, Zuberer DA (1991) Use of chrome azurol S reagents to evaluate siderophore production by rhizosphere bacteria. Biol Fertil Soils 12:39–45. doi:10.1007/BF00369386

Antoun H, Beauchamp CJ, Goussard N et al (1998) Potential of *Rhizobium* and *Bradyrhizobium* species as growth promoting rhizobacteria on non-legumes: effect on radishes (*Raphanus sativus* L.). Plant Soil 204:57–67

Avis TJ et al (2008) Multifaceted beneficial effects of rhizosphere microorganisms on plant health and productivity. Soil Biol Biochem 40:1733–1740. doi:10.1016/j.soilbio.2008.02.013

Bhardwaj D et al (2014) Biofertilizers function as key player in sustainable agriculture by improving soil fertility, plant tolerance and crop productivity. Microb Cell Fact 13:66. doi:10.1186/1475-2859-13-66

Bonaldi M, Chen X, Kunova A, Pizzatti C, Saracchi M, Cortesi P (2015) Colonization of lettuce rhizosphere and roots by tagged *Streptomyces*. Front Microbiol 6:25. doi:10.3389/fmicb.2015.00025

Cao Y, Zhang Z, Ling N, Yuan Y, Zheng X, Shen B et al (2011) *Bacillus subtilis* SQR 9 can control *Fusarium* wilt in cucumber by colonizing plant roots. Biol Fertil Soils 47:495–506. doi:10.1007/s00374-011-05562

Chabot R, Antoun H, Cescas MP (1996) Growth promotion of maize and lettuce by phosphate-solubilizing *Rhizobium leguminosarum* biovar phaseoli. Plant Soil 184:311–321. doi:10.1007/BF00010460

Cheng HP, Walker GC (1998) Succinoglycan is required for initiation and elongation of infection threads during nodulation of alfalfa by *Rhizobium meliloti*. J Bacteriol 180:5183–5191. PMCID: PMC107556

Chun J, Lee JH, Jung Y, Kim M, Kim S, Kim BK, Lim YW (2007) EzTaxon: a web-based tool for the identification of prokaryotes based on 16S ribosomal RNA gene sequences. Int J Syst Evol Microbiol 57:2259–2261

Compant S, Clément C, Sessitsch A (2010) Plant growth-promoting bacteria in the rhizo- and endosphere of plants: their role, colonization, mechanisms involved and prospects for utilization. Soil Biol Biochem 42:669–678. doi:10.1016/j.soilbio.2009.11.024

Dutta S, Podile AR (2010) Plant growth promoting rhizobacteria (PGPR): the bugs to debug the root zone. Crit Rev Microbiol 36(3):232–244. doi:10.3109/10408411003766806

Flores-Félix JD, Menéndez E, Rivera LP et al (2013) Use of *Rhizobium leguminosarum* as a potential biofertilizer for *Lactuca sativa* and *Daucus carota* crops. J Plant Nutr Soil Sci:1–7. doi:10.1002/jpln.2013001161

Fujishige NA, Kapadia NN, De Hoff PL, Hirsch AM (2006) Investigations of *Rhizobium* biofilm formation. FEMS Microbiol Ecol 56(2):195–206

Gamalero E, Lingua G, Berta G, Lemanceau P (2003) Methods for studying root colonization by introduced beneficial bacteria. Agronomie 23:407–418. doi:10.1051/Agro:2003014

García-Fraile P, Carro L, Robledo M et al (2012) *Rhizobium* promotes non-legumes growth and quality in several production steps: towards a biofertilization of edible raw vegetables healthy for humans. PLoS ONE 7(5):e38122. doi:10.1371/journal.pone.0038122

García-Fraile P, Menéndez E, Rivas R (2015) Role of bacterial biofertilizers in agriculture and forestry. AIMS Bioeng 2(3):183–205. doi:10.3934/bioeng.2015.3.183

Janczarek M, Kutkowska J, Piersiak T, Skorupska A (2010) *Rhizobium leguminosarum* bv. *Trifolii* rosR is required for interaction with clover, biofilm formation and adaptation to the environment. BMC Microbiol 10:284. doi:10.1186/1471-2180-10-284

Kamilova F, Kravchenko LV, Al Shaposhnikov et al (2006) Organic acids, sugars, and L-tryptophane in exudates of vegetables growing on stonewool and their effects on activities of rhizosphere bacteria. Mol Plant Microbe Interact 9:250–256. doi:10.1094/MPMI-19-0250

Khalid A, Arshad M, Zahir ZA (2004) Screening plant growth-promoting rhizobacteria for improving growth and yield of wheat. J Appl Microbiol 96:473–480. doi:10.1046/j.1365-2672.2003.02161.x

Khurana R, Uversky VN, Nielsen L, Fink AL (2001) Is Congo red an amyloid-specific dye? J Biol Chem 276(25):22715–22721

Krzyzanowska D, Obuchowski M, Bikowski M, Rychlowski M, Jafra S (2012) Colonization of potato rhizosphere by GFP-tagged *Bacillus subtilis* MB73/2, *Pseudomonas* sp. P482 and *Ochrobactrum* sp. A44 shown on large sections of roots using enrichment sample preparation and confocal laser scanning microscopy. Sensors 12:17608–17619. doi:10.3390/S121217608

Mateos PF, Baker B, Philip-Hollingsworth S, Squartini A, Peruffo A, Nuti M, Dazzo F (1995) Direct in situ identification of cellulose microfibrils associated with *Rhizobium leguminosarum* biovar. *Trifolii* attached to the root epidermis of white clover. Can J Microbiol 41:202–207

Menéndez E, García-Fraile P, Rivas R (2015) Biotechnological applications of bacterial cellulases. AIMS Bioengineering 2(3):163–182. doi:10.3934/bioeng.2015.3.163

Mishra RP, Singh RK, Jaiswal HK, Kumar V, Maurya S (2006) Rhizobium-mediated induction of phenolics and plant growth promotion in rice (*Oryza sativa* L.). Curr Microbiol 52:383–389

Mulas D, García-Fraile P, Carro L et al (2011) Distribution and efficiency of *Rhizobium leguminosarum* strains nodulating *Phaseolus vulgaris* in Northern Spanish soils: selection of native strains that replace conventional N fertilization. Soil Biol Biochem 43:2283–2293. doi:10.1016/j.soilbio.2011.07.018

Nihorimbere V, Ongena M, Smargiassi M, Thonart P (2011) Beneficial effect of the rhizosphere microbial community for plant growth and health. Biotechnol Agron Soc 15(2):327–337

O'Hara GW, Goss TJ, Dilworth MJ, Glenn AR (1989) Maintenance of intracellular pH and acid-tolerance in *Rhizobium meliloti*. Appl Environ Microb 55:1870–1876

O'Toole GA (1999) Microbial biofilms: from ecology to molecular genetics. Microbiol Mol Biol R 64:847–867

Peix A, Rivas-Boyero AA, Mateos PF, Rodriguez-Barrueco C, Martınez-Molina E, Velazquez E (2001) Growth promotion of chickpea and barley by a phosphate solubilizing strain of *Mesorhizobium mediterraneum* under growth chamber conditions. Soil Biol Biochem 33 (1):103–110

Peix A, Rivas R, Santa-Regina I, Mateos PF, Martínez-Molina E, Rodriguez-Barrueco C, Velázquez E (2004) *Pseudomonas lutea* sp. nov., a novel phosphate-solubilizing bacterium isolated from the rhizosphere of grasses. Int J Syst Evol Microbiol 54:847–850. doi:10.1099/ijs.0.02966-0

Perrine-Walker FM, Prayitno J, Rolfe BG et al (2007) Infection process and the interaction of rice roots with rhizobia. J Exp Bot 58(12):3343–3350. doi:10.1093/jxb/erm181

Perrine-Walker FM, Hynes MF, Rolfe BG et al (2009) Strain competition and agar affect the interaction of rhizobia with rice. Can J Microbiol 55(10):1217–1223. doi:10.1139/W09-077

Ramey BE, Koutsoudis M, Von Bodman SB, Fuqua C (2004) Biofilm formation in plant–microbe associations. Curr Opin Microbiol 7:602–609

Robledo M, Rivera LP, Jiménez-Zurdo JI, Rivas R, Dazzo FB, Velázquez E, Martínez-Molina E, Hirsch AM, Mateos PF (2012) Role of *Rhizobium* endoglucanase CelC2 in cellulose biosynthesis and biofilm formation on plant roots and abiotic surfaces. Microb Cell Fact 11:125

Schwyn B, Neilands JB (1987) Universal chemical assay for the detection and determination of siderophores. Anal Biochem 160:47–56

Teather RM, Wood PJ (1982) Use of Congo red-polysaccharide interactions in enumeration and characterization of cellulolytic bacteria from the bovine rumen. App Environ Microb 43 (4):777–780

Vacheron J, Desbrosses G, Boffaud ML et al (2013) Plant growth-promoting rhizobacteria and root system functioning. Front Plant Sci 4:356. doi:10.3389/fpls.2013.00356

Veliz-Vallejos DF, Van Noorden GE, Yuan M et al (2014) A Sinorhizobium meliloti-specific N-acyl homoserine lactone quorum-sensing signal increases nodule numbers in *Medicago truncatula* independent of autoregulation. Plant Sci 5:551. doi:10.3389/fpls.2014.00551

Vincent JM (1970) A manual for the practical study of root nodule bacteria. Blackwell Scientific Publications, Oxford, pp 73–79. doi:10.1002/jobm.19720120524

Wang ET, Chen WX (2004) Estimation of biotic and abiotic factors that affect migration of rhizobia. In: Biological resources and migration. Springer, pp 167–161. doi:10.1007/978-3-662-06083-4

Wielvo J, Podlesna A, Kidal D et al (2015) The diversity of pea microsymbionts in various types of soils and their effects on plant host productivity. Microbes Environ 30(3):254–261. doi:10.1264/jsme2.ME14141

Williams A, Wilkinson A, Krehenbrink M et al (2008) Glucomannan-mediated attachment of *Rhizobium leguminosarum* to pea root hairs is required for competitive nodule infection. J Bacteriol 190(13):4706–4715

Xie SS, Wu HJ, Zang HY et al (2014) Plant growth promotion by spermidine-producing *Bacillus subtilis* OKB105. Mol Plant Microbe Interact 27(7):655–663. doi:10.1094/MPMI-01-14-0010-R

Yuan J, Zhang N, Huang Q et al (2015) Organic acids from root exudates of banana help root colonization of PGPR strain *Bacillus amyloliquefaciens* NJN-6. Sci Rep 5:13438. doi:10.1038/srep13438

Chapter 11
Interactions Between *Micromonospora* and Arbuscular Mycorrhizal Fungi

Pilar Martínez-Hidalgo, Inmaculada García-Romera and Juan Antonio Ocampo

Abstract *Micromonospora* is a Gram-positive bacterium that can be isolated from nitrogen-fixing root nodules from healthy leguminous plants and promotes plant growth. Selected strains of *Micromonospora* isolated in our laboratory from alfalfa root nodules showed in vitro antifungal activity against several pathogenic fungi. Moreover, root inoculation of tomato plants with these *Micromonospora* strains effectively reduced leaf infection by the fungal pathogen *Botrytis cinerea*, despite temporal and spatial separation between the inoculation of both microorganisms. Although *Micromonospora* improves plant growth and has antifungal activity against plant pathogens, the interaction between *Micromonospora* and AM fungi has not been studied before. Thus, the aim of the study was to work on the understanding of this interaction. The results obtained allow us to conclude that *Micromonospora* strains ALFpr18c and ALFb5 did not adversely affect the colonization and development of the AM fungi *R. irregularis* and *F. mosseae* on plant root. *Micromonospora* strain ALFpr18c increased the beneficial effect of the AM fungi on shoot dry weight of alfalfa and tomato when the AM fungi were established first on the roots. Therefore, it is possible to use *Micromonospora* ALFpr18c combined with AM fungi to improve plant growth. *Micromonospora* strains should be considered excellent plant probiotic bacteria as they combine plant growth-promoting activity with both direct antifungal activity against plant pathogens and the ability to prime plant immunity and did not adversely affect the colonization and development of the AM fungi.

Keywords Actinobacteria · *Micromonospora* · Arbuscular mycorrhizal fungi · *Rhizophagus irregularis* · *Funneliformis mosseae* · Symbiosis

---

P. Martínez-Hidalgo (✉)
Department of Microbiology and Genetics, University of Salamanca, Salamanca, Spain
e-mail: martinezhp@usal.es

I. García-Romera · J.A. Ocampo
Department of Soil Microbiology, Estación Experimental del Zaidín, CSIC, Granada, Spain

© Springer International Publishing Switzerland 2016

F. González-Andrés and E. James (eds.), *Biological Nitrogen Fixation and Beneficial Plant–Microbe Interactions*, DOI 10.1007/978-3-319-32528-6_11

11.1 Introduction

Some soil microorganisms are necessary for plant development and growth as they improve nutrient uptake, produce plant growth-stimulating substances, and prevent pathogen attacks (Turner et al. 2013). Arbuscular mycorrhizal (AM) fungi are an important component of the soil microbial biomass; they are obligate symbiotic soil fungi that colonize the roots of the majority of terrestrial plants (Smith and Read 2008). This symbiosis is mutualistic based on bidirectional nutrient transfer between the symbionts. The plant benefits particularly through enhanced phosphorus, water, and mineral nutrient uptake which often results in better growth (Smith and Smith 2012). However, AM fungi not only improve plant growth but it can also enhance soil structure, interact with other microbes, and alleviate biotic and abiotic plant stress (Gianinazzi et al. 2010).

Other soil microorganisms are known to promote the AM symbiosis (Adriano-Anaya et al. 2006) and have been named "mycorrhizal helper microorganisms" (McAllister et al. 1997). Actinobacteria are interesting and common components of the soil rhizosphere (McCarthy and Williams 1992), but studies on the interaction between actinobacteria and AM fungi are scarce and mainly centered on the *Streptomyces* genus and were inconclusive, since both antagonistic and synergic effects of *Streptomyces* on AM fungi have been described (Franco-Correa et al. 2010).

Micromonospora is an actinobacterial genus which is widely distributed in nature, inhabiting soil and plant tissues even in root nodules of *Ensifer* colonized plants. Some *Micromonospora* strains are able to promote plant growth (Martínez-Hidalgo et al. 2014; Solans et al. 2015). However, *Micromonospora* has been reported to have antifungal activity against certain pathogenic fungi (Martínez-Hidalgo et al. 2015).

Although *Micromonospora* has been found to be an interesting possibility for new formulation of inoculants, the interaction between *Micromonospora* and AM fungi has not been studied before. The aim of the study is to assess the interaction between *Micromonospora*, AM fungi, and the plant.

11.2 Materials and Methods

11.2.1 Greenhouse Experiments

The experiments were carried out in 0.3-l pots of soil obtained from Granada Spain (see Siles et al. 2014 for soil characteristics). The soil was steam-sterilized and mixed with sterilized sand to a proportion of 2:3 (V:V). Seeds of tomato (*Solanum lycopersicum* cv. Muchamiel) and alfalfa (*Medicago sativa* ecotype Aragón) were surface-sterilized with $HgCl_2$ for 10 min, thoroughly rinsed with sterilized water and sown in moistened sand. After germination, uniform seedlings were planted and grown in a greenhouse with supplementary light under controlled conditions. The *Micromonospora* sp. pr18c and b5 strains used in this study were isolated from surface-sterilized root nodules of naturally occurring alfalfa plants from Castilla y León (Spain) (Martínez-Hidalgo et al. 2014). For inoculation of plants, an aqueous suspension in sterile distilled water was prepared from cultures grown in SA1 medium (Trujillo et al. 2005). The AM fungi used were *Rhizophagus irregularis* (DAOM 197198) and *Funneliformis mosseae* (BEG 12) which were inoculated as described by Martín-Rodríguez et al. (2015) and García-Sánchez et al. (2014). Four treatments were used: (1) uninoculated controls (2) soil pots inoculated with *R. irregularis* or *F. mosseae* (3) soil pots inoculated with *Micromonospora* sp. pr18c or *Micromonospora* sp. b5, and (4) soil pots inoculated with *R. irregularis* or *F. mosseae* and *Micromonospora* sp. pr18c or *Micromonospora* sp. b5. Plants were inoculated with the AM fungi at the time of transplanting (after 3 weeks of growth). The *Micromonospora* sp. strains were inoculated at the same time as *R. irregularis* or *F. mosseae* or after a fortnight. Five replicate pots per treatment were used. Plants were harvested after 8 weeks, and shoot dry mass was determined. The root system was cleared and stained for microscope examination (Phillips and Hayman 1970), and the percentage of total root length colonized by AM fungi was measured (Giovannetti and Mosse 1980).

11.2.2 Statistical Analysis

The data were analyzed by one-way analysis of variance (ANOVA), and treatment means were compared using Duncan's multiple range test.

11.3 Results

The inoculation of *Micromonospora* sp. ALFpr18c and ALFb5 did not affect the percentage of AM root length colonization reached either in alfalfa or in tomato inoculated with *R. irregularis* or *F. mosseae* (Tables 11.1, 11.2, 11.3, and 11.4).

A significant increase of the shoot dry weight of alfalfa inoculated with *Micromonospora* sp. ALFpr18c a fortnight after *R. irregularis* was observed. However, the joint inoculation of *R. irregularis* and *Micromonospora* sp. ALFb5 inoculated at the same time or after fortnight of the AM fungus, decreased the shoot dry weight of alfalfa (Table 11.1).

The co-inoculation of *Micromonospora* sp. ALFpr18c and *F. mosseae* decreased the shoot dry weight of alfalfa when the bacterial strain was inoculated either at the same time or a fortnight after the AM fungus. However, there was a significant increase of the shoot dry weight of alfalfa when *Micromonospora* sp. ALFb5 was inoculated a fortnight after *F. mosseae* (Table 11.2).

Table 11.1 Shoot dry weight and percentage of root length colonization of alfalfa (*Medicago sativa*) by *Rhizophagus irregularis* in the presence of the *Micromonospora* sp. pr18c and b5 strains inoculated at the same time (T0) or fortnight after (T1) the arbuscular fungus

Treatments	T0		T1	
	Dry weight (mg)	% root length mycorrhization	Dry weight (mg)	% root length mycorrhization
Non-inoculated	10.2 a	0.0	10.2 a	0.0
R. irregularis	63.3 c	18.4 a	63.5 c	19.2 a
M. pr18c and *R. irregularis*	63.3 c	12.3 a	150.1 d	13.5 a
M. b5 and *R. irregularis*	33.3 b	19.6 a	26.6 b	15.5 a

Each value is the mean for five pots. Column values followed by the same letter are not significantly different according to Duncan's multiple range test ($P = 0.05$)

Table 11.2 Shoot dry weight and percentage of root length colonization of alfalfa (*Medicago sativa*) by *Funneliformis mosseae* in the presence of the *Micromonospora* sp. pr18c and b5 strains inoculated at the same time (T0) or fortnight after (T1) the arbuscular fungus

Treatments	T0		T1	
	Dry weight (mg)	% root length mycorrhization	Dry weight (mg)	% root length mycorrhization
Non-inoculated	10.1 a	0.0	10.1 a	0.0
F. mosseae	46.6 c	22.5 a	46.5 c	25.2 a
M. pr18c and *F. mosseae*	30.2 b	28.4 a	33.3 b	22.6 a
M. b5 and *F. mosseae*	43.3 c	26.5 a	133.3 d	38.3 a

Each value is the mean for five pots. Column values followed by the same letter are not significantly different according to Duncan's multiple range test ($P = 0.05$)

Table 11.3 Shoot dry weight and percentage of root length colonization of tomato (*Solanum lycopersicum*) by *Rhizophagus irregularis* in the presence of the *Micromonospora* sp. pr18c and b5 strains inoculated at the same time (T0) or fortnight after (T1) the arbuscular fungus

Treatments	T0		T1	
	Dry weight (mg)	% root length mycorrhization	Dry weight (mg)	% root length mycorrhization
Non-inoculated	26.3 a	0.0	22.3 a	0.0
R. irregularis	113.2 c	34.2 a	120.3 c	39.2 a
M. pr18c and *R. irregularis*	154.5 c	43.3 a	220.3 d	43.5 a
M. b5 and *R. irregularis*	72.1 b	30.5 a	53.2 b	35.5 a

Each value is the mean for five pots. Column values followed by the same letter are not significantly different according to Duncan's multiple range test ($P = 0.05$)

Table 11.4 Shoot dry weight and percentage of root length colonization of tomato (*Solanum lycopersicum*) by *Funneliformis mosseae* in the presence of the *Micromonospora* sp. pr18c and b5 strains inoculated at the same time (T0) or fortnight after (T1) the arbuscular fungus

Treatments	T0		T1	
	Dry weight (mg)	% root length mycorrhization	Dry weight (mg)	% root length mycorrhization
Non-inoculated	22.1 a	0.0	24.2 a	0.0
F. mosseae	216.5 c	38.1 a	205.6 c	37.2 a
M. pr18c and *F. mosseae*	194.5 c	42.4 a	112.3 c	39.4 a
M. b5 and *F. mosseae*	52.2 b	33.5 a	72.2 b	34.5 a

Each value is the mean for five pots. Column values followed by the same letter are not significantly different according to Duncan's multiple range test ($P = 0.05$)

The shoot dry weight of tomato increased when *Micromonospora* sp. ALFpr18c was inoculated a fortnight after *R. irregularis*, but the inoculation of *R. irregularis* together *Micromonospora* sp. ALFb5 decreased the shoot dry weight of tomato (Table 11.3).

The co-inoculation of *Micromonospora* sp. ALFb5 and *F. mosseae* decreased the shoot dry weight of tomato when the bacterial strain was inoculated either at the same time or fortnight after the AM fungus. However, no effect was found on the shoot dry weight of tomato when *Micromonospora* sp. ALFb5 and *F. mosseae* were inoculated together (Table 11.4).

11.4 Discussion

As was expected, the colonization of alfalfa and tomato by the AM fungi *R. irregularis* and *F. mosseae* increased the shoot dry weight of both plants.

The interaction between soil rhizosphere microorganisms is important for plant development and growth. The different effect of soil rhizosphere microorganisms on AM fungi has been described previously (Adriano-Anaya et al. 2006; Arriagada et al. 2010). *Micromonospora* strains are able to increase the development of the symbiotic bacteria *Ensifer meliloti* in plant root and increase plant growth (Martínez-Hidalgo et al. 2015). *Micromonospora* is also capable of producing secondary metabolites and enzymes involved in the inhibition of fungal spore germination and fungal growth (Martínez-Hidalgo et al. 2015). We found that the development of AM fungi into the root of alfalfa and tomato was not affected by *Micromonospora* sp. ALFpr18c and ALF5b strains. However, the effect of the AM fungi *R. irregularis* and *F. mosseae* on plant growth was affected by both strains of *Micromonospora* sp. It is known that a correlation between the percentage of AM root colonization and their effect on plant growth is not common (Adriano-Anaya et al. 2006). Moreover, the existence of soil microorganisms that increased the effect of the AM symbiosis on plant growth but did not increase the AM colonization of root has also been reported (Arriagada et al. 2010).

Most of the AM helper microorganisms mainly influence the AM fungi when the latter are in the presymbiotic phase of the symbiosis development (Frey-Klett et al. 2007). However, *Micromonospora* sp. ALFpr18c increased the beneficial effect of *R. irregularis* on the shoot dry weight of alfalfa and tomato when the actinobacteria was inoculated a fortnight after *R. irregularis*, i.e., when the colonization of the root by the AM fungus is taking place inside the roots of these plants. Is it known that the AM symbiosis form a mycorrhizosphere which can improve the growth and metabolic function of their microbial inhabitants which provide nutrients that can be incorporated to the plant through the AM mycelium avoiding the competition with other rhizosphere microorganisms for nutrients (Barea et al. 2013). The great capacity of secondary metabolites and exoenzymes production by *Micromonospora* sp. ALFpr18c (Martínez-Hidalgo 2014) may explain its beneficial effect on the dry weight of AM colonized plants. However, when *Micromonospora* sp. ALFpr18c was inoculated together with *F. mosseae,* the growth of alfalfa decreased but not the growth of tomato.

Otherwise, we found that *Micromonospora* ALFb5 decreased the beneficial effect of *R. irregularis* on the shoot dry weight of both plants, but when this *Micromonospora* strain was co inoculated with the AM fungus *F. mosseae* a positive effect on alfalfa plant shoot dry weights was observed. Beneficial and antagonistic action of different *Micromonospora* strains against the effect of other microorganisms on plant development has been found (Martínez-Hidalgo 2014, 2015) in this study; it has been shown that the actinobacteria only benefits the effect of the AM fungi on plant, when a certain combination of a plant and AM endophyte and *Micromonospora* strain converge, indicating that compatibility studies between the microorganisms studied and the plant used are needed.

11.5 Conclusions

The *Micromonospora* strains ALFpr18c and ALFb5 did not affect the colonization and development of the AM fungi *R. irregularis* and *F. mosseae* on plant root. *Micromonospora* strains ALFpr18c increased the beneficial effect of the AM fungi on shoot dry weight of alfalfa and tomato when the AM fungi were established on the roots. Therefore, *Micromonospora* strain ALFpr18c should be considered excellent plant probiotic bacteria as they combine plant growth-promoting activity with both direct antifungal activity against plant pathogens and the ability to prime plant immunity and did not adversely affect the colonization and development of the AM fungi. In conclusion, it is possible to use *Micromonospora* ALFpr18c combined with AM fungi to improve plant growth.

Acknowledgements This work was supported by Junta de Castilla y León and by MICINN Grant, Spain. P.M.-H. was supported by a fellowship from CSIC JAE-PRE.

References

Adriano-Anaya ML, Salvador-Figueroa M, Ocampo JA, García-Romera I (2006) Hydrolytic enzymes production in maize (*Zea mays*) and sorghum (*Sorghum bicolor*) roots inoculated with *Gluconacetobacter diazotrophicus* and *Glomus intraradices*. Soil Biol Biochem 38:879–886

Arriagada C, Pereira G, García-Romera I, Ocampo JA (2010) Improved zinc tolerance in *Eucalyptus globulus* inoculated with *Glomus deserticola* and *Trametes versicolor* or *Coriolopsis rigida*. Soil Biol Biochem 42:118–124

Barea JM, Pozo MJ, Azcón R, Azcón-Aguilar C (2013) Microbial interactions in the rhizosphere. In: Molecular microbial ecology of the rhizosphere. Wiley, pp 29–44

Franco-Correa M, Quintana A, Duque C, Suarez C, Rodríguez MX, Barea JM (2010) Evaluation of actinomycete strains for key traits related with plant growth promotion and mycorrhiza helping activities. Appl Soil Ecol 45:209–217

Frey-Klett P, Garbaye J, Tarkka M (2007) The mycorrhiza helper bacteria revisited. New Phytol 176:22–36

García-Sánchez M, Palma JM, Ocampo JA, García-Romera I, Aranda E (2014) Arbuscular mycorrhizal fungi alleviate oxidative stress induced by ADOR and enhance antioxidant responses of tomato plants. J Plant Physiol 171:421–428

Gianinazzi S, Gollotte A, Binet MN, van Tuinen D, Redecker D, Wipf D (2010) Agroecology: the key role of arbuscular mycorrhizas in ecosystem services. Mycorrhiza 20:519–530

Giovannetti M, Mosse B (1980) An evaluation of techniques for measuring vesicular-arbuscular mycorrhizal infection in roots. New Phytol 84:489–500

Martín-Rodríguez JA, Ocampo JA, Molinero-Rosales N, Tarkowská D, Ruíz-Rivero O, García-Garrido JM (2015) Role of gibberellins during arbuscular mycorrhizal formation in tomato: New insights revealed by endogenous quantification and genetic analysis of their metabolism in mycorrhizal roots. Physiol Plant 154:66–81

Martínez-Hidalgo P, Galindo-Villardón P, Igual JM, Martínez-Molina E (2014) *Micromonospora* from nitrogen fixing nodules of alfalfa (*Medicago sativa* L.). A new promising Plant Probiotic Bacteria. Sci Rep 4:6389. doi:10.1038/srep06389

Martínez-Hidalgo P, García JM, Pozo MJ (2015) Induced systemic resistance against *Botrytis cinerea* by *Micromonospora* strains isolated from root nodules. Front Microbiol 6:922

McAllister CB, García-Garrido JM, García-Romera I, Godeas A, Ocampo JA (1997) Interactions between *Alternaria alternata*, *Fusarium equiseti* and *Glomus mosseae* and its effects on plant growth. Biol Fertil Soils 24:301–305

McCarthy AJ, Williams ST (1992) Actinomycetes as agents of biodegradation in the environment. Gene 115:189–192

Phillips JM, Hayman DS (1970) Improved procedures for clearing roots and staining parasitic and versicular-arbuscular mycorrhizal fungi for rapid assessment of infection. Trans Br Mycol Soc 55:158–161

Siles JA, Pérez-Mendoza D, Ibáñez JA, Scervino JM, Ocampo JA, García-Romera I, Sampedro I (2014) Assessing the impact of biotransformed dry olive residue application to soil: effects on enzyme activities and fungal community. Int Biodeterior Biodegradation 89:15–22

Smith SE, Read R (2008) Mycorrhizal symbiosis, 3rd edn. Academic Press, Amsterdam, Boston, 787 p

Smith SE, Smith FA (2012) Fresh perspective on the roles of arbuscular mycorrhizal fungi in plant nutrition and growth. Mycologia 104:1–13

Solans M, Ruiz OA, Wall LG (2015) Effect of actinobacteria on Lotus tenuis – Mesorhizobium loti symbiosis: preliminary study. Symbiosis 65(1):33–37

Trujillo ME, Fernández-Molinero C, Velázquez E, Kroppenstedt RM, Schumann P, Mateos PF et al (2005) *Micromonospora mirobrigensis* sp. nov. Int J Syst Evol Microbiol 55:877–880. doi:10.1099/ijs.0.63361-63360

Turner TR, James EK, Poole PS (2013) The plant microbiome. Genome Biol 14:209

Chapter 12
Different Effects on *Vigna unguiculata* Plants After the Inoculation with Strains from Two *Bradyrhizobium* Symbiovars

Ana Bejarano, Luís R. Silva, Paula Andrade, Encarna Velázquez and Raúl Rivas

Abstract The species *Vigna unguiculata* (cowpea) from the Tribe Phaseolae is indigenous to Africa and was introduced from this continent into some European countries. In a previous work, we analysed the core and symbiotic genes of several slow-growing rhizobial strains nodulating *V. unguiculata* in a soil from Extremadura. Most of them belong to the genus *Bradyrhizobium* group I on the basis of the 16S rRNA gene analysis with only one strain belonging to the group II. These two groups also correspond to two different symbiovars on the basis of the *nodC* gene analysis, genisteae and vignae, respectively. In the present work, we analysed the effect of the inoculation of strains VUPME29 and VUPME10 representative from symbiovars genisteae and vignae, respectively, on *V. unguiculata* yield and bioactive compounds production. The results showed that plants inoculated with the strain VUPME10 from symbiovar vignae have higher number of nodules per plant, dry shoot weight and N content than those inoculated with the strain VUPME29 from symbiovar genistearum. Also, differences were found in the concentration of some bioactive compounds in leaves of *V. unguiculata* inoculated with these two strains. This is the first study about the effect of inoculation with strains from different symbiovars in the yield and the bioactive profile of a legume.

Keywords *Vigna unguiculata* · Legumes · *Bradyrhizobium* · Symbiovars · Effectiveness · Bioactive compounds

A. Bejarano · E. Velázquez (✉) · R. Rivas
Departamento de Microbiología y Genética, Universidad de Salamanca, Salamanca, Spain
e-mail: evp@usal.es

L.R. Silva
CICS-UBI—Centro de Investigação em Ciências da Saúde, Universidade da Beira Interior, Covilhã, Portugal

P. Andrade
3REQUIMTE/Laboratório de Farmacognosia, Departamento de Química, Faculdade de Farmácia, Universidade do Porto, Porto, Portugal

© Springer International Publishing Switzerland 2016

F. González-Andrés and E. James (eds.), *Biological Nitrogen Fixation and Beneficial Plant–Microbe Interactions*, DOI 10.1007/978-3-319-32528-6_12

12.1 Introduction

Vigna unguiculata L. Walp (cowpea) is a legume from tribe Phaseolae cultivated in America, Asia, and Africa (Ehlers and Hall 1997). Cowpea is the most important food legume in Africa (Belane and Dakora 2012) since their seeds contain a high content of proteins (23 %), carbohydrates (56 %), fibre (4 %), lipids (1 %), vitamins, and several micronutrients (Iqbal et al. 2006; Belane and Dakora 2012). They are rich in essential amino acids such as lysine, methionine, threonine, tryptophan, and minerals such as potassium, magnesium, calcium, and phosphorus (Iqbal et al. 2006; Belane and Dakora 2012).

In addition to seeds, young leaves of cowpea are consumed in several African and Asian countries (Nielsen et al. 1997), making this legume one of the most important leaf vegetables in African countries (Barrett 1990) where fresh leaves are directly consumed from home gardens and fields (Bittenbender 1992). They are rich in proteins, minerals such as phosphorous, iron, and calcium, and vitamins such as ascorbic acid, niacin, and β-carotene (Nielsen et al. 1997), but no data are available about the phytochemicals present in the leaves of cowpea.

The species *V. unguiculata* was introduced in Europe from North-eastern Africa around 300 BC (Paludosi and Ng 1997), and it is cultivated in Mediterranean countries. In Spain, it is cultivated in Extremadura, a warm region with acidic soils suitable for the cropping of this legume, where their seeds are very appreciated by the consumers being leaves intended for animal consumption. The interest of cowpea cultivation in Spain could be increased since it can be included in greening practices as a nitrogen-fixing bean due its ability to form nitrogen-fixing symbioses with rhizobia, mainly belonging to genus *Bradyrhizobium* (Bejarano et al. 2014).

In Africa, the cowpea production is strongly dependent on symbiotic N_2 fixation mainly provided by *Bradyrhizobium* strains (Krasova-Wade et al. 2003; Steenkamp et al. 2008; Pule-Meulenberg et al. 2010, Grönemeyer et al. 2014). Some of these African strains belong to the recently described symbiovar vignae together with one strain isolated in Spain where cowpea was mostly nodulated by the symbiovar genistearum (Bejarano et al. 2014). Although differences in the effectiveness of nodules occupied by different *Bradyrhizobium* ITS genotypes were observed in African countries (Pule-Meulenberg et al. 2010), there are no data about the effectiveness on *V. unguiculata* of different *Bradyrhizobium* symbiovars nodulating this legume.

Therefore, the first aim of this study was to analyse the nodulation and nitrogen fixation abilities of representative strains from the two different *Bradyrhizobium* symbiovars found in Spain, one of them present so far only in cowpea nodules in Europe (genistearum) and the other present in both Europe and Africa (vignae).

In addition to the importance of cowpea as food, in the last years, it has gained interest as a nutraceutical legume or pulse due to its proteins (Duranti 2006) and antioxidant activity (Das et al. 2012). Cowpea seeds contain significant levels of phenolic compounds (Cai et al. 2003; Gutiérrez-Uribe et al. 2011), flavonols (Dueñas et al. 2005; Gutiérrez-Uribe et al. 2011), anthocyanins (Ha et al. 2010; Ojwang et al. 2012), and proanthocyanidins (Ojwang et al. 2013), whereas no data about the cowpea leaves composition are available.

In two previous works, we have shown that the inoculation of soybean, a widely know nutraceutical legume (Carbonaro et al. 2015) with *B. japonicum,* leads to an increase in the content of phenolic compounds and organic acids in leaves (Couto et al. 2011) and the organic acids on seeds (Silva et al. 2013). Therefore, the second aim of this study was to analyse the content of these compounds in *V. unguiculata* leaves and seeds after the inoculation of the representative strains from the two different *Bradyrhizobium* symbiovars. As far as we know, this is the first study about the effect of different *Bradyrhizobium* symbiovars in the yield and chemical composition of leaves and seeds from *V. unguiculata.*

12.2 Materials and Methods

12.2.1 Effectiveness Assays

V. unguiculata seeds were surface-sterilized for 20 min in sodium hypochlorite solution (7 %) and then washed ten times with sterile distilled water. Surface-sterilized seeds were placed in pots of 2.5 l capacity containing a 50 % (v/v) mixture of peat and vermiculite autoclaved during 1 h in three consecutive days.

Thirty plants, one per pot, were used for each treatment: (i) inoculation with strain VUPME29 from symbiovar genistearum, (ii) inoculation with strain VUMPE10 from symbiovar vignae, and (iii) an uninoculated control. Five days after sowing, each plant was inoculated with 100 ml of a bacterial suspension containing 10^6 CFU/ml that was obtained after seven days of incubation on YMA plates (Vincent 1970) at 28 °C. Both inoculated and uninoculated control plants were irrigated with water from a bottom reservoir every 48 h along 3 months in a laboratory greenhouse under temperatures of 15–20 °C at night and 25–35 °C in the day, humidity being set at 70 %. After this time, 15 plants per treatment were harvested, the number of nodules per plant was counted, and the plants were dried at 40 °C in a stove vent for chemical analyses. The remaining 15 plants were maintained in the greenhouse, and the seeds were harvested 130 days after inoculation; then, they were dried at room temperature. The N content was analysed in the Ionomic Service from IRNASA-CSIC (Spain). Statistical analyses were carried out using StatView program for Macintosh computers. Data were analysed by one-way analysis of variance, and mean values were compared by Fisher's protected LSD (least significant difference) test ($P \leq 0.05$).

12.2.2 Phenolic Composition Analysis

All chemicals used were of analytical grade. The standards *p*-coumaric acid, ferulic acid, sinapic acid, chlorogenic acid, protocatechuic acid, *p*-hydroxybenzoic acid, catechin, and kaempferol-3-*O*-glucoside were obtained from Sigma-Aldrich (St. Louis, MO, USA). Vitexin, isovitexin, quercetin-3-*O*-glucoside, and quercetin-3-*O*-rutinoside were from Extrasynthese (Genay, France). Methanol and formic and hydrochloric acids were obtained from Merck (Darmstadt, Germany). The water was treated in a Milli-Q water purification system (Millipore, Bedford, MA).

Dried seeds and leaves were weighed and milled (mean particle size lower than 910 μm). The samples corresponded to the mixture of the dried seeds and lyophilized leaves of fifteen replicates of each treatment. Each was further divided into three aliquots and analysed separately.

The phenolic compounds were analysed on an analytical HPLC unit (Gilson), using a previously described procedure (Silva et al. 2013). Twenty microliters of each methanolic extract was analysed on an analytical HPLC unit (Gilson), using Spherisorb ODS2 (25.0 × 0.46 cm; 5 μm, particle size; Waters, Milford, MA) column. The solvent system used was a gradient of water–formic acid (19:1) (A) and methanol (B), starting with 5 % methanol and installing a gradient to obtain 15 % B at 3 min, 25 % B at 13 min, 30 % B at 25 min, 35 % B at 35 min, 45 % B at 39 min, 45 % B at 42 min, 50 % B at 44 min, 55 % B at 47 min, 70 % B at 50 min, 75 % B at 56 min, and 80 % B at 60 min, at a solvent flow rate of 0.9 mL/min. Detection was achieved with a Gilson diode array detector (DAD). Spectral data from peaks were accumulated in the range 200–400 nm, and chromatograms were recorded at 280, 320, and 350 nm. The data were processed on Unipoint System software (Gilson Medical Electronics, Villiers-le-Bel, France). Phenolic compound quantification was achieved by measuring the absorbance recorded in the chromatograms relative to external standards. *p*-coumaric derivatives were quantified as *p*-coumaric acid. Phenolic acid derivatives were quantified as ferulic acid. Unidentified flavonoids were quantified as quercetin. Hydroxybenzoic acids and flavan-3-ols were quantified at 280 nm, hydroxycinnamic acids at 320, and the other flavonoids at 350 nm. This procedure was performed in triplicate.

All data were recorded as mean ± standard deviation of triplicate determinations. Mean values were compared using one-way analysis of variance (one-way ANOVA) (GraphPad Prism version 5.00, GraphPad Software, Inc., San Diego, CA). Differences were considered significant at $p < 0.05$.

12.3 Results and Discussion

12.3.1 Plant Assays

In several studies, the nitrogen fixation effectiveness of cowpea bradyrhizobia has been analysed (Awonaike et al. 1990; Stamford et al. 1995; Fening and Danso 2002; Pule-Meulenberg et al. 2010; Mathu et al. 2012; Nyoki and Ndakidemi 2014). From these studies, it can be concluded that the inoculation with *Bradyrhizobium* strains increased the N uptake by the cowpea plants, although differences in the effectiveness of different *Bradyrhizobium* strains were found in some of them (Awonaike et al. 1990). Several strains used in the inoculation experiments were not identified (Awonaike et al. 1990; Stamford et al. 1995; Fening and Danso 2002). Some of them were commercial strains (Mathu et al. 2012; Nyoki and Ndakidemi 2014) sometimes identified as *B. japonicum* (Nyoki and Ndakidemi 2014) or were identified on the basis of the IGS region (Pule-Meulenberg et al. 2010). In none of these studies, the strains were identified at symbiovar level that is determinant of the effectiveness in nodulation and nitrogen fixation of rhizobial strains. In the present work, we compared the effectiveness of two strains representative of two symbiovars, genistearum nodulating cowpea in Spain (VUPME29) and vignae nodulating cowpea in Spain and in Africa (VUPME10) where *V. unguiculata* has its distribution centre and from where it was distributed to Europe (Paludosi and Ng 1997).

The results of plant assays showed significant differences between the two strains analysed, VUPME29 and VUPME10, in nodule numbers, shoot dry weight, plant N content, and dry weight of seeds (Table 12.1). The inoculation with the strain VUPME29 from the symbiovar genistearum nodulated cowpea but showed no significant differences in the shoot dry weight and N content per plant with respect to the uninoculated control at the time of the leaves harvest. However, at the final harvest, the dry weight of seeds was significantly higher than that of seeds from uninoculated plants. This suggests a delay in the N fixation process in nodules from the plants inoculated with the strain VUPME29 with respect to the strain VUMPE10 (Table 12.1). The inoculation with the strain VUPME10 from the symbiovar vignae yields plants with high number of nodules that were very effective in the production of shoot dry biomass, N content per plant, and seed dry weight (Table 12.1).

The results of this work showed that the symbiovar vignae is significantly more effective than the symbiovar genistearum which is in agreement with the geo graphical distribution of both symbiovars. The symbiovar vignae is probably indigenous to Africa where they coexist for a long time with *V. ungiculata* before the introduction of this legume in other continents together with its seeds (Bejarano et al. 2014). Therefore, this symbiovar is well adapted to *V. unguiculata* symbiosis, and its scarce presence in Spanish soils could be due to lower competitiveness with respect to the strains from symbiovar genistearum present in these soils. This

Table 12.1 Results from microcosm experiment in *Vigna unguiculata*

Treatment	Nodules per plant	Shoot dry weight (g) (±S.E.)	N (%) (±S.E.)	N content per plant (mg) (±S.E.)	Dry weight per seed (g) (±S.E.)
Control	0	7.0 (± 0.6)[a]	1.64 (± 0.02)[a]	0.11 (± 0.02)[a]	0.09 (± 0.01)[a]
VUPME29	98.3 (± 3.6)[a]	6.1 (± 0.4)[a]	1.57 (± 0.01)[a]	0.10 (± 0.01)[a]	0.14 (± 0.01)[b]
VUPME10	289.7 (± 17.6)[b]	9.1 (± 0.2)[b]	3.32 (± 0.17)[b]	0.30 (± 0.01)[b]	0.17 (± 0.01)[c]

Values followed by the same letter in each treatment are not significantly different from each other at $P = 0.05$ according to Fisher's protected LSD (least significant difference). S.E. = standard error symbiovar, probably indigenous to Europe (Stepkowski et al. 2005), presents a good adaptation to the Spanish soils, but it is not the best partner of *V. unguiculata* that establish symbiosis with this symbiovar due to its high promiscuity degree. Therefore, biofertilization of *V. unguiculata* with strains from the symbiovar vignae in European soils will enhance the quality and the yield of this legume facilitating its use in greening practices.

12.3.2 Phenolic Compounds

Phenolic compounds are products from the secondary metabolism in plants; they play vital roles during plant development and contribute to the colour of plants (Liu 2013). Phenolics have been reported to possess health benefits, like prevention of chronic diseases, such as cancer and cardiovascular diseases (Cai et al. 2003). They have been analysed in previous works in cowpea seeds (Cai et al. 2003; Dueñas et al. 2005; Ojwang et al. 2012), but there are no data about those present in cowpea leaves.

Concerning the seeds, it has been reported the presence of protocatechuic, *p*-hydroxybenzoic, caffeic, *p*-coumaric, ferulic, 2,4-dimethoxybenzoic, gallic, and cinnamic acids (Cai et al. 2003; Gutiérrez-Uribe et al. 2011) and flavonols such as myricetin, quercetin, kaempferol, myricetin-3-*O*-glucoside, quercetin-3-*O*-galactoside, quercetin-3-*O*-glucoside, or quercetin feruloyl-diglycoside (Dueñas et al. 2005; Gutiérrez-Uribe et al. 2011). Some varieties contain anthocyanins (Ha et al. 2010; Ojwang et al. 2012) and proanthocyanidins such as catechin-3-*O*-glucoside (Ojwang et al. 2013).

The analysis of *V. unguiculata* seeds in this study revealed the presence of 13 phenolic compounds from which only protocatechuic, *p*-hydroxybenzoic, caffeic, *p*-coumaric, and ferulic acids were previously reported (Cai et al. 2003; Dueñas et al. 2005; Gutiérrez-Uribe et al. 2011). No changes in the phenol profile were found after the inoculation with the *Bradyrhizobium* strains. In agreement with our previous results on seeds of soybean inoculated with *Bradyrhizobium* (Silva et al. 2013), the inoculation decreased the contents of phenol compounds in cowpea seeds (Table 12.2). Nevertheless, some differences were found in the amounts of some phenols such as catechin and kaempferol-3-*O*-glucoside depending on the inoculated symbiovar, being higher in seeds from plants inoculated with the strain VUPME29 (symbiovar genistearum).

The analysis of phenolic compounds performed by first time in this study in cowpea leaves allowed the identification of 15 phenolic compounds that were also found in inoculation treatments (Table 12.3). Within them, we identified in cowpea leaves quercetin-3-*O*-rutinoside, vitexin, and isovitexin that were reported for the first time in *V. unguiculata* seeds in previous works (Cai et al. 2003; Dueñas et al. 2005; Gutiérrez-Uribe et al. 2011). The total content of phenolics in leaves was similar in all treatments with 1389, 1091, and 1341 mg/kg for control, VUPME10, and VUPME29, respectively (Table 12.3). The *p*-coumaric acid derivative 2 was the main phenolic compound found in all samples (43 %). No significant

Table 12.2 Phenolic compound profile of *Vigna unguiculata* dried seeds (mg/kg)[a]

Phenolic compounds (mg/kg)	Control	VUPME10	VUPME29
Protocatechuic acid	1.3 ± 0.0[a]	0.4 ± 0.0[b,a]	0.4 ± 0.0[b,a]
Catechin	2.7 ± 0.4[a]	0.8 ± 0.1[b,a]	1.7 ± 0.1[b,b]
p-hydroxybenzoic acid	0.5 ± 0.1[a]	0.1 ± 0.0[b,a]	0.8 ± 0.1[b,b]
Chlorogenic acid	0.7 ± 0.1[a]	0.3 ± 0.0[b,a]	nd
p-coumaric acid	0.1 ± 0.0[a]	nd	nd
Ferulic acid	0.1 ± 0.0[a]	nd	0.1 ± 0.0[a,a]
Vitexin	0.1 ± 0.0[a]	0.1 ± 0.0[a,a]	0.1 ± 0.0[a,a]
Unknown flavonoid 1	0.4 ± 0.0[a]	0.2 ± 0.0[b,a]	0.2 ± 0.0[b,a]
Unknown flavonoid 2	1.0 ± 0.0[a]	0.4 ± 0.0[b,a]	0.2 ± 0.0[b,b]
Kaempferol-3-*O*-glucoside	1.9 ± 0.1[a]	0.5 ± 0.0[b,a]	1.7 ± 0.0[b,b]
Quercetin-3-*O*-glucoside	0.3 ± 0.0[a]	0.1 ± 0.0[b,a]	0.2 ± 0.0[a,a]
Unknown flavonoid 3	0.1 ± 0.0[a]	0.1 ± 0.0[a,a]	nd
Unknown flavonoid 4	0.1 ± 0.0[a]	nd	0.1 ± 0.0[a,a]
Σ	9.3	3.0	5.5

[a] Values are expressed as mean ± standard deviation of three assays; Σ, sum of the determined phenolic compounds; [a] in the same line, compared the samples, the same letters represents no significant differences (p>0.05); [b] in the same line, compared the samples, the presence of different letters represents no significant differences ($p \leq 0.05$)

differences were found among treatments although the inoculation with both strains showed a trend to increase the contents on vitexin, sinapic acid, and the unknown flavonoid 5 and to decrease the contents of isovitexin, quercetin-3-*O*-rutinoside, and kaempferol-3-*O*-glucoside (Table 12.3). Newly, differences were found in the amounts of several phenolic compounds depending on the inoculated symbiovar, highlighting the higher amounts of the kaempferol-3-*O*-glucoside and *p*-coumaric acid derivative 2 in leaves after the inoculation of the strain VUPME29 (symbiovar genistearum). In turn, the amount of the unknown flavonoid 3 was higher in leaves of plants inoculated with the strain VUPME10 (Table 12.3).

The results obtained from the present work for cowpea are in agreement with those previously obtained from soybean where the inoculation with *B. japonicum* led to an increase in some phenolic compounds in the leaves (Couto et al. 2011), but led to a decrease in the seeds (Silva et al. 2013). It is surprising that despite the increase of the yield found after inoculation was higher in the case of vignae symbiovar (Table 12.1), the inoculation with this strain led to higher decreases in some phenolic compounds in both leaves and seeds. This can be related to the dilution effect already reported by Silva et al. (2013) who showed that plants with higher biomass showed a reduction in phenolic compounds.

In summary, the results of this work showed that the strain VUPME10 from symbiovar vignae present in Africa and Spain has a significant better effect on the yield of *V. unguiculata* in both leaves and seeds than the strain VUPME29 from symbiovar genisteae indigenous to Europe. This showed that the inoculation with strains of symbiovar vignae scarcely present in Spain can be a good biofertilization

Table 12.3 Phenolic compounds profile of *Vigna unguiculata* dried leaves (mg/kg)[a]

Phenolic compounds (mg/kg)		Control	VUPME10	VUPME29
1	*p*-Coumaric acid derivative 1	59.6±1.4[a]	55.4±1.4[b,a]	58.2±1.2[a,a]
2	*p*-Coumaric acid derivative 2	584.8±29.5[a]	455.8±14.5[b,a]	598.9±32.4[a,b]
3	*p*-Coumaric	14.4±0.3[a]	14.1±1.4[a,a]	10.5±0.3[b,b]
4	Phenolic acid derivative	26.7±0.9[a]	24.2±1.1[b,a]	32.1±0.3[b,b]
5	Ferulic acid	8.0±0.4[a]	10.4±0.5[b,a]	8.3±0.4[a,b]
6	Sinapic acid	5.6±0.1[a]	15.5±2.2[b,a]	16.8±1.6[b,a]
7	Vitexin	22.4±0.4[a]	26.8±0.1[b,a]	27.1±1.3[b,a]
8	Quercetin-3-*O*-rutinoside	56.7±7.0[a]	37.1±0.5[b,a]	43.3±1.2[b,a]
9	Isovitexin	37.7±3.5[a]	15.8±1.3[b,a]	20.7±4.7[b,a]
10	Unknown flavonoid 1	83.9±1.4[a]	55.9±1.6[b,a]	80.5±1.3[a,b]
11	Unknown flavonoid 2	66.5±4.7[a]	55.2±1.3[b,a]	76.1±3.5[b,b]
12	Unknown flavonoid 3	98.7±0.4[a]	123.7±7.3[b,a]	73.0±9.0[b,b]
13	Unknown flavonoid 4	96.7±2.3[a]	55.4±1.8[b,a]	82.5±3.8[b,b]
14	Unknown flavonoid 5	39.1±1.3[a]	41.0±4.2[a,a]	62.3±9.9[b,b]
15	Kaempferol-3-*O*-glucoside	187.7±2.3[a]	104.3±3.8[b,a]	150.6±7.2[b,b]
	$\sum$	1388.5	1090.8	1340.7

[a] Values are expressed as mean ± standard deviation of three assays; $\sum$, sum of the determined phenolic compounds; [a] in the same line, compared the samples, the same letters represents no significant differences ($p>0.05$); [b] in the same line, compared the samples, the presence of different letters represents no significant differences ($p \leq 0.05$)

practice to increase the production of this legume in this country allowing its introduction to the greening schemes for European farmers. The inoculation with both symbiovars does not change the phenolic profile of leaves or seeds although differences were found in the amounts of several phenolic compounds between inoculation treatments with the symbiovars genistearum and vignae. This is the first report about differences found in the bioactive compounds of a legume depending on the rhizobial symbiovar inoculated.

Acknowledgements This work was supported by the "Fundação para a Ciência e a Tecnologia"—FCT (PEst-C/SAU/UI0709/2014) cofunded by Fundo Europeu de Desenvolvimento Regional—FEDER via Programa Operacional Factores de Competitividade—COMPETE/QREN. Luís R. Silva acknowledges FCT for the financial support for the post-doc grant (SFRH/BPD/105263/2014).

References

Awonaike KO, Kumarasinghe KS, Danso SKA (1990) Nitrogen fixation and yield of cowpea (*Vigna unguiculata*) as influenced by cultivar and *Bradyrhizobium* strain. Field Crops Res 24:163–171

Barrett RP (1990) Legume species as leaf vegetables. In: Janick J, Simon JE (eds) Advances in new crops. Timber Press, Portland, OR, pp 391–396

Bejarano A, Ramirez-Bahena MH, Velázquez E et al (2014) *Vigna unguiculata* is nodulated in Spain by endosymbionts of Genisteae legumes and by a new symbiovar (vignae) of the genus *Bradyrhizobium*. Syst Appl Microbiol 37:533–540

Belane AK, Dakora FD (2012) Elevated concentrations of dietarily-important trace elements and macronutrients in edible leaves and grain of 27 cowpea (*Vigna unguiculata* L. Walp.) genotypes: implications for human nutrition and health. Food Nutr Sci 3:377–386

Bittenbender HC (1992) Handling and storage of cowpea *Vigna unguiculata* (L) Walp as a leaf vegetable. Trop Agric 69:197–199

Cai R, Hettiarachchy NS, Jalaluddin M (2003) High-performance liquid chromatography determination of phenolic constituents in 17 varieties of cowpeas. J Agric Food Chem 51:1623–1627

Carbonaro M, Maselli P, Nucara A (2015) Structural aspects of legume proteins and nutraceutical properties. Food Res Int 76:19–30. doi:10.1016/j.foodres.2014.11.007

Couto C, Silva LR, Valentão P et al (2011) Effects induced by the nodulation with *Bradyrhizobium japonicum* on *Glycine max* (soybean) metabolism and antioxidant potential. Food Chem 127:1487–1495

Das L, Bhaumik E, Raychaudhuri U et al (2012) Role of nutraceuticals in human health. J Food Sci Technol 49:173–183

Dueñas M, Fernández D, Hernández T et al (2005) Bioactive phenolic compounds of cowpeas (*Vigna sinensis* L). Modifications by fermentation with natural microflora and with *Lactobacillus plantarum* ATCC 14917. J Sci Food Agric 85:297–304

Duranti M (2006) Grain legume proteins and nutraceutical properties. Fitoterapia 77:67–82

Ehlers JD, Hall AE (1997) Cowpea (*Vigna unguiculata* L. Walp.). Field Crops Res 53:187–204

Fening JO, Danso SKA (2002) Variation in symbiotic effectiveness of cowpea bradyrhizobia indigenous to Ghanaian soils. Appl Soil Ecol 21:23–29

Grönemeyer JL, Kulkarni A, Berkelmann D et al (2014) Rhizobia indigenous to the Okavango region in Sub-Saharan Africa: diversity, adaptations, and host specificity. Appl Environ Microbiol 80:7244–7257

Gutiérrez-Uribe JA, Romo-López I, Serna-Saldívar SO (2011) Phenolic composition and mammary cancer cell inhibition of extracts of whole cowpeas (*Vigna unguiculata*) and its anatomical parts. J Funct Foods 3:290–297. doi:10.1016/j.jff.2011.05.004

Ha TJ, Lee MH, Park CH et al (2010) Identification and characterization of anthocyanins in yard-long beans (*Vigna unguiculata* ssp. *sesquipedalis* L.) by high-performance liquid chromatography with diode array detection and electrospray ionization/mass spectrometry (HPLC-DAD-ESI/MS) analysis. J Agric Food Chem 58:2571–2576

Iqbal A, Khalil IA, Ateeq N et al (2006) Nutritional quality of important food legumes. Food Chem 97:331–335

Krasova-Wade T, Ndoye I, Braconnier S et al (2003) Diversity of indigenus bradyrhizobia associated with three cowpea cultivars (*Vigna unguiculata* (L.) Walp.) grown under limited and favorable water conditions in Senegal (West Africa). African J Biotech 2:13–22

Liu RH (2013) Health-promoting components of fruits and vegetables in the diet. Adv Nutr 4:384s–392s

Mathu S, Herrmann L, Pypers P et al (2012) Potential of indigenous bradyrhizobia versus commercial inoculants to improve cowpea (*Vigna unguiculata* L. walp.) and green gram (*Vigna radiata* L. wilczek.) yields in Kenya. Soil Sci Plant Nut 58:750–763

Nielsen SS, Ohler TA, Mitchell CA (1997) Cowpea leaves for human consumption: production, utilization, and nutrient composition. In: Singh BB, Mohan Raj DR, Dashiell KE, Jackai LEN (eds) Advances in Cowpea Research. International Institute of Tropical Agriculture, Ibadan, Nigeria. Japan International Research Center for Agricultural SciencesTsukuba Ibaraki, Japan, pp 326–332

Nyoki D, Ndakidemi PA (2014) Effects of *Bradyrhizobium japonicum* inoculation and supplementation with phosphorus on macronutrients uptake in cowpea (*Vigna unguiculata* (L.) Walp). Am J Plant Sci 5:442–451

Ojwang LO, Dykes L, Awika JM (2012) Ultra performance liquid chromatography-tandem quadrupole mass spectrometry profiling of anthocyanins and flavonols in cowpea (*Vigna unguiculata*) of varying genotypes. J Agric Food Chem 60:3735–3744

Ojwang LO, Yang L, Dykes L, Awika J (2013) Proanthocyanidin profile of cowpea (*Vigna unguiculata*) reveals catechin-O-glucoside as the dominant compound. Food Chem 139:35–43

Paludosi S, Ng NQ (1997) Origin, taxonomy and morphology of *Vigna unguiculata* (L.) Walp. In: Singh BB, Mohan Raj DR, Dashiell KE, Jackai LEN (eds) Advances in Cowpea Research. Copublication of International Insitute of Troplical Agriculture (IITA) and Japan International Research Center of Agricultural Siences (JIRCAS), Nigeria, pp 1–12

Pule-Meulenberg F, Belane AK, Krasova-Wade T et al (2010) Symbiotic functioning and bradyrhizobial biodiversity of cowpea (*Vigna unguiculata* L. Walp.) in Africa. BMC Microbiol 10:89

Silva LR, Pereira MJ, Azevedo J et al (2013) Inoculation with *Bradyrhizobium japonicum* enhances the organic and fatty acids content of soybean (*Glycine max* (L.) Merrill) seeds. Food Chem 141:3636–3648

Stamford NP, Chamber-Perez M, Camacho-Martínez M (1995) Symbiotic effectiveness of several tropical *Bradyrhizobium* strains on cowpea under a long-term exposure to nitrate: relationships between nitrogen fixation and nitrate reduction activities.]. Plant Physiol 147:378–382

Steenkamp ET, Stepkowski T, Przymusiak A et al (2008) Cowpea and peanut in southern Africa are nodulated by diverse *Bradyrhizobium* strains harboring nodulation genes that belong to the large pantropical clade common in Africa. Mol Phylogenet Evol 48:1131–1144

Stepkowski T, Moulin L, Krzyzańska A et al (2005) European origin of *Bradyrhizobium* populations infecting lupins and serradella in soils of Western Australia and South Africa. Appl Environ Microbiol 71:7041–7052

Vincent JM (1970) The cultivation, isolation and maintenance of rhizobia. In: Vincent JM (ed) A manual for the practical study of root-nodule. Blackwell Scientific, Oxford, pp 1–13

Part IV
Physiology and Biochemistry of Beneficial Microorganisms and Associated Plants

Chapter 13
Salt Stress Tolerance in *Casuarina glauca* and Its Relation with Nitrogen-Fixing *Frankia* Bacteria

Nuno Duro, Mário da Costa, Paula Batista-Santos, Paula Scotti-Campos, Isabel P. Pais, Filipe Colwell, Ana P. Rodrigues, José N. Semedo, Fernando C. Lidon, José C. Ramalho, Katharina Pawlowski and Ana I. Ribeiro-Barros

Abstract Salinity is one of the most widespread abiotic stresses. It is estimated that salt stress will cause the loss of more than 50 % of arable land by the year 2050. A promising solution for the recovery of saline soils encompasses the use of actinorhizal plants, a group of perennial dicotyledonous angiosperms highly resilient to extreme environmental conditions. These plants are also able to establish a root nodule symbiosis with N_2-fixing actinobacteria of the genus *Frankia*. *Casuarina glauca*, the model actinorhizal species, tolerates NaCl concentrations above seawater levels. Such ability seems to be innate and independent of the symbiotic relationship with N_2-fixing *Frankia*. In this work, we present a mini review of the basic mechanisms underlying salt tolerance in *C. glauca* focusing on the impact of salt on the photosynthesis, redox status, and membrane integrity.

N. Duro · M. da Costa · P. Batista-Santos · F. Colwell
A.P. Rodrigues · J.C. Ramalho · A.I. Ribeiro-Barros (✉)
Plant Stress and Biodiversity Group, Instituto Superior de Agronomia (ISA), Universidade de Lisboa, Av. República, Quinta do Marquês, 2784-505, Oeiras Lisbon, Portugal
e-mail: aribeiro@isa.ulisboa.pt

N. Duro · M. da Costa · A.I. Ribeiro-Barros
Instituto de Tecnologia Química e Biológica, Universidade Nova de Lisboa, Av. República, Quinta do Marquês, Oeiras Lisbon, Portugal

P. Scotti-Campos · I.P. Pais · J.N. Semedo
Unidade de Biotecnologia e Recursos Genéticos, Instituto Nacional de Investigação Agrária e Veterinária, I.P. (INIAV), Av. República, Quinta do Marquês, 2784-505, Oeiras Lisbon, Portugal

F.C. Lidon · J.C. Ramalho · A.I. Ribeiro-Barros
GeoBioTec, Faculdade de Ciências e Tecnologia, Universidade Nova de Lisboa, Caparica, Portugal

K. Pawlowski
Department of Ecology, Environment and Plant Sciences, Stockholm University, 106 91 Stockholm, Sweden

© Springer International Publishing Switzerland 2016

F. González-Andrés and E. James (eds.), *Biological Nitrogen Fixation and Beneficial Plant–Microbe Interactions*, DOI 10.1007/978-3-319-32528-6_13

Keywords *Casuarina glauca* · *Frankia* · Salt stress · N_2-fixing symbiosis

13.1 Introduction

Casuarina glauca Sieb. Ex Spreng. (fam. Casuarinaceae) (swamp oak) is a model actinorhizal plant that establishes N_2-fixing root nodule symbiosis with bacteria of the genus *Frankia* (Zhong et al. 2010). This species shows an outstanding capacity to endure extreme environments, particularly saline conditions, as it tolerates sea-water levels of salt and is commonly planted in areas with a shallow, saline water table (Diagne et al. 2013). Thus, *C. glauca* is a promising species for the prevention of desertification, the stabilization of coastal dunes, and the remediation of degraded lands (Diem and Dommergues 1990; Ribeiro et al. 2011; Sayed 2011). This aspect is particularly relevant under the context of global climate changes which increasingly impose longer periods of drought and soil salinization (Wang et al. 2003; Chaves et al. 2009). Indeed, salt constraints are expected to have destructive global effects, resulting in up to 50 % loss of cultivable lands by the middle of the twenty-first century (Mahajan and Tuteja 2005). Moreover, the spreading of agriculture to arid and semi-arid regions under intensive irrigation management will promote secondary soil salinization as a result of changes in the hydrologic balance between water input (irrigation or rainfall) and water use (transpiration) (Chaves et al. 2009). In this paper, we describe the main mechanisms underlying salt stress tolerance in symbiotic and non-symbiotic *C. glauca*. In the first part, a general overview of the tolerance limits based on the analysis of plant biomass and water and mineral contents will be shortly addressed. The second part includes the discussion of major physiological adaptations related to photosynthesis, antioxidative status, and membrane integrity at the leaf level.

13.2 Salt Tolerance Limits

Tolerant plants are able to adopt a variety of strategies to cope with saline environments, using specific mechanisms of salt elimination or compartmentalization in the vacuoles, avoiding the toxic effects on photosynthesis and other key metabolic processes (Munns and Tester 2008). Clemens et al. (1983) ranked *Casuarina* spp. according to their salt tolerance ability: *C. equisetifolia* appears in the top as a highly tolerant species, followed by *C. cunninghamiana*, *C. glauca*, and *C. littoralis* (moderately tolerant), *C. cristata, C. decaisneana*, *C. stricta*, and *C. tolurosa* (moderately sensitive), and *C. inophloia* (sensitive). However, salt tolerance in the Casuarinaceae is dependent on the plant ecotype (Carter et al. 2006).

El-Lakany and Luard (1982) reported that *C. glauca* from Australia was able to tolerate up to 500 mM NaCl. Such tolerance may be further enhanced when the plant is in symbiosis with particular *Frankia* strains. However, *Frankia* is generally scarce in stressed soils, and thus, preinoculation of the host plants is often required in order to form effective N_2-fixing associations before field transplanting (Hafeez et al. 1999).

Batista-Santos et al. (2015) reported an outstanding salt tolerance in a *C. glauca* ecotype from New Zealand, exhibiting the first stress symptoms (leaf chlorosis and necrosis, accompanied by a decrease in growth) only at 600 mM NaCl. This phenotype was independent of the presence of symbiotic *Frankia* sp. strain Thr (Duro et al. 2016). The symptoms were associated with a moderate reduction in relative water content (RWC), which is a common effect of salinity stress associated with limited water availability for cell extension processes and plant vigour (Katerji et al. 1997) as well as with a strong osmotic adjustment, essential to maintain a potential gradient of water influx and to sustain metabolic activity (Chaves et al. 2009). Indeed, the synthesis of compatible solutes for osmotic adjustment is an acclimation response to salinity (Kozlowski 1997), which also facilitates adjustments in ion exchange (uptake, extrusion, and sequestration) and promotes the re-equilibrium of cellular homeostasis, detoxification, and survival under stress (Chaves et al. 2009). On the other hand, by reducing growth (an thereby metabolism), *C. glauca* is able to maintain the contents of key nutrients (nitrogen and carbon) in branchlets up to 600 mM NaCl (Batista-Santos et al. 2015), reinforcing its capability to control the basal metabolism and survive in harsh environments.

13.3 Photosynthetic Apparatus

The effects of salinity on photosynthesis range from the restriction of CO_2 diffusion to the carboxylation sites on chloroplasts, *via* limitation of stomatal opening, to modifications of photochemistry and carbon metabolism in the mesophyll (Sudhir and Murthy 2004; Chaves et al. 2009). By inducing stomatal closure, salt stress reduces CO_2 availability to Rubisco, thereby inhibiting C-fixation. As photon capture and electron transport are relatively insensitive to salt stress, the impact of this stress is particularly associated with the reduction of Calvin cycle activity which leads to a reduction of ATP use. The difference in sensitivities will cause an imbalance between the processes engaged in energy absorption, transfer, and transformation, i.e. the amount of light energy absorbed will exceed the amount of energy that can be used in CO_2 fixation. Under these circumstances, excited chlorophyll molecules accumulate, leading to the overproduction of reactive oxygen species (ROS) (e.g. singlet oxygen, $^1O_2$, superoxide, $O_2^{\cdot-}$, hydrogen peroxide, H_2O_2, and hydroxyl radical, $OH\cdot$) at PSI and PSII. These highly reactive molecules will cause oxidative stress conditions in the cell, affecting photosynthetic electron transport and metabolism, as well as membrane integrity (Logan 2005; Chaves et al. 2009; Fortunato et al. 2010). The extent of these negative effects varies according to the intensity and duration of the stress and is also affected by leaf age and plant genotype (Munns and Tester 2008; Chaves et al. 2009).

Under salt stress conditions, the reduction of plant photosynthetic rates has been associated with dehydration and high concentrations of Na^+ and/or Cl^- inside chloroplasts, the latter inhibiting Rubisco (Seeman and Critchley 1985). Nevertheless, the maintenance of the potential functioning of PSs and Rubisco at a considerably higher level than net photosynthesis (P_n) suggests that salt stress may suppress the carboxylase activity of this enzyme, promoting its oxygenase activity (Sivakumar et al. 2000). According to Batista-Santos et al. (2015), in *C. glauca*, P_n and stomatal conductance (G_s) were gradually affected with increasing salt concentrations. This reduction seems to be associated with metabolic changes rather than stomatal closure, as internal CO_2 concentration was slightly altered. In agreement with these observations, Duro et al. (2016) reported the enrichment of $\delta^{13}C$ promoted by salt in *C. glauca* branchlets, which is normally associated with stomatal closure and concomitant decrease of CO_2 availability in the chloroplasts. Under these conditions, Rubisco, which usually prefers ^{12}C, is forced to use all available carbon, enriching the biomass in ^{13}C (therefore increasing $\delta^{13}C$) (Brugnoli and Lauteri 1991; Van Groenigen and Van Kessel 2002; Amor and Cuadra-Crespo 2011). Indeed, under saline conditions, stomatal regulation is among the first events to occur, facilitating the maintenance of RWC within limits where photosynthetic capacity and quantum yield are barely affected, although the photosynthetic pathway will be downregulated (Chaves et al. 2009; Jaoudé et al. 2013).

Batista-Santos et al. (2015) also reported that despite the reduction of P_n to almost negligible values, the photosynthetic assimilation potential (A_{max}) was maintained high, reflecting the maintenance of the global photosynthetic potential. Additionally, the majority of fluorescence parameters were only slightly affected by salt exposure. In general, although the yield of photochemistry (Y_{II}) decreased, the *C. glauca* branchlets presented a reinforcement of downregulated processes ($Y_{(NPQ)}$), likely associated with an increase in energy dissipation by zeaxanthin, instead of non-photochemical losses associated with photodamaging effects ($Y_{(NO)}$). On the other hand, the maintenance of PSII centres in the open state, given by the low impact on the photochemical quenchings and the much lower impact on YII than in Pn, suggests the presence of photorespiration. In salt-tolerant species, leaf photosynthetic capacity might be reinforced under salinity conditions, e.g. through higher maximum quantum efficiency of PSII photochemistry (Redondo-Gomez et al. 2007) or carboxylation activity (Debez et al. 2006). In *C. glauca*, such increases were not observed, but non-stomatal limitations associated with PSII, electron transport, and enzyme activity were observed mostly from 400 mM NaCl upwards, which was consistent with the small A_{max} decrease (Batista-Santos et al. 2015).

Another important aspect to be highlighted is that salt stress damage in *C. glauca* was more prominent at the biochemical level with severe effects in key enzymes from the photosynthetic (total and initial Rubisco) and respiratory (pyruvate kinase and malate dehydrogenase) pathways, particularly at 600 mM NaCl. Such reduction in enzyme activity seemed to be related to enhanced degradation and/or reduced synthesis (Batista-Santos et al. 2015). The presence of symbiotic *Frankia* Thr did not affect the photosynthetic performance of *C. glauca* submitted to increasing salt concentrations (Batista-Santos et al. 2015).

13.4 Antioxidant System

In all biological systems, there is a constant production of ROS. In fact, the production of ROS is an inevitable consequence of photosynthetic activity and under normal metabolic conditions 10–30 % of the thylakoid electron transport might lead to O_2 photoreduction (Asada 1994; Logan 2005). Moreover, salinity increases the natural production of ROS, which may result in membrane lipoperoxidation and membrane injury (Leshem 1987; Zhu 2001; Campos et al. 2003). In defence against ROS, plants have developed a complex antioxidative system, which provides a crucial protection against oxidative damage by regulating the lifetime of ROS within the cellular environment. This system comprises numerous enzymes that catalyse ROS scavenging, such as ascorbate peroxidase (APX), superoxide dismutase (SOD), glutathione reductase (GR) (all involved in the glutathione–ascorbate cycle), and catalase (CAT), as well as compounds of low molecular weight, such as ascorbate and glutathione (Noctor and Foyer 1998; Fortunato et al. 2010). SOD will scavenge $O_2^{\cdot -}$, producing H_2O_2, after which APX (in the chloroplast) and CAT (extrachloroplast) will promote the breakdown of H_2O_2 into oxygen and water. GR performs the reduction of glutathione, an important antioxidant molecule which reduces ascorbate radicals in the ascorbate–glutathione cycle (Smirnoff 1995; Ramalho et al. 1998; Logan 2005). The correlation between the reinforcement of the antioxidant system and salinity tolerance has been observed in a growing number of species (Türkan and Demiral 2009). In many cases, it has been proposed that salt stress tolerance is related to a higher activity of antioxidant enzymes such as APX, CAT, and SOD (Shalata et al. 2001). In *C. glauca*, salt stress enhances the antioxidant response, as shown by the activities of SOD, APX, GR, and CAT (Table 13.1). In line with the observations on Batista-Santos et al. (2015), the symbiosis between *Frankia* Thr and *C. glauca* did not interfere with the plant tolerance to salt stress.

Table 13.1 Changes in malondialdehyde (MDA), superoxide dismutase (SOD), catalase (CAT), ascorbate peroxidase (APX), and glutathione reductase (GR) content in branchlets of non-symbiotic (KNO_3+) and symbiotic (NOD+) *Casuarina glauca* plants exposed to four [NaCl]

Parameter	N-source	0 mM NaCl	200 mM NaCl	400 mM NaCl	600 mM NaCl
MDA	KNO_3+	25.3 aA	21.8 aB	32.6 aA	25.9 aB
	NOD+	29.4 bA	48.5 aA	39.8 abA	42.5 abA
SOD	KNO_3+	1577 cA	4229 bA	5369 aA	4980 aA
	NOD+	1422 cA	3940 bA	4819 aA	4701 aA
CAT	KNO_3+	365 cA	1797 bA	1896 aA	2003 aA
	NOD+	303 cA	1433 bB	1428 bB	1508 aB
APX	KNO_3+	4.0 cA	13.6 bA	17.4 aA	17.3 aA
	NOD+	5.3 bA	12.4 aA	17.5 aA	13.6 aA
GR	KNO_3+	5.2 bA	19.8 aA	21.5 aA	22.4 aA
	NOD+	6.2 cA	19.7 abA	20.3 aA	14.7 bB

Each value represents the mean ± SE ($n = 5$). Different letters express significant differences between salt concentrations for KNO_3+ or NOD+ (a, b, c) or between KNO_3+ and NOD+ for the same salt concentration (A, B), with a and A representing the highest values

13.5 Lipid Analysis and Membrane Integrity

Salt stress can induce changes in the structure and function of cell membranes that result from modification of membrane lipids and/or altered permeability (Kuiper 1984; Surjus and Durand 1996). Thus, these modifications have been widely used to evaluate protoplasmic tolerance to abiotic stresses (Campos et al. 2003; Matos et al. 2010; Scotti-Campos et al. 2014). The ability to maintain membrane integrity, and hence cell compartmentalization and metabolism, is crucial for the survival of plants under environmental stress conditions. Such capacity mostly relies on the dynamics of the lipid matrix of cellular membranes, which includes quantitative and qualitative lipid modifications that allow the plants to cope with stressful conditions (Dias et al. 2010; Matos et al. 2010; Partelli et al. 2011; Scotti-Campos et al. 2013, 2014). Malondialdehyde (MDA), the end product of lipoperoxidation, is commonly used as marker of environmental stresses (Dias et al. 2010; Campos et al. 2003), including salinity (Khan and Panda 2008).

In *C. glauca*, membrane permeability was markedly altered only at 600 mM NaCl (Fig. 13.1; Scotti-Campos et al. 2016), which is an indicator of high salinity tolerance (Mansour and Salama 2004). These observations agree with the moderate impact of salt observed in the photosynthetic structures of *C. glauca*, as well as with their functional performance at high salt concentrations (Batista-Santos et al. 2015). Symbiotic *C. glauca* plants presented higher MDA levels, which reflects a higher damage (Table 13.1). This might be due to the fact that in this group of plants N-supply was restricted as the symbiosis with *Frankia* strain Thr was no longer effective from 200 mM NaCl upward, therefore limiting, e.g. protein repair and *de novo* protein synthesis (Duro et al. 2016).

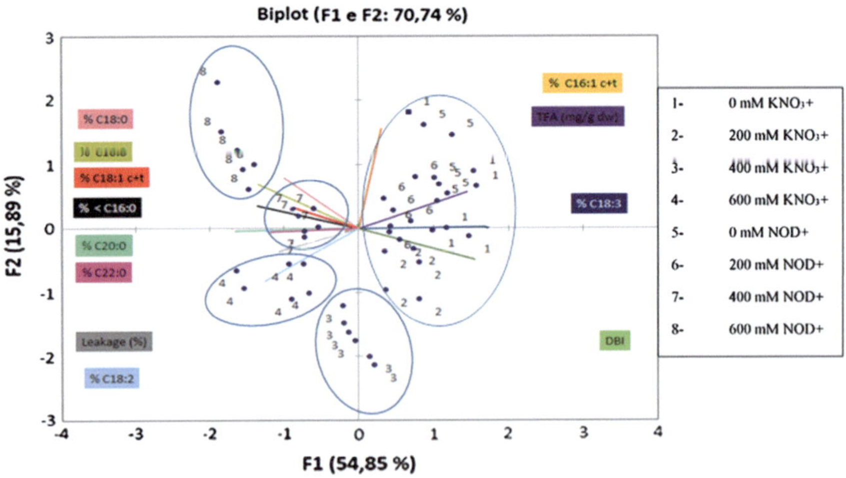

Fig. 13.1 Principal component analysis of *C. glauca* lipid metabolism. Fatty acid composition (mol%), double bond index (DBI), and electrolyte leakage (%) in branchlets of symbiotic (KNO_3+) and non-symbiotic (NOD+) plants exposed to four [NaCl]

Another marker of salt stress in plants is the reduction of the biosynthesis of unsaturated fatty acids that decreases membrane fluidity and integrity (Ellouze et al. 1995). In line with previous results, the lipid profiles of *C. glauca* exposed to increasing salt concentrations (Fig. 13.1) seem to be related to the stabilization of cellular membranes and to the protection of cells from salinity injury, as they would help to rigidify the membranes and to reduce their permeability to salt ions (Huang 2006). On the other hand, Allakhverdiev et al. (2001) showed that unsaturation of fatty acids stimulates the antiporter system Na^+/H^+ (that removes Na^+ from the cytosol with energy consumption) by increasing membrane fluidity. This protects the activity of PSII and PSI from the accumulation of Na^+. Thus, in *C. glauca*, unsaturated fatty acids seem to play an important role in the maintenance of photosynthetic mechanisms under salt stress.

13.6 Conclusion

C. glauca is able to tolerate moderate to high levels of salt. Such capacity seems to be dependent on the plant ecotype and, at least in some cases, is innate and independent of its symbiotic capacity with *Frankia* bacteria. Salt tolerance in *C. glauca* is connected with photosynthetic adjustments and membrane preservation as well to a controlled oxidative state environment. However, the available literature is still scarce, and further complementary studies are needed so that the best *C. glauca* genotypes and *Frankia* strain combinations are selected.

Acknowledgements This work was supported by Fundação para a Ciência e Tecnologia under the scope of the project PTDC/AGR-FOR/4218/2012 and grant SFRH/BPD/78619/2011 (P. Batista-Santos).

References

Allakhverdiev SI, Kinoshita M, Inaba M et al (2001) Unsaturated fatty acids in membrane lipids protect the photosynthetic machinery against salt induced damage in *Synechococcus*. Plant Physiol 125:1842–1853

Amor FM, Cuadra-Crespo P (2011) Alleviation of salinity stress in broccoli using foliar urea or methyl-jasmonate: analysis of growth, gas exchange, and isotope composition. Plant Growth Regul 63:55–62

Asada K (1994) Mechanisms for scavenging reactive molecules generated in chloroplasts under light stress. In: Baker NR, Bowyer JR (eds) Photoinhibition of photosynthesis—from molecular mechanisms to the field. BIOS Sci Pub, Oxford, pp 129–142

Batista-Santos P, Duro N, Rodrigues AP et al (2015) Is salt stress tolerance in *Casuarina glauca* Sieb. ex Spreng. associated with its nitrogen-fixing root-nodule symbiosis? An analysis at the photosynthetic level. Plant Physiol Biochem 96:97–109

Brugnoli E, Lauteri M (1991) Effects of salinity on stomatal conductance, photosynthetic capacity, and carbon isotope discrimination of salt-tolerant (*Gossypium hirsutum* L.) and salt-sensitive (*Phaseolus vulgaris* L.) C3 non-halophytes. Plant Physiol 95:628–635

Campos PS, Quartin VL, Ramalho JC et al (2003) Electrolyte leakage and lipid degradation account for cold sensitivity in leaves of *Coffea* sp. plants. J Plant Physiol 160:283–292
Carter JL, Colmer TD, Veneklaas EJ (2006) Variable tolerance of wetland tree species to combined salinity and waterlogging is related to regulation of ion uptake and production of organic solutes. New Phytol 169:123–133
Chaves MM, Flexas J, Pinheiro C (2009) Photosynthesis under drought and salt stress: regulation mechanisms from whole plant to cell. Annals Bot 103:551–560
Clemens J, Campbell LC, Nurisjah S (1983) Germination, growth and mineral ion concentrations of Casuarina species under saline conditions. Aust J Bot 31:1–9
Debez A, Saadaoui D, Ramani B et al (2006) Leaf H+-ATPase activity and photosynthetic capacity of *Cakile maritima* under increasing salinity. Environ Exp Bot 57:285–295
Diagne N, Diouf D, Svistoonoff S et al (2013) Casuarina in Africa: distribution, role and importance of arbuscular mycorrhizal, ectomycorrhizal fungi and *Frankia* on plant development. J Environ Manage 128:204–209
Dias AS, Barreiro MG, Campos PS et al (2010) Wheat cellular membrane thermotolerance under heat stress. J Agron Crop Sci 196:100–108
Diem HG, Dommergues YR (1990) Current and potential uses and management of Casuarinaceae in tropics and subtropics. In: Schwintzer CR, Tjepkma JD (eds) The biology of *Frankia* and actinorhizal plants. Academic Press, San Diego, pp 317–342
Duro N, Batista-Santos P, Maia R et al (2016) The impact of salinity on the symbiosis between *Casuarina glauca* Sieb. ex Spreng. and N_2-fixing *Frankia* bacteria based on the analysis of nitrogen and carbon metabolism. Plant Soil 398:327–337
El-Lakany MH, Luard EJ (1982) Comparative salt tolerance of selected Casuarina species. Aust For Res 13:11–20
Ellouze C, Takahashi M, Wittung P et al (1995) Elongation of helical pitch of RecA filament upon ATP and ADP binding evidenced from small angle neutron scattering data. EUK J Biochem 233:579–583
Fortunato A, Lidon FC, Batista-Santos P et al (2010) Biochemical and molecular characterization of the antioxidative system of *Coffea* sp. under cold conditions in genotypes with contrasting tolerance. J Plant Physiol 167:333–342
Hafeez FY, Hameed S, Malik KA (1999) *Frankia* and *Rhizobium* strains as inoculum for fast growing trees in saline environment. Pak J Bot 31:173 182
Huang B (2006) Cellular membranes in stress sensing and regulation of plant adaptation to abiotic stresses. In: Huang B (ed) Plant environment interactions, 3rd edn. CRC Press, New York, pp 1–25
Jaoudé RA, De Dato G, Palmegiani M et al (2013) Impact of fresh and saline water flooding on leaf gas exchange in two Italian provenances of *Tamarix africana* Poiret. Plant Biol 15:109–117
Katerji N, van Hoorn JW, Hamdy A et al (1997) Osmotic adjustment of sugar beets in response to soil salinity and its influence on stomatal conductance, growth and yield, growth and yield. Agric Water Manag 34:57–69
Khan MH, Panda SK (2008) Alterations in root lipid peroxidation and antioxidative responses in two rice cultivars under NaCl-salinity stress. Acta Physiol Plant 30:81–89
Kozlowski TT (1997) Responses of woody plants to flooding and salinity. Tree Physiol 1:1–29
Kuiper PJC (1984) Functioning of plant cell membrane under saline conditions: membrane lipid composition and ATPase. In: Staples RC, Toenniessen GH (eds) Salinity tolerance in plants. Wiley, New York, pp 77–91
Leshem Y (1987) Membrane phospholipid catabolism and Ca^{2+} activity in control of senescence. Physiol Plant 69:551–559
Logan BA (2005) Reactive oxygen species and photosynthesis. In: Smirnoff N (ed) Antioxidants and reactive oxygen in plants. Blackwell Publishing, Oxford, pp 250–267
Mahajan S, Tuteja N (2005) Cold, salinity and drought stresses: an overview. Arch Biochem Biophys 444:139–158

Mansour MMF, Salama KHA (2004) Cellular basis of salinity tolerance in plants. Environ Exp Bot 52:113–122
Matos MC, Campos PS, Passarinho JA et al (2010) Drought effect on photosynthetic activity, osmolyte accumulation and membrane integrity of two *Cicer arietinum* genotypes. Photosynthetica 48:303–312
Munns R, Tester M (2008) Mechanisms of salinity tolerance. Annu Rev Plant Physiol 59:651–681
Noctor G, Foyer C (1998) Ascorbate and glutathione: keeping active oxygen under control. Annu Rev Plant Physiol Plant Mol Biol 49:249–279
Partelli FL, Batista-Santos P, Campos PS et al (2011) Characterization of the main lipid components of chloroplast membranes and cold induced changes in *Coffea* sp. Environ Exp Bot 74:194–204
Ramalho JC, Campos PS, Teixeira M et al (1998) Nitrogen dependent changes in antioxidant systems and in fatty acid composition of chloroplast membranes from *Coffea arabica* L. plants submitted to high irradiance. Plant Sci 135:115–124
Redondo-Gomez S, Mateos-Naranjo E, Davy AJ et al (2007) Growth and photosynthetic responses to salinity of the salt-marsh shrub *Atriplex portulacoides*. Ann Bot 100:555–563
Ribeiro A, Graça I, Pawlosky C et al (2011) Actinorhizal plant defence-related genes in response to symbiotic *Frankia*. Funct Plant Biol 38:639–644
Sayed WF (2011) Improving Casuarina growth and symbiosis with *Frankia* under different soil and environmental conditions-review. Folia Microbiol Prague 56:1–9
Scotti-Campos P, Pham-Thi AT, Semedo JN et al (2013) Physiological responses and membrane integrity in three *Vigna* genotypes with contrasting drought tolerance. Emir J Food Agric 25:1002–1013
Scotti-Campos P, Pais I, Partelli FL et al (2014) Phospholipids profile in chloroplasts of *Coffea* spp. genotypes differing in cold acclimation ability. J Plant Physiol 171:243–249
Scotti-Campos P, Duro N, da Costa M et al (2016) Antioxidative ability and membrane integrity in salt-induced responses of *Casuarina glauca* Sieber ex Spreng. in symbiosis with N_2-fixing *Frankia* Thr or supplemented with mineral nitrogen. J Plant Physiol 196:60–69
Seeman JR, Critchley C (1985) Effects of salt stress on the growth, ion content, stomatal behaviour and photosynthetic capacity of salt-sensitive species, *Phaseolus vulgaris* (L). Planta 164:151–162
Shalata A, Mittova V, Volokita M et al (2001) Response of cultivated tomato and its wild salt-tolerant relative *Lycopersicon pennelli* to salt-dependent oxidative stress: the root antioxidative system. Physiol Plant 122:487–494
Sivakumar P, Sharmila P, Pardha Saradhi P (2000) Proline alleviates salt-stress-induced enhancement in ribulose-1,5-bisphosphate oxygenase activity. Biochem Biophys Res Commun 279:512–515
Smirnoff N (1995) Metabolic flexibility in relation to environment. In: Smirnoff N (ed) Environment and plant metabolism. Flexibility and acclimation. BIOS Scientific Publishers, Oxford, pp 1–16
Sudhir P, Murthy SDS (2004) Effects of salt stress on basic processes of photosynthesis. Photosynthetica 42:481–486
Surjus A, Durand M (1996) Lipid changes in soybean root membranes in response to salt treatment. J Exp Bot 47:17–23
Türkan I, Demiral T (2009) Recent developments in understanding salinity tolerance. Environ Exp Bot 67:2–9
Van Groenigen JW, Van Kessel C (2002) Salinity-induced patterns of natural abundance carbon-13 and nitrogen-15 in plant and soil. Soil Sci Soc Am J 66:489–498
Wang W, Vinocur B, Altman A (2003) Plant responses to drought, salinity and extreme temperatures: towards genetic engineering for stress tolerance. Planta 218:1–14
Zhong C, Zhang Y, Chen Y et al (2010) Casuarina research in China. Symbiosis 50:107–114
Zhu JK (2001) Plant salt tolerance. Trends Plant Sci 6:66–71

Part V
Inoculants For Agriculture and Environmental Science

Chapter 14
Induced Systemic Resistance Could Explain the Reduction in the Incidence of Black Sigatoka (*Mycosphaerella fijiensis*) in Banana Plants Inoculated with Bacteria Isolated from Banana Tree Roots in the Dominican Republic

Iris-Esther Marcano, César-Antonio Díaz-Alcántara, Victoria Seco, Beatriz Urbano and Fernando González-Andrés

Abstract Black sigatoka caused by *Mycosphaerella fijiensis* (Morelet) is one of the main diseases affecting banana trees worldwide, resulting in a reduced fruit yield. Banana is a major export crop in many tropical and subtropical countries, and many farmers have decided to shift to an organic production system for produce that is destined for the international market. Induced systemic resistance (ISR) protects plants against disease-causing pathogens, and this mechanism can be triggered by some bacteria associated with plant roots. The inoculation of plant roots with bacteria able to induce systemic resistance in plants has been proposed for disease control in organic agriculture as an alternative to chemicals. Fifty-six non-pathogenic and endophytic strains isolated from roots of banana trees cv. Cavendish in organic systems from four regions of the Dominican Republic were the starting point for the present work. We used 26 of them to inoculate the root systems of banana plants, the leaves of which were inoculated with *M. fijiensis* inoculum, in order to analyse their effect on the control of black sigatoka under growth chamber

I.-E. Marcano · C.-A. Díaz-Alcántara
Facultad de Ciencias Agronómicas y Veterinarias, Universidad Autónoma de Santo Domingo, Santo Domingo, Dominican Republic

V. Seco
Departamento de Ingeniería y Ciencias Agrarias, Universidad de León, León, Spain

B. Urbano
Departamento de Ingeniería Agrícola y Forestal, Universidad de Valladolid, Valladolid, Spain

F. González-Andrés (✉)
Instituto de Medio Ambiente, Recursos Naturales y Biodiversidad, Universidad de León, León, Spain
e-mail: fgona@unileon.es

© Springer International Publishing Switzerland 2016

F. González-Andrés and E. James (eds.), *Biological Nitrogen Fixation and Beneficial Plant–Microbe Interactions*, DOI 10.1007/978-3-319-32528-6_14

conditions. The selection of the 26 strains to be tested in plants was based on the following plant growth-promoting (PGP) traits: production of siderophore (17 strains), production of IAA (two strains) and ACC deaminase activity (two strains); the remaining five showed low values for all of the PGP traits. The best controllers of sigatoka were five strains from the genus *Bacillus* (related to *B. licheniformis*, *B. siamensis* and *B. subtilis* ssp. *inaquosorum*) and one *Rhizobium massiliae*. With the six strains, the severity index (SI) of the plants coinoculated with the bacteria and the pathogen did not differ from the noninoculated control. The sigatoka control observed has been tentatively assigned to ISR phenomena, which is discussed in the chapter. The six selected strains could be used in the short term to control black sigatoka in organic banana production systems in the Dominican Republic.

Keywords Organic agriculture · *Mycosphaerella fijiensis* · Black sigatoka · Banana tree · Induced systemic resistance · *Bacillus*

14.1 Introduction

Countries located in tropical and subtropical Latin America plus the Philippines, the majority of them considered developing countries, are the leading banana exporters worldwide (Arias et al. 2004). In the Caribbean Islands, the Dominican Republic is the leading producer of bananas (FAO 2012), which is this country's most exported agricultural product (CEI-RD 2013). The standards of maximum residue levels of pesticides pose a threat for the exportation of agrarian products from developing countries (Ferro et al. 2015); for this reason, organic systems are frequently adopted by producers to avoid pesticide residue-related problems. In the Dominican Republic, 54 % of the total acreage of banana is grown organically (FAO 2013), and 85 % of the organic bananas produced in this country are for exportation (CEI-RD 2013).

However, organic farming dramatically reduces or even suppresses the use of chemical inputs, and in such a situation, alternative products or processes must be used for the control of plagues or diseases (Sinha et al. 2010). As such an alternative, the exploitation of the intrinsic biological potential of microbial rhizosphere processes has been generally proposed to improve plant health by various mechanisms of action (Babalola 2010; Lugtenberg and Kamilova 2009; Maheshwari 2011). In agriculture, the most direct way to take advantage of microbial rhizosphere processes is the use of seeds or soil inoculants based on selected soil beneficial microorganisms (Bhardwaj et al. 2014). Such microorganisms interact with plants to improve their photosynthetic activity and growth, as a consequence of a complex mix of different modes of action; on the one hand, it is the control of disease by several mechanisms and, on the other, the improvement of plant nutrition through better nutrient uptake, nitrogen fixation and phosphate, potassium and iron solubilization or mineralization (Glick et al. 2007; Esitken et al. 2010; Adesemoye et al. 2009; De Souza et al. 2013; Sinha et al. 2010, Singh et al. 2011).

One of the most promising mechanisms of diseases control for agriculture is the development by plants of a state of enhanced defensive capacity upon appropriate stimulation, known as induced systemic resistance (ISR). Different soil beneficial microorganisms have been shown to trigger ISR in plants, and therefore, they have been proposed as root inoculants for the control of crop diseases (van Loon et al. 2006; Van Wees et al. 2008; Pieterse et al. 2014).

Black sigatoka caused by the fungi *Mycosphaerella fijiensis* Morelet is one of the main diseases affecting banana trees worldwide (Etebu et al. 2003; Chuc-Uc et al. 2011; Vishnevetsky et al. 2011; Portal et al. 2012). Sometimes, it is mixed with yellow sigatoka caused by the fungi *Mycosphaerella musicola* Leach (Ferreira et al. 2004). Such pathogens reduce the functional leaf area, reducing, as a consequence, fruit yield (Lhomme and Franciso Jiménez 1992; Castaño Zapata 2006). They also affect the physiology of the plant, triggering a premature ripening of the fruit and causing a reduction in economic yield (Valerio et al. 2002; Ceballos et al. 2012).

In the previous work (Marcano 2014, unpublished Ph.D. Thesis), we created a collection of bacterial strains associated with banana tree roots, including endophytic and rizospheric ones, which were isolated from organic farms located in each one of the four provinces of the Dominican Republic (Fig. 14.1) where banana is cultivated in commercial fields. Eighty-three endophytic strains were identified by means of 16S rRNA gene sequencing (unpublished), and they were also analysed for several plant growth-promoting (PGP) activities in vitro. Marcano (2014, unpublished Ph.D. Thesis) tested the general agronomic performance of five of such strains, which were preselected on the basis of their in vitro PGP activity.

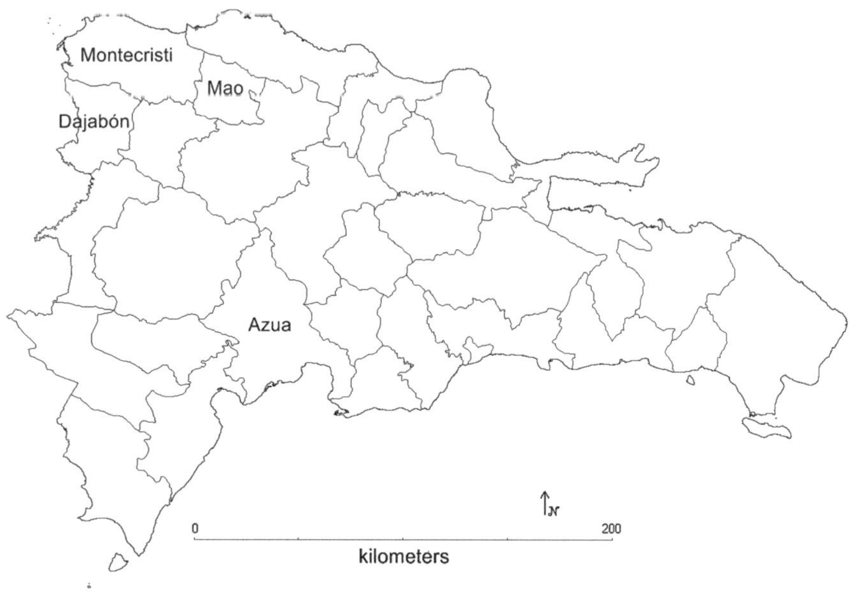

Fig. 14.1 Location of the provinces from which the bacterial strains were isolated

They were tested in a field trial in a commercial farm in Dominican Republic, along with an uninoculated control, and unexpectedly one of them clearly controlled the incidence of black sigatoka, and two other slightly controlled the disease. The black sigatoka attack is generalized throughout the country, and it is naturally present in all the commercial farms of banana.

On the basis of this result, we suspected that other bacterial strains from the collection could have similar or greater effects on the control of black sigatoka. Therefore, the objective of the present work was to analyse the biocontrol effect against black sigatoka of 26 bacterial strains from the collection, in banana plants inoculated with black sigatoka inoculum. The experiment was carried out in a growth chamber with controlled temperature, photoperiod and humidity, in 10-l containers. The bacteria used in the experiment were selected on the basis of their in vitro PGP activity. The inoculation was made in the substratum, at the moment when the plants were transplanted to the containers.

14.2 Materials and Methods

14.2.1 Bacterial Strains

The endophytic bacteria used for this work were isolated by Marcano (2014, unpublished Ph.D. Thesis). The bacteria were isolated from surface-disinfected banana roots in organic banana plantations in four different provinces of the Dominican Republic (Fig. 14.1) and identified on the basis of their 16S rRNA gene sequences (sequences not shown). Briefly, feeder roots up to 1 mm in diameter were surface-disinfected with 70 % ethanol for 1 min and 7 % NaClO for 5 min, and three successive washings with sterile distilled water. Following roots were crushed and plated onto petri dishes with TSA medium supplemented with 1 mg l^{-1} cycloheximide to prevent fungal growth. Plates were evaluated daily for strains selection and purification in the same medium, until the obtention of pure cultures. For the identification of each strain, the 16S rRNA gene was sequenced by Macrogen (the Netherlands), yielding products of 1400 bp or more. Briefly, the amplification of the 16S rRNA gene was performed using the primer pair 27F and 1492R, and the sequencing of the gene was carried out using the primers 518F and 800R (Lane 1991; Weisburg et al. 1991). The obtained sequences (not shown) were compared with those from EzTaxon-e server (Kim et al. 2012). The strains with a similarity higher than 99.5 % with a type strain were considered of the same species that the type strain; for similarities lower than 99.5 %, the isolate was considered phylogenetically close to such species. Out of the 83 endophytic strains identified by means of the 16S rRNA sequencing by Marcano (2014, unpublished Ph.D. Thesis), 56 of them were accepted and the rest discarded because they showed either pathogenic activity against humans or plants; from them, 26 were selected on the basis of their in vitro PGP activity for the present work (Table 14.1).

Table 14.1 Bacterial strains selected for the black sigatoka control experiment, in growth chamber

Province of origin	Strain code	Identification on the basis of 16S rRNA gene sequencing	*In vitro* PGP activities				Sigatoka control *in vitro*[d]		
		Accession from the EzTaxon database, with the most similar 16S rRNA sequence	IAA (μg/ml)	Ca_3PO_4 solubilizing[a]	Siderophore production[b]	ACC deaminase[c]	6 dafi	10 dafi	14 dafi
Azua	**DR_A_04**	*Pseudomonas koreensis* Ps 9-14(T)	5.10	1	**2**	0.00	1	1	1
Azua	**DR_A_06**	*Pseudomonas taiwanensis* BCRC 17751(T)	1.93	0	**2**	109.32	1	1	1
Azua	**DR_A_07**	*Bacillus amyloliquefaciens* subsp. *plantarum* FZB42(T)	1.90	0	1	0.00	1	1	1
Azua	**DR_A_08**	*Bacillus siamensis* KCTC 13613(T)	2.26	0	1	21.08	1	1	1
Azua	DR_A_09	close to *Bacillus sonorensis* NRRL B-23154(T)	1.68	0	**3**	0.00	1	1	1
Azua	**DR_A_12**	*Pseudomonas plecoglossicida* FPC951(T)	4.46	1	**2**	1.77	1	1	1
Azua	DR_A_13	close to *Bacillus sonorensis* NRRL B-23154(T)	0.95	0	1	**133.70**	1	1	1
Azua	**DR_A_14**	*Bacillus subtilis* subsp. *inaquosorum* KCTC 13429(T)	1.31	0	**2**	8.51	1	1	1
Dajabón	DR_D_02	close to *Bacillus licheniformis* ATCC 14580(T)	2.01	0	**2**	0.00	1	1	1
Dajabón	**DR_D_03**	*Pseudomonas taiwanensis* BCRC 17751(T)	0.81	1	**2**	0.00	1	1	1
Dajabón	DR_D_06	close to *Leclercia adecarboxylata* GTC 1267(T)	**5.84**	0	1	2.93	1	1	1
Montecristi	DR_M_02	close to *Bacillus sonorensis* NRRL B-23154(T)	1.17	0	**3**	0.00	1	1	1
Montecristi	**DR_M_04**	*Bacillus aryabhattai* B8W22(T)	1.23	0	1	20.38	1	1	1
Montecristi	**DR_M_05**	*Pseudomonas mosselii* CIP 105259	5.13	1	**2**	4.14	1	1	1
Montecristi	**DR_M_07**	*Pseudomonas plecoglossicida* FPC951(T)	2.38	1	1	6.08	1	1	1
Montecristi	DR_M_09	Close to *Bacillus licheniformis* ATCC 14580(T)	2.24	0	**3**	4.55	1	1	1
Montecristi	**DR_M_11**	*Rhizobium massiliae* 90A	3.50	0	**2**	0.00	1	0	0
Montecristi	**DR_M_12**	*Bacillus subtilis* subsp. *inaquosorum* KCTC 13429(T)	1.52	0	**2**	6.59	1	1	1

(continued)

Table 14.1 (continued)

Province of origin	Strain code	Identification on the basis of 16S rRNA gene sequencing	*In vitro* PGP activities				Sigatoka control *in vitro*[d]		
		Accession from the EzTaxon database, with the most similar 16S rRNA sequence	IAA (μg/ml)	Ca_3PO_4 solubilizing[a]	Siderophore production[b]	ACC deaminase[c]	6 dafi	10 dafi	14 dafi
Mao	**DR_O_01**	*Pseudomonas plecoglossicida* FPC951(T)	3.09	1	**2**	0.15	1	1	1
Mao	DR_O_02	close to *Pseudomonas nitroreducens* DSM 14399(T)	1.30	0	**3**	0.62	1	1	1
Mao	**DR_O_06**	*Bacillus aerophilus* 28 K(T)	2.60	0	1	**127.15**	1	0	0
Mao	**DR_O_08**	*Bacillus amyloliquefaciens* subsp. *plantarum* FZB42(T)	1.42	0	1	7.93	1	1	1
Mao	**DR_O_09**	*Pseudomonas taiwanensis* BCRC 17751(T)	2.15	1	**2**	16.19	1	1	1
Mao	**DR_O_10**	*Bacillus subtilis* subsp. *inaquosorum* KCTC 13429(T)	1.91	0	**3**	15.34	1	1	1
Mao	**DR_O_11**	*Bacillus safensis* FO-036b(T)	2.19	0	**3**	0.72	1	0	0
Mao	**DR_O_15**	*Bacillus aerophilus* 28 K(T)	**7.63**	0	0	0.18	1	1	1

The bacteria were isolated by Marcano (201[illegible], unpublished Ph.D. Thesis). Those strains with the strain code in bold characters have been identified based on 16S rRNA gene sequencing; for the rest, we indicate the strain with the most similar 16S rRNA sequence in the EzTaxon database (see text for identification criteria). For each strain, the value corresponding to the PGP trait used for the selection of the strain is indicated in bold characters. Five strains (DR_A_07; DR_A_08; DR_M_04; DR_M_07; DR_O_08) with low expression of all the PGP traits were also selected (see text for details). As complementary information, the direct control of the fungus *Mycosphaerella fijiensis* in vitro is also shown

[a] 0 absence of halo; 1 presence of halo

[b] 0 absence of halo; 1 halo <5 mm; 2 halo between 5 and 10 mm; 3 halo >10 mm

[c] μM α-Ketobutyrate mg of protein^{-1} h^{-1}

[d] Inhibition observed x days after fungus inoculation (dafi): 0 no inhibition of fungal growth; 1 inhibition halo <110 % of the diameter of the bacterial colony; 2 strong inhibition, greater halo

14.2.2 Determination of PGP Traits In Vitro and Statistical Analysis

The 56 nonpathogenic strains isolated by Marcano (2014, unpublished Ph.D. Thesis) were analysed for four PGP traits in vitro: the ability to solubilize tricalcium phosphate was tested in YED-P media according to Peix et al. (2004); the strains were incubated at 28 °C for 5 days; and the presence or absence of halo was observed. Siderophore production was determined using M9-CAS agar media (Schwyn and Neilands 1987; Alexander and Zuberer 1991); strains were incubated in this media at 28 °C for at least three days; yellowish orange halos around the colonies on M9-CAS agar were indicative of siderophore secretion. IAA production was measured in JMM media (O'Hara et al. 1989) supplemented with 0.17 g/l tryptophan, as described (Khalid et al. 2004). The ACC deaminase activity was determined according to Penrose and Glick (2003), measuring the released amounts of α-ketobutyrate (μmol α-ketobutyrate mg protein^{-1} h^{-1}).

The values observed in the 56 strains (data not shown) were used to select 26 strains that were used for the in vivo analysis of black sigatoka control with banana plants in growth chamber (Table 14.1).

14.2.3 Experimental Design in the Growth Chamber

The experimental design was completely randomized with three replications. Each experimental unit consisted of a 10-l container with a professional substratum (70 % blonde peat, 30 % black peat; N-P-K 12-14-24, pH 6) from Pindstrup Mosebrug SAE, with one banana plant from the variety Cavendish (*Musa acuminata* group AAA), produced *in vitro* by the Dominican Institute of Biotechnology and Industry (IIBI), in the Dominican Republic. When the plants were transplanted to the containers, they had between 4 and 6 developed leaves. The treatments consisted of the inoculation of the plants with pathogenic fungi (*M. fijiensis* Morelet) plus one of each of 26 bacteria strains. There were also two controls without bacterial inoculation; one of these controls was inoculated with pathogenic fungi, and the other was not inoculated at all. At transplantation, the plants were inoculated in the root system with 10 ml of the corresponding bacterial suspension in NaCl (0.90 %) to an approximate concentration of 9 × 10^8 CFU/ml. Each strain had previously been grown in TSA medium at 28 °C.

14.2.4 Pathogen Production and Inoculation

The strain of *M. fijiensis* Morelet used for the experiment was CBS_121358 (Centraalbureau Voor Schimmelcultures CBS, Fungal Biodiversity Centre, Institute of the Royal Netherlands Academy of Arts and Sciences KNAW, The Netherlands). In order to propagate the fungus inoculum, we used the procedure described by Leiva-Mora et al. (2010), consisting of inoculation with a mycelial suspension of the fungus (10^5 mycelial fragments per millilitre). To prepare the suspension, the fungus had been previously grown in Malt Extract Agar (MEA) and inoculated in Erlenmeyer flasks with 100 ml of potato dextrose broth (PDB). The flasks were shaken at 28 °C for 14 days, following homogenization with a BRAUN homogenizer. One per cent gelatin was added to enhance fixation with the plant leaf. All of the fully opened leaves on each banana plant were inoculated as described by Leiva-Mora et al. (2010); the backside of the leaves was slightly damaged with a metal brush and subsequently inoculated with the inoculum using a paint brush, followed by air-drying for a period of 2 h. Next, the growth chamber was set for 72 h at 90–100 % relative humidity (RH). After this period, the climatic conditions were 28 °C for 13 h and 22 °C for 11 h at 80–90 % RH.

To estimate the severity of black sigatoka attack on banana leaves, we used the Stover classification system, modified by Gauhl (Orjeda 1998; Morales Romero et al. 2011). Each leaf received a value corresponding to the necrotic area according to Table 14.2.

The severity index (SI) for all plants in each treatment was calculated according to the formula of Towsend and Heuberguer (Orjeda 1998; Morales Romero et al. 2011)

$$\mathrm{SI} = \Sigma nb / [(N - 1)\,T]$$

where n is the number of leaves in each stage according to Table 14.2, b is the value corresponding to the stage, N is the number of stages considered (7), and T is the total number of evaluated leaves.

The incidence of black sigatoka in the plants was estimated 140 days after transplantation, measuring SI, fresh and dry aerial biomass and plant height. The data were subjected to ANOVA appropriate for a completely randomized design.

Table 14.2 System for the classification of black sigatoka symptoms in banana leaves according to Stover, modified by Gauhl (Orjeda 1998; Morales Romero et al. 2011)

Stage	Description
Stage 0	No damages
Stage 1	Less than 1 % of the leaf blade with damages (line or less than 10 spots)
Stage 2	1 a 5 of the leaf blade with damages
Stage 3	6 a 15 of the leaf blade with damages
Stage 4	16 a 33 % of the leaf blade with damages
Stage 5	34 a 50 % de la lámina con síntomas
Stage 6	51 a 100 % de la lámina con síntomas

14.2.5 *In Vitro Antagonistic Effect of Bacterial Strains Against* **Mycosphaerella fijiensis** *in a Dual Culture System*

The fungal strain used was the same as indicated for the banana plant experiment, and the fungal inoculum was produced by the same procedure except for the addition of gelatin, which was not used in the present experiment. The inoculum (1.5 ml) was plated onto 90 mm dia. Petri dishes with MEA growth medium, extended uniformly and air-dried in a sterile flow chamber. Subsequently, a growth mass of the corresponding bacteria, previously cultured on TSA medium, was placed in the centre of the dish, to serve as the inoculum. The plates were incubated at 28 °C for 14 days, and the growth of both microorganisms was scored at 6, 10 and 14 days after the initiation of the experiment. The plate was coded as a 0 when there was no inhibition of fungal growth, as a 1 when there was slight inhibition (the radius of the inhibition halo was less than 110 % of the diameter of the bacterial colony) and a 2 when there was strong inhibition, observed by a larger halo.

14.3 Results and Discussion

The development of alternatives to chemical products for the control of disease on organic banana plantations is a key aspect of the profitability of this crop for exportation in the producing countries. As the two most serious problems with organically produced crops in the Dominican Republic are the black sigatoka and nematode infections (Heslop-Harrison and Schwarzacher 2007; Chuc-Uc et al. 2011; Vishnevetsky et al. 2011; Portal et al. 2012; Bellamy 2013), the present work attempts to search for a microorganism-based solution to control black sigatoka on organic banana plantations.

14.3.1 **In Vitro** *Antagonistic Effect of Bacterial Strains Against* **Mycosphaerella fijiensis** *in a Dual Culture System*

Eighty-eight per cent of the 56 nonpathogenic bacterial strains analysed showed a slight antagonistic effect in vitro against the fungi 6 days after fungus inoculation (dafi), but the effect persisted after 14 dafi in only for 75 % of cases. However, none of the strains included in the present work showed a strong antagonistic effect, like the one shown by other strains from the bacterial collection of Marcano (2014, unpublished Ph.D. Thesis). Table 14.1 shows the results of the in vitro antagonistic effect against the fungi for the 26 strains that were tested in plants (the results of the rest are not shown).

14.3.2 Selection of 26 Strains Based on Their PGP Traits, to Be Included in the Experiment with Plants

For the selection of bacteria, the direct effect of biocontrol of *Mycosphaerella fijiensis* was excluded as a selection criterion. This was because our objective was to enhance the plant's defence against the pathogen by means of root inoculation, rather than to spray the biocontroller on the plant's leaves, which are the plant organs in which the whole pathogen cycle occurs. Therefore, for strain selection, we prioritized siderophore production because, according to García-Gutiérrez et al. (2012) and Djavaheri et al. (2012), siderophore assimilation is linked to the ISR processes. With these criteria, we selected 17 strains (Table 14.1). Moreover, we also selected four other isolates with a low expression of the trait siderophore production, two of them with high IAA production (Table 14.1) and two others with high ACC deaminase activity (Table 14.1). On the other hand, we selected five other isolates with a low expression of all the PGP traits analysed. The reason to include strains with different expression levels of PGP traits in vitro was to shed light on the relationship between PGP traits and the stimulation of plant resistance against black sigatoka.

14.3.3 Different Incidence of Black Sigatoka in Banana Plants Inoculated with the Selected Bacterial Strains

In the growth chamber experiment, there were significant differences between treatment for SI and plant height ($p < 0.001$), but not for aerial biomass (Table 14.3). As expected, there was a negative correlation between SI and growth parameters; in the case of plant height, the negative correlation was highly significant ($p < 0.001$) with a Pearson correlation coefficient of −0.722. For fresh and dry biomass, the Pearson correlation coefficients were −0.572 ($p = 0.001$) and −0.488 ($p < 0.01$), respectively.

For six strains, the SI in the plants coinoculated with the corresponding strain plus black sigatoka inoculum did not differ from that of the negative control (the control inoculated neither with the pathogen nor the bacteria), and therefore, it can be deduced that inoculation of the root system with such strains controls the pathogen. Such strains were DR_D_02 *Bacillus* sp. philogenetically close to *B. licheniformis*; DR_A_08 *Bacillus siamensis*; DR_M_12, DR_A_14 and DR_O_10 the three *Bacillus subtilis* subsp. *inaquosorum*; *and* DR_M_11 *Rhizobium massiliae* (Table 14.3). The treatments inoculated with two strains belonging to the species *Pseudomonas plecoglossicida* showed a SI of 0.42 and 0.46, respectively, significantly higher than that of the negative control (0.18). This is interesting because Marcano (2014, unpublished Ph.D. Thesis) found a good control of black sigatoka in a field experiment in a commercial plantation with one strain of *P. plecoglossicida*. From this result, it can be deduced that there are bacterial

Table 14.3 Mean values of the parameters measured 140 days after transplantation, in banana plants grown in 10-l containers in a growth chamber and inoculated with the indicated bacteria (treatments) and *Mycosphaerella fijiensis* (except the negative control)

Inoculation treatment	Severity index (SI)		Aerial biomass (fresh weight) (g)	Aerial biomass (dry weight) (g)	Plant height (cm)	
Negative control (no inoculation)	0.18	a	152.12	19.43	23.90	abcde
DR_D_02	0.31	ab	135.93	15.50	27.63	e
DR_A_08	0.32	ab	114.82	13.17	25.51	cde
DR_M_12	0.34	abc	85.95	9.86	23.40	abcde
DR_M_11	0.34	abc	102.75	14.51	22.43	abcde
DR_A_14	0.38	abcd	101.91	11.03	23.40	abcde
DR_O_10	0.38	abcd	99.89	11.95	23.20	abcde
DR_A_04	0.42	bcde	141.93	16.52	24.90	bcde
DR_M_07	0.42	bcde	133.07	15.77	25.50	cde
DR_O_09	0.46	bcdef	141.15	16.69	26.40	de
DR_O_01	0.46	bcdef	152.39	19.69	27.60	e
DR_O_15	0.49	bcdefg	92.20	10.07	22.00	abcde
DR_D_06	0.49	bcdefg	114.65	14.31	25.20	cde
DR_O_11	0.53	cdefgh	70.12	7.94	18.88	abcde
DR_M_09	0.53	cdefgh	113.91	18.44	25.85	cde
DR_D_03	0.53	cdefgh	111.98	13.86	22.18	abcde
DR_O_08	0.57	defgh	112.05	13.70	22.83	abcde
DR_A_13	0.57	defgh	121.64	14.68	23.16	abcde
DR_O_06	0.57	defgh	92.13	10.63	20.72	abcde
DR_A_07	0.60	efgh	142.74	17.37	25.30	cde
DR_M_05	0.61	efgh	109.31	13.44	22.00	abcde
Sigatoka inoculated control	0.64	fgh	75.77	9.31	16.15	ab
DR_A_12	0.66	fgh	104.14	12.70	18.06	abcd
DR_M_04	0.66	fgh	69.94	8.44	17.21	abc
DR_M_02	0.68	gh	78.98	10.21	15.51	a
DR_O_02	0.69	gh	61.13	7.57	15.73	a
DR_A_09	0.70	gh	86.09	11.85	18.06	abcd
DR_A_06	0.73	h	71.10	8.31	15.17	a
Degrees of freedom	27		27	27	27	
Mean squares	0.060		2113.522	36.923	43.329	
F	13.760***		2.087 ns	2.383 ns	5.550***	

For those parameters in which significant differences exist, values followed by the same letter did not show significant differences (Tukey's test, $p < 0.05$) (significance level: ***$p < 0.001$; *ns* not significant) species with higher potential to control black sigatoka than *P. plecoglossicida*. Therefore, the results obtained in the present work are very promising, because the strains with the best results in the growth chamber could improve the good results obtained in the field by Marcano (2014, unpublished Ph.D. Thesis).

14.3.4 ISR as the Mechanisms Potentially Involved in the Control of Black Sigatoka in the Growth Chamber Experiment

Five out of the six bacterial strains that better reduced the incidence of black sigatoka in the plant experiment also showed a slight antagonistic effect in vitro against *M. fijiensis*; the exception was *R. massiliae,* which only showed a slight inhibitory halo and only up to six days after the establishment of the experiment. We tentatively assigned the observed control in plants to ISR processes, and not to a direct control of the pathogen by the bacteria, because of the spatial separation between the two microorganisms. The inoculation was performed in the roots, while the fungus *M. fijiensis* was inoculated in the leaves, where it is known to complete its entire life cycle (Bennett and Arneson 2003). Thus, a direct antagonism of the bacteria on the fungus in our experimental conditions is very unlikely. The protection observed was, therefore, likely related to an effect on plant defence mechanisms. However, in order to unravel the phenomena involved in the protection observed, in a future work, a further molecular study on the enzymes and signalling pathways related to the ISR in this particular crop and with this particular pathogen would be necessary. It is well known that upon appropriate stimulation, plants can develop a state of enhanced defensive capacity, known as ISR and also that different soil beneficial microorganisms have been shown to trigger ISR in plants (Martínez-Hidalgo et al. 2015). Although in general, ISR mechanisms rely on jasmonic acid signalling (Pozo and Azcón-Aguilar 2007; Van Wees et al. 2008; Pieterse et al. 2014), there are no investigations into the molecular aspects of ISR-based control of black sigatoka in banana trees; Marcano (2014, unpublished Ph.D. Thesis) and Oye Anda et al. (2015) observed a reduction in the black sigatoka attack triggered by rhizospheric bacteria and arbuscular mycorrhizal fungus, respectively, but not from a molecular standpoint.

Other researchers have mentioned the control of other banana diseases with ISR mechanisms mediated by root-associated bacteria; for example, Fishal et al. (2010) observed the control of *Fusarium oxysporum* with ISR triggered by one strain of *Pseudomonas* sp. and one of *Burkholderia* sp.; Peeran et al. (2014) worked with *Colletotrichum musae* in banana trees and showed evidence of an increase in the activity of defence-related enzymes after inoculation with the strain FP7 (*Pseudomonas fluorescens*) in field conditions and Harish et al. (2008, 2009) controlled the *Banana Bunch Top Virus* (BBTW) with endophytic *P. fluorescens* and *Bacillus* sp., also attributed to ISR processes.

From the observed results, there is no correlation between the PGP traits of the bacterium and its control of disease in banana plants. As an example of this, the second best controller (DR_A_08 *Bacillus siamensis*) did not show any PGP trait. Conversely, other strains without PGP traits did not show ISR against the disease (DR_A_07 *Bacillus amyloliquefaciens* subsp. *plantarum* and DR_M_04 *Bacillus aryabhattai*).

14.4 Conclusion

The present work attempted to search for safe-for-humans bacterial strains to control black sigatoka disease on banana plantations. Twenty-six out of 56 non-pathogenic strains isolated in the previous work were inoculated into the root systems of banana plants grown in a growth chamber, and the leaves of plants were also inoculated with *M. fijiensis* inoculum. Compared with controls, the SI of the disease in plants inoculated with six strains (five of the genus *Bacillus* including *B. licheniformis, B. siamensis* and *B. subtilis* subsp. *inaquosorum*, and one of the genus *R. massiliae*) did not differ from that of the negative control, inoculated neither with the pathogen nor with bacteria. Therefore, these strains are potentially useful to control disease by root inoculation on commercial plantations, being more effective under the conditions present in a growth chamber than the strain from *P. plecoglossicida,* which produced significant disease control in afield experiment. In this work, it was discussed that the observed control of disease can be tentatively assigned to ISR processes. This work opens the door to a promising future of commercial products for the biocontrol of black sigatoka, based on endophytic bacteria isolated from banana roots. Prior to their introduction into the market, it is necessary to carry out field trials with the strains showing the best performance in the growth chamber.

Acknowledgements This work has been financially supported by the Spanish Ministry of Foreign Affairs and Cooperation (Projects PCI-AECID A/023132/09, PCI-AECID A/030020/10 and PCI-AECID A1/035364/11). I.-E. M. was granted by the Spanish Ministry of Foreign Affairs and Cooperation (MAEC-AECID grant 2010-2014). The manuscript has been professionally proof read by PRS.

References

Adesemoye AO, Torbert HA, Kloepper JW (2009) Plant growth promoting rhizobacteria allow reduced application rates of chemical fertilizers. Microb Ecol 58:921–929. doi:10.1007/s00248-009-9531-y

Alexander DB, Zuberer DA (1991) Use of Chrome Azurol S reagents to evaluate siderophore production by rhizosphere bacteria. Biol Fertil Soils 12:39–45

Arias P, Dankers C, Liu P, Pilkauskas P (2004) La economía mundial del banano 1985–2002

Babalola OO (2010) Beneficial bacteria of agricultural importance. Biotechnol Lett 32:1559–1570. doi:10.1007/s10529-010-0347-0

Bellamy A (2013) Banana production systems: identification of alternative systems for more sustainable production. Ambio 42:334–343. doi:10.1007/s13280-012-0341-y

Bennett RS, Arneson PA (2003) Sigatoka negra. Spanish translation by Knight RJ Jr (2005). The Plant Health Instructor

Bhardwaj D, Ansari MW, Sahoo RK, Tuteja N (2014) Biofertilizers function as key player in sustainable agriculture by improving soil fertility, plant tolerance and crop productivity. Microb Cell Fact 13:66. doi:10.1186/1475-2859-13-66

Castaño Zapata J (2006) Evaluación in vitro de extractos vegetales sobre *Mycosphaerella fijiensis* Morelet. Agronomia 14:37–50

Ceballos I, Mosquera S, Angulo M et al (2012) Cultivable bacteria populations associated with leaves of banana and plantain plants and their antagonistic activity against *Mycosphaerella fijiensis*. Microb Ecol 64:641–653. doi:10.1007/s00248-012-0052-8

CEI-RD (2013) Perfil Económico del Banano. Centro de Exportación e Inversión de la República Dominicana. Santo Domingo República Dominicana

Chuc-Uc J, Brito-Argáez L, Canto-Canché B (2011) The in vitro secretome of *Mycosphaerella fijiensis* induces cell death in banana leaves. Plant Physiol Biochem 49:572–578. doi:10.1016/j.plaphy.2011.02.006

De Souza R, Beneduzi A, Ambrosini A et al (2013) The effect of plant growth-promoting Rhizobacteria on the growth of rice (*Oryza sativa* L.) cropped in Southern Brazilian fields. Plant Soil 366:585–603. doi:10.1007/s11104-012-1430-1

Djavaheri M, Mercado-Blanco J, Versluis C et al (2012) Iron-regulated metabolites produced by *Pseudomonas fluorescens* WCS374r are not required for eliciting induced systemic resistance against *Pseudomonas* syringae pv. tomato in *Arabidopsis*. Microbiol Open 1:311–325. doi:10.1002/mbo3.32

Esitken A, Yildiz HE, Ercisli S, Donmez MF, Turan M, Gunes A (2010) Effects of plant growth promoting bacteria (PGPB) on yield, growth and nutrient contents of organically grown strawberry. Sci Hortic 124:62–66. doi:10.1016/j.scienta.2009.12.012

Etebu E, Pasberg-Gauhl C, Gauhl F, Daniel-Kalio L (2003) Preliminary studies of in vitro stimulation of sexual mating among isolates of *Mycosphaerella fijiensis*, causal agent of black sigatoka disease in bananas and plantains. Phytoparasitica 31:69–75. doi:10.1007/BF02979768

FAO (2012) Crece la producción y la exportación de banano en la República Dominicana. http://www.fao.org/agronoticias/agro-noticias/detalle/es/c/127976/. Accessed 08 Feb 2014

FAO (2013) República Dominicana y la Unión Europea firman convenio para fortalecer industria bananera. http://www.fao.org/agronoticias/agro-noticias/detalle/es/c/179989/. Consultada el 10 Feb 2013

Ferreira CF, Oliveira Silva S, Damasco Sobrinho NP et al (2004) Molecular characterization of banana (AA) diploids with contrasting levels of black and yellow sigatoka resistance. Am J Appl Sci 1:276–278

Ferro E, Wilson JS, Otsuki T (2015) The effect of product standards on agricultural exports. Food Policy 50:68–79. doi:10.1016/j.foodpol.2014.10.016

Fishal EMM, Meon S, Yun WM (2010) Induction of tolerance to Fusarium wilt and defense-related mechanisms in the plantlets of susceptible berangan banana pre-inoculated with *Pseudomonas* sp. (UPMP3) and *Burkholderia* sp. (UPMB3). Agric Sci China 9:1140–1149. doi:10.1016/S1671-2927(09)60201-7

García-Gutiérrez L, Romero D, Zeriouh H et al (2012) Isolation and selection of plant growth-promoting rhizobacteria as inducers of systemic resistance in melon. Plant Soil 358:201–212. doi:10.1007/s11104-012-1173-z

Glick BR, Cheng Z, Czarny J, Duan J (2007) Promotion of plant growth by ACC deaminase-producing soil bacteria. In: Bakker PAHM, Raaijmakers JM, Bloemberg G, Höfte M, Lemanceau P, Cooke BM (eds) New perspectives and approaches in plant growth-promoting Rhizobacteria research, pp 329–339. doi:10.1007/s10658-007-9162-4

Harish S, Kavino M, Kumar N, Saravanakumar D, Soorianathasundaram K, Samiyappan R (2008) Biohardening with plant growth promoting rhizosphere and endophytic bacteria induces systemic resistance against Banana *Bunchy Top Virus*. Appl Soil Ecol 39:187–200. doi:10.1016/j.apsoil.2007.12.006

Harish S, Kavino M, Kumar N, Balasubramanian P, Samiyappan R (2009) Induction of defense-related proteins by mixtures of plant growth promoting endophytic bacteria against *Banana bunchy top virus*. Biol Control 51:16–25. doi:10.1016/j.biocontrol.2009.06.002

Heslop-Harrison JS, Schwarzacher T (2007) Domestication, genomics and the future for banana. Ann Bot 100:1073–1084. doi:10.1093/aob/mcm191

Khalid A, Arshad M, Zahir ZA (2004) Screening plant growth-promoting Rhizobacteria for improving growth and yield of wheat. J Appl Microbiol 96(3):473–480

Kim OS, Cho YJ, Lee K et al (2012) Introducing EzTaxon-e: a prokaryotic 16S rRNA gene sequence database with phylotypes that represent uncultured species. Int J Syst Evol Micr 62:716–721. doi:10.1099/ijs.0.038075-0

Lane DJ (1991) 16S/23S rRNA sequencing. In: Stackebrandt E, Goodfellow M (eds) Nucleic acid techniques in bacterial systematics. Wiley, New York, pp 115–175

Leiva-Mora M, Alvarado-Capó Y, Acosta-Suárez M et al (2010) Protocolo para la inoculación artificial de plantas de *Musa* spp. con *Mycosphaerella fijiensis* y evaluación de su respuesta mediante variables epifitiológicas y componentes de la resistencia. Biotecnología Vegetal 10:79–88

Lhomme J, Francisco Jiménez O (1992) Estimating dew duration on banana and plantain leaves from standard meteorological observations. Agric For Meteorol 62:263–274. doi:10.1016/0168-1923(92)90018-Y

Lugtenberg B, Kamilova F (2009) Plant-growth-promoting Rhizobacteria. Annu Rev Microbiol 63:541–556. doi:10.1146/annurev.micro.62.081307.162918

Maheshwari DK (2011) Plant growth and health promoting bacteria. Microbiology monographs. Springer, Munster. doi:10.1007/978-3-642-13612-2

Marcano IE (2014) Aislamiento y caracterización de bacterias de la rizosfera de banano (*Musa* sp.) en República Dominicana y selección de cepas para el desarrollo de biofertilizantes. Tesis Doctoral

Martínez-Hidalgo P, García JM, Pozo MJ (2015) Induced systemic resistance against Botrytis cinerea by Micromonospora strains isolated from root nodules. Front Microbiol 6:922. doi:10.3389/fmicb.2015.00922

Morales Romero L, Ullauri Espinoza MA, Dávila Martínez A, Folgueras Montiel M (2011) Respuesta de genotipos mejorados de plátanos (*Musa* spp.) a *Mycosphaerella fijiensis* Morelet 1–20

O'Hara GW, Goss TJ, Dilworth MJ et al (1989) Maintenance of intracellular pH and acid tolerance in *Rhizobium meliloti*. Appl Environ Microbiol 55(8):1870–1876

Orjeda G (1998) Evaluación de la resistencia de los bananos a las enfermedades de Sigatoka negra y marchitamiento por *Fusarium*. Guías técnicas INIBAP 3. IPGRI, Roma, Italia; Red Internacional para el mejoramiento del banano y el plátano, Montpellier, Francia

Oye Anda CC, de Boulois HD, Declerck S (2015) The arbuscular mycorrhiza fungus Rhizophagus Irregularis MUCL 41833 decreases disease severity of Black Sigatoka on banana c.v. Grande name, under in vitro culture conditions. Fruits 70:37–46. doi:10.1051/fruits/2014041

Peeran MF, Nagendran K, Gandhi K, Raguchander T, Prabakar K (2014) Water in oil based PGPR formulation of *Pseudomonas fluorescens* (FP7) showed enhanced resistance against *Colletotrichum musae*. Crop Protection 65:186–193. doi:10.1016/j.cropro.2014.07.010

Peix A, Rivas R, Santa-Regina I et al (2004) Pseudomonas lutea sp. nov., a novel phosphate-solubilizing bacterium isolated from the rhizosphere of grasses. Int J Syst Evol Micr 54:847–850

Penrose DM, Glick BR (2003) Methods for isolating and characterizing ACC deaminase-containing plant growth-promoting rhizobacteria. Physiol Plant 118:10–15

Pieterse CMJ, Zamioudis C, Berendsen RL et al (2014) Induced systemic resistance by beneficial microbes. Annu Rev Phytopathol 52:347–375. doi:10.1146/annurev-phyto-082712-102340

Portal O, Acosta-Suárez M, Ocaña B et al (2012) A green fluorescent protein-transformed Mycosphaerella fijiensis strain shows increased aggressiveness on banana. Australas Plant Path 41:645–647. doi:10.1007/s13313-012-0155-1

Pozo MJ, Azcón-Aguilar C (2007) Unraveling mycorrhiza-induced resistance. Curr Opin Plant Biol 10:393–398. doi:10.1016/j.pbi.2007.05.004

Schwyn B, Neilands JB (1987) Universal chemical assay for the detection and determination of siderophores. Ann Biochem 160:47–56

Singh JS, Pandey VC, Singh DP (2011) Efficient Soil microorganisms: a new dimension for sustainable agriculture and environmental development. Agric Ecosyst Environ 140:339–353. doi:10.1016/j.agee.2011.01.017

Sinha RK, Valani D, Chauhan K, Agarwal S (2010) Embarking on a second green revolution for sustainable agriculture by vermiculture biotechnology using earthworms: reviving the dreams of Sir Charles Darwin. J. Agric Biotech Sustain Dev 2:113–128

Valerio R, Lindorf H, García de García E (2002) Anatomía foliar comparada de ocho cultivares de banano con relación a la resistencia o susceptibilidad a la sigatoka (amarilla y negra). Agronomia Trop 52:507–521

Van Loon LC, Geraats BPJ, Linthorst HJM (2006) Ethylene as a modulator of disease resistance in plants. Trends Plant Sci 11:184–191. doi:10.1016/j.tplants.2006.02.005

Van Wees SCM, Van der Ent S, Pieterse CMJ (2008) Plant immune responses triggered by beneficial microbes. Curr Opin Plant Biol 11:443–448. doi:10.1016/j.pbi.2008.05.005

Vishnevetsky J, White T Jr, Palmateer A et al (2011) Improved tolerance toward fungal diseases in transgenic Cavendish banana (*Musa* spp. AAA group) cv. grand nain. Transgenic Res 20: 61–72. doi:10.1007/s11248-010-9392-

Weisburg WG, Barns SM, Pelletier DA, Lane DJ (1991) 16S Ribosomal DNA amplification for phylogenetic study. J Bacteriol 173:697–703

Chapter 15
Microbial Inoculants with Autochthonous Bacteria for Biodiverse Legume Pastures in Portuguese Agro-Forestry Ecosystems

Ricardo Soares, Eva Arcos, Eugénio Ferreira and Isabel Videira e Castro

Abstract One key point to achieve the success of microorganisms like rhizobia as biofertilizers (inoculants) is the characterization and selection of autochthonous strains. The aim of this work was to study the nitrogen-fixing efficiency of autochthonous *Rhizobium* strains, using as hosts several agronomic important annual *Trifolium* species (*Trifolium subterraneum*, *Trifolium incarnatum*, *Trifolium suaveolens*, and *Trifolium vesiculosum*), and compare them with a peat commercial inoculant. For this purpose, three preselected strains (89Ts2a, 123Ts2a, and 149Ts2) isolated from several *Trifolium* spp. grown in the south of Portugal were characterized beyond their symbiotic performance, also for their "in vitro" PPB activities (solubilization of mineral phosphate and production of siderophores). In addition, molecular identification was performed using the sequence of 16S rRNA and recA genes. Results of shoot dry weight indicated that autochthonous *Rhizobium* strains were highly efficient in nitrogen fixation. For *T. subterraneum* and *T. incarnatum*, strain 123TS2a had the best performance, and for *T. suaveolens* and *T. vesiculosum*, the best performance was accomplished by strain 89TS2a. These two strains were combined in a mixed inoculum that was shown to promote higher yields than the commercial peat inoculant in the majority of *Trifolium* species tested. Results of "in vitro" activities indicated that two strains (89Ts2 and 123Ts2a) could solubilize phosphate, and the three strains could produce siderophores. All these strains had strong phylogenetic relationship with *Rhizobium leguminosarum* and are closely related to biovar *trifolii* and *viciae*.

Keywords Biological nitrogen fixation · Rhizobia · Peat commercial inoculant · *Trifolium* spp. · PPB activities

R. Soares · E. Arcos · E. Ferreira · I. Videira e Castro (✉)
Laboratório de Microbiologia do Solo, Instituto Nacional
de Investigação Agrária e Veterinária, I.P. (INIAV), Oeiras, Portugal
e-mail: isabel.castro@iniav.pt

© Springer International Publishing Switzerland 2016

F. González-Andrés and E. James (eds.), *Biological Nitrogen Fixation and Beneficial Plant–Microbe Interactions*, DOI 10.1007/978-3-319-32528-6_15

15.1 Introduction

Nitrogen fixation comprises a crucial aspect for crop development throughout the world in all agricultural systems. The Haber–Bosch process (industrial nitrogen fixation) developed in the last century has provided a large amount of N to agriculture, which raised human population as it was never seen before. Despite its high productivity, the energy consumption and ecological effects of this process are too dramatic for a sustainable agriculture. One possible, more ecological, and low-cost alternative seems to be the biological nitrogen fixation (BNF), which has been intensively studied over the years (Ferguson et al. 2010; Huo et al. 2012; Okon and Labandera-Gonzalez 1994; Santi et al. 2013). In particular, legume crops can associate symbiotically with a group of soil bacteria called rhizobia that, once established in specialized structures designated by root nodules, are able to fix the di-nitrogen present in the air and provide it to the plant in a usable form (ammonia) (Kamboj et al. 2008; van Rhijn and Vanderleyden 1995).

In Portugal, a strategy for improving pastures in a Mediterranean agro-forestry system, characterized by a large existence of cork oak trees, named as "montado" (or "dehesa" in Spain), has been developed (Ferreira et al. 2010; Ferreira and Castro 2011). This strategy is based on the establishment of biodiverse permanent pastures, including selected and improved plant species, in which inoculated legumes with highly efficient rhizobia have an important role. One key point to achieve the success of microorganisms, like rhizobia, as biofertilizers (inoculants) is the characterization and selection of the autochthonous strains by optimizing the interaction between legume species and rhizobia strains, which highlight the high efficiency in nitrogen fixation. In such biodiverse pastures, the legumes are composed mainly of annual and perennial clovers. *Trifolium* sp. (clover) is one of the largest genera within the *Fabaceae* family. It contains annual and perennial species with an important value for improving pastures. This genus is very specific in its association with rhizobia, interacting efficiently with *Rhizobium leguminosarum* bv. *trifolii* strains. Generally, the symbiotic effectiveness can vary in a wide range (Drew et al. 2011; Ferreira and Marques 1992; Slattery and Coventry 1995), especially between annual and perennial clovers (Howieson et al. 2005; Vincent 1954). Thus, to reach the maximum potential of crop growth, it is very important to select and combine correctly the strains to be used as inoculants.

Besides nitrogen fixation, rhizobia can also promote plant growth throughout other mechanisms, such as phosphate solubilization and siderophore production. These and other activities can be present in bacteria designated as plant probiotic bacteria (PPB) (Glick 1995; Kloepper et al. 1989). Unlike nitrogen, which is available in large quantities in a reservoir (atmosphere), phosphorous is mostly immobilized in the soil and often poorly present in soluble forms (Ezawa et al. 2004; He et al. 2007). Siderophores are low-weight iron chelators that are prone to restrain the growth of plant phytopathogens, through iron deprivation (Lemanceau and Alabouvette 1993).

The aim of this work was to study the behavior of different annual clover species, usually present in biodiverse legume pastures (Crespo 2006) inoculated with selected autochthonous *Rhizobium* strains, and compare their nitrogen-fixing efficiency with a commercial peat inoculant. Genetic identification and PPB activities were also aimed for further characterization of the strains.

15.2 Materials and Methods

15.2.1 Rhizobia Strains

Autochthonous *Rhizobium* strains were isolated from the root nodules of several native clovers (mainly *Trifolium subterraneum*) growing in different fields across southern Portugal (Ferreira and Marques 1992), following the methodology described by Vincent (1970). Each isolate (pure culture) is stored in yeast mannitol agar media (YMA) at 4 °C and belongs to the "INIAV Rhizobia Bacteria Collection". For selecting the most effective strains to be used in a future biofertilizer for annual clovers, a set of 10 strains were inoculated in *T. subterraneum* cv. Clare and their symbiotic effectiveness was compared and used as a preliminary evaluation. Using this criterion, strains 89TS2a, 123TS2a, and 149TS2 were preselected for further characterization.

15.2.2 Symbiotic Effectiveness

The three preselected rhizobial strains were evaluated in four annual *Trifolium* species: *T. subterraneum* L., *Trifolium incarnatum* L., *Trifolium suaveolens* L., and *Trifolium vesiculosum* Savi. *Trifolium* sp. seeds were rinsed with ethanol, surface-sterilized using 0.1 % of $HgCl_2$ for 4 min, washed extensively with sterile distilled water, left in sterile water to hydrate for 1–2 h, and placed in water agar (0.8 %) plates, at 28 °C, in the dark, during 1–2 days. The pregerminated seeds were transferred to flasks (one seed per flask) containing 50 ml agar slants of Jensen medium (Jensen 1941). Six replicate seedlings of each clover species were inoculated with 1 ml suspension containing approximately 10^8 cells of each rhizobial strain, either individually or in mixture, or with 1 ml of the peat inoculant suspended in liquid Jensen medium (¼ diluted). Uninoculated (T0) and nitrogen (TN) controls were also included for each clover species, by adding 1 ml liquid Jensen medium (¼ diluted) and 1 ml 1.75 % KNO_3, respectively. Plants were grown during 8 weeks in a controlled-environment chamber with 16-h light/8-h dark cycle at 23 °C (day)/18 °C (night). After this period, plants were harvested and examined for nodulation and the shoots were dried in an oven at 80 °C for 2 days.

The values of shoot dry weight (*X*) were used to calculate the index of effectiveness (Es) according to the formula (Ferreira and Marques 1992)

$$\mathrm{Es}(\%) = \frac{\mathbf{Xs} - \mathbf{XT0}}{\mathbf{XTN} - \mathbf{XT0}} \times 100,$$

where **X**s represents the mean dry weight of inoculated shoots; **X**TN the mean dry weight of plants with nitrogen control; and **X**T0 the mean dry weight of uninoculated plants.

15.2.3 In Vitro *PPB Activities*

The ability to solubilize mineral phosphate was assayed on yeast extract dextrose (YED) agar medium plates supplemented with 0.2 % of Ca_3PO_4 (Peix et al. 2001). After growth, strains with halos around the colonies were considered as phosphate solubilizers. Siderophore production was screened by growing each strain in tryptone yeast (TY) medium (Beringer 1974) and thereafter overlaying with 15 ml CAS (chrome azurol S) medium (Pérez-Miranda et al. 2007). The formation of orange halos around colonies revealed the presence of these low-weight iron chelation molecules.

15.2.4 Molecular Identification

Rhizobium strains were grown overnight in TY liquid medium at 28 °C. Total genomic DNA was extracted using Bio-Rad Aqua Pure Genomic DNA extraction kit. 16S rRNA and *rec*A genes were amplified using Taq PCR Master Mix Kit with approximately 20–40 ng of total genomic DNA of each strain. For 16S rRNA gene sequencing, 1.5-kb (approximately) fragment was amplified using 41f/1488r primers (Weisburg et al. 1991). For *rec*A gene, a fragment with about 500 bp was amplified using the primer pair *rec*A63/*rec*A555 (Gaunt et al. 2001). Nucleotide sequences were edited, using Chromas Lite software (version 2.1.1), and homologies were searched in NCBI BLASTn database. Phylogenetic analysis was performed using Mega (version 5) software (Tamura et al. 2011) and neighbor-joining (NJ) (Saitou and Nei 1987) and bootstrap (Felsenstein 1985) methods. Phylogenetic distances were calculated using p-distance nucleotide substitution model and the "pairwise deletion" option. Confidence values were based on 1000 bootstrap replications.

15.3 Results

15.3.1 Single-Strain Inoculation

Results obtained with the individual inoculation of the three strains showed differences in the shoot dry weight among the tested species of *Trifolium*. For *T. suaveolens* and *T. vesiculosum*, the highest values were obtained with strain 89Ts2a; for *T. subterraneum* and *T. incarnatum*, the highest values were obtained when inoculated with strain 123Ts2a. On the other hand, the inoculation of strain 149Ts2, in the four tested species of clover, did not promote, in a general way, an increase in the plant shoot dry weight (Fig. 15.1).

Figure 15.2 shows the indices of effectiveness of rhizobial strains when inoculated in the different *Trifolium* species. Strain 89Ts2a showed the highest effectiveness in *T. suaveolens*, with an Es of 120 %. In the remaining *Trifolium* sp., the Es value of this strain was below 80 %. However, in the case of *T. vesiculosum*, strain 89Ts2a had the highest Es value of the three strains (with 77 %). Strain 123Ts2a had the highest effectiveness found in this study (124 %) when inoculated in *T. subterraneum*. For *T. incarnatum*, the Es was 80 % when inoculated with 123Ts2a. Less prominent, strain 149Ts2 produced lower Es than strain 89Ts2a or 123Ts2a in all tested *Trifolium* sp.

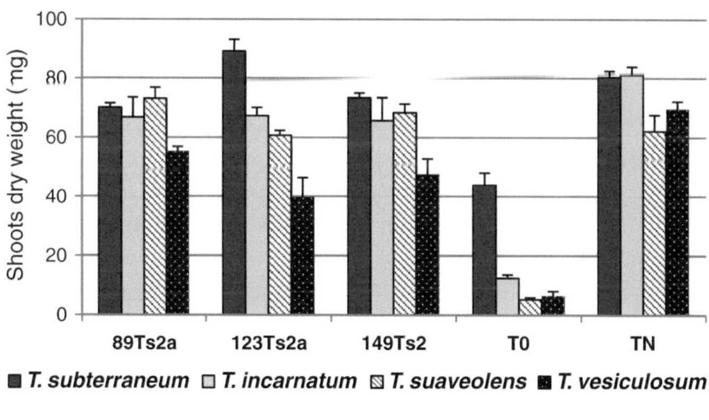

Fig. 15.1 Shoot dry weight of four different *Trifolium* species inoculated with autochthonous rhizobia strains. It was also included TN (plants only with nitrogen) and T0 (uninoculated plants and without mineral N) controls

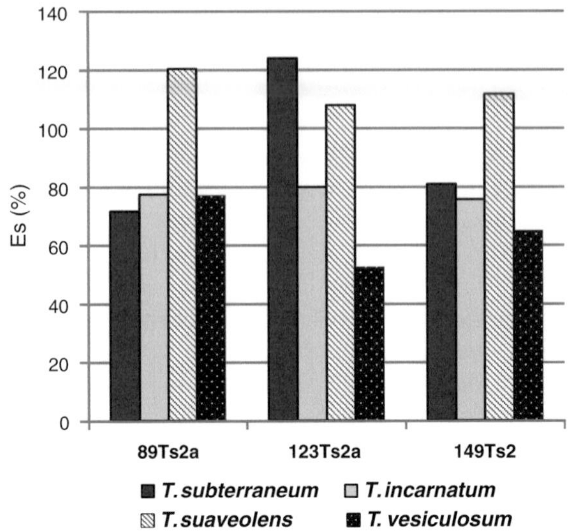

Fig. 15.2 Indices of effectiveness (Es) of rhizobia autochthonous strains, inoculated in four different *Trifolium* species

15.3.2 Mixture of Strains and Commercial Peat Inoculation

Given the results obtained with the single-strain inoculation, mixed inoculation with strains 89Ts2a and 123Ts2a was also attempted. The results obtained with the inoculation of this mixture in *T. subterraneum*, *T. suaveolens*, and *T. vesiculosum* revealed higher values of shoot dry weight when compared with those obtained by the inoculation of the commercial peat. On the contrary, the highest values of shoot dry weight in *T. incarnatum* were accomplished by the peat inoculant (Fig. 15.3).

The indices of effectiveness (Fig. 15.4) showed that the mixture of strains was more effective than the commercial peat inoculant in *T. subterraneum*, with an Es of 106 %, in *T. suaveolens*, with an Es of 96 %, and also in *T. vesiculosum*, with an Es of 47 %, which is much higher than the Es value (26 %) found with the peat inoculant. On the contrary, in *T. incarnatum*, the peat inoculant had higher Es (93 %) than the mixture of strains (59 %).

15.3.3 In Vitro *PPB Activities*

All strains were screened for the presence of in vitro activities related to the promotion of plant growth, other than nitrogen fixation. The ability to solubilize tricalcium phosphate was evaluated by the formation of clearance halos around the growth areas in agar plates containing insoluble tricalcium phosphate. Positive

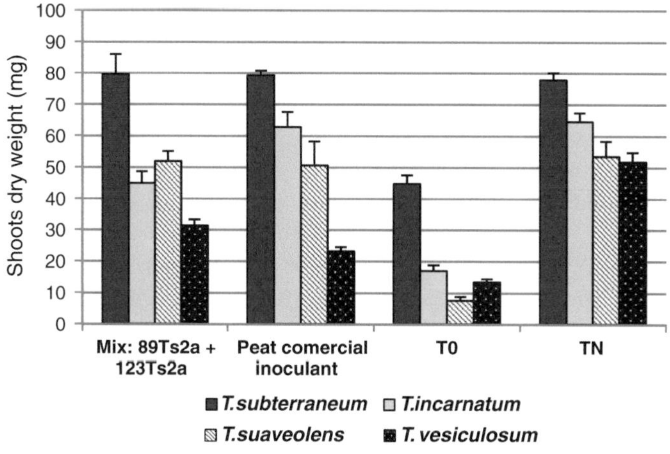

Fig. 15.3 Shoot dry weight of four different *Trifolium* species, inoculated with a mixture of autochthonous rhizobia strains (89TS2a and 123TS2a) and with a commercial peat inoculant. It was also included TN (plants only with nitrogen) and T0 (uninoculated plants and without mineral N) controls

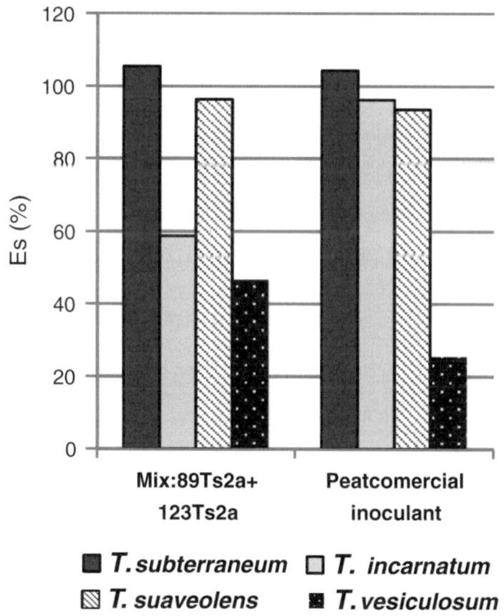

Fig. 15.4 Indices of effectiveness of the mixture of rhizobia autochthonous strains, 89Ts2a and 123Ts2a, and of the commercial peat inoculant, when inoculated in four different *Trifolium* species

activity was observed in all strains tested. On the other hand, the presence of orange halos, around the colonies, indicative of siderophore production, was detected only in strains 89Ts2a and 123Ts2a.

15.3.4 Molecular Identification

Phylogenetic analysis of 16S rRNA and recA of the rhizobial isolates (Figs. 15.5 and 15.6) indicated a strong similarity of all strains with the species *R. leguminosarum*. The 16S rRNA gene sequence (along a stretch of 1137 nt) of strains 123Ts2a and 149Ts2 was 100 % identical and displayed 99.9 % sequence identity with representative strains of *R. leguminosarum* bv. *trifolii* and bv. *viciae*. Strain 89Ts2a was clustered with *R. leguminosarum* bv. *trifolii* Rtl E12, showing 99.4 % of sequence identity. On the other hand, using the sequence of *rec*A gene (along a stretch of 405 nt), strain 89Ts2a showed 100 % sequence identity with *R. leguminosarum* bv. *viciae* CCBAU 03100 and strain 123Ts2a displayed 99.8 % of identical nucleotides with *R. leguminosarum* R458702248, which was isolated from *Trifolium arvense*. Additionally, strain 149Ts2a showed 100 % sequence identity with *R. leguminosarum* bv. *trifolii* WSM 1325 (Table 15.1).

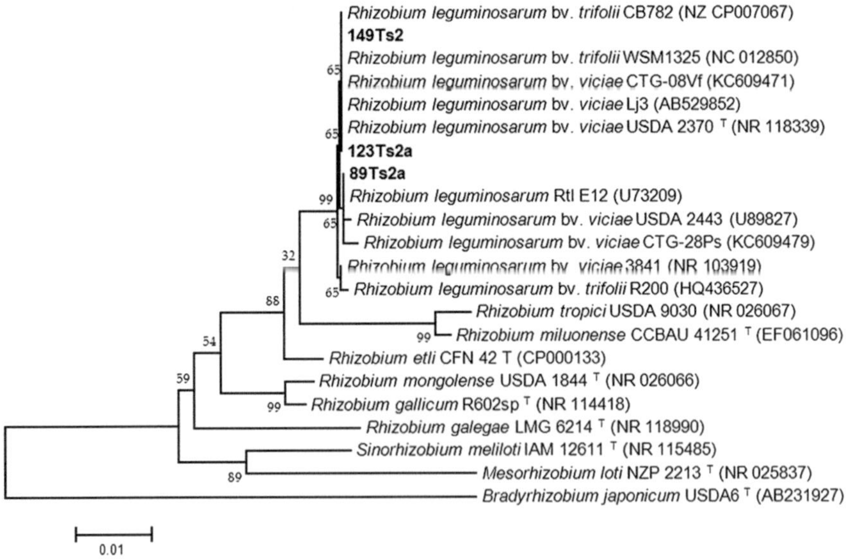

Fig. 15.5 16S rRNA gene phylogeny based on 1137 positions. In each branch is presented the confidence levels of bootstrap. GenBank accession codes are presented next to each strain

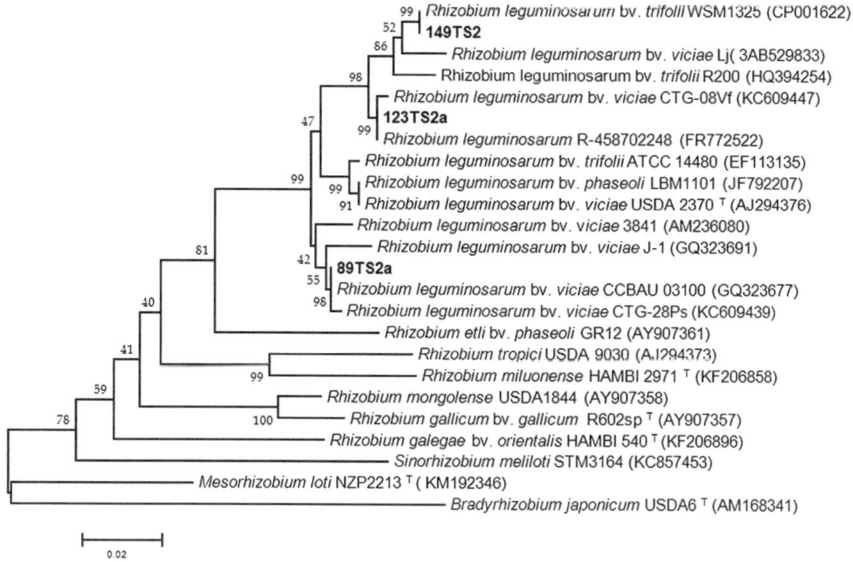

Fig. 15.6 *rec*A gene phylogeny based on 405 positions. In each branch is presented the confidence levels of bootstrap. GenBank accession codes are presented next to each strain

Table 15.1 Autochthonous strains best match, using NCBI GenBank database, of 16S rRNA and *rec*A gene

Strain	16S rRNA gene		*rec*A gene	
	Identification	%	Identification	%
89TS2a	*R. leguminosarum* bv. *trifolii* Rtl E12	99.4	*R. leguminosarum* bv. *viciae* CCBAU 03100	100
123Ts2a	*R. leguminosarum* bv. *trifolii* GTG-14-Ps, Lj3; *R. leguminosarum* bv. *viciae* CB782, WSM1325	99.9	*R. leguminosarum* R458702248	99.8
149Ts2a	*R. leguminosarum* bv. *trifolii* GTG-14-Ps, Lj3; *R. leguminosarum* bv. *viciae* CB782, WSM1325	99.9	*R. leguminosarum* bv. *trifolii* WSM 1325	100

15.4 Discussion

There are many studies in the literature reporting the isolation of *R. leguminosarum* bv. *trifolii* strains with equal or even superior symbiotic performance than the commercial in-use (Bergersen et al. 1971; Collins et al. 2002; Ferreira and Marques 1992; Gibson et al. 1975; Slattery and Coventry 1995). In this study, the combination of two autochthonous strains, 89TS2a and 123Ts2a, showed to be a strong alternative to the commercial peat inoculant, presenting higher indices of effectiveness in all annual clover species tested with the exception of *T. incarnatum*. The results obtained with this last host could be due to the fact that the autochthonous strains of *R. leguminosarum* were not isolated from *T. incarnatum* but from *T. subterraneum* and for the reason that the maximum symbiotic performance could not be reached, although they are all annual *Trifolium*. This disadvantage over the peat inoculant could be possibly overcome by the isolation and selection of *R. leguminosarum* bv. *trifolii* strains from the more prominent nodules of *T. incarnatum* naturally growing plants. So, new data concerning the effect of the Es, once in mixture with the other two strains, would be needed.

Strain 149Ts2a was not chosen to integrate this combination of strains, as it presented lower indices of effectiveness in comparison with the other two tested strains, when inoculated separately or in mixture (data not shown).

The use of inoculants with autochthonous strains can represent an advantage over inoculants made out of strains with different origins, since native strains are usually better adapted to survive and persist in the local ecosystem conditions (Aguilar et al. 2001; Brockwell et al. 1982; Chatel et al. 1973). Furthermore, as it is known, Portuguese soils are generally deficient in soluble phosphorous, particularly in the "montado" ecosystem (Tóth et al. 2014). Strains 89Ts2a and 123Ts2a could solubilize and provide this important macronutrient to plants. The production of siderophores can also be of great importance to an inoculant, since it could prevent different types of diseases induced by phytopathogens. However, further work will be required to accurately evaluate this effect against phytopathogens, such as those which have a drastic impact on agriculture and economy, like the *Phytophthora cinnamomi* pathogen that affects the cork oaks in the "montado" ecosystem (Brasier 1996). All these activities can enhance the value of an inoculant and may have a clear advantage in plant growth promotion over other commercial inoculants where these activities are not present.

The molecular characterization showed that the three autochthonous strains had strong phylogenetic relationships with the species *Rhizobium leguminosarum* as it was expected since the *Trifolium* genus is one of the best examples of specificity within the legume–rhizobia symbiosis. Strain 89Ts2a showed 100 % sequence identity with a biovar *viciae* strain in the *rec*A gene, but shared also a high sequence identity (99.4 %) with a biovar *trifolii* strain in the 16S rRNA gene. On the other hand, strains 123Ts2a and 149Ts2a displayed high identity in 16S rRNA gene with biovar *trifolii* and biovar *viciae* strains. In order to determine to which biovar or symbiovar (Rogel et al. 2011) our strains belong, more information would be needed, such as the phylogeny of symbiotic genes and symbiotic performance information about the corresponding plants of the two genera discussed.

15.5 Conclusion

The selection of autochthonous *R. leguminosarum* strains has allowed us to develop an alternative to a peat base commercial inoculant for annual clovers. Our option of using autochthonous strains instead of commercial peat inoculants could bring higher yields in most species of annual clovers used in biodiverse pastures in Portuguese agro-forestry ecosystems, such as *T. subterraneum*, *T. suaveolens* and *T. vesiculosum*. These selected autochthonous strains have also the capacity for undergoing phosphate solubilization and siderophore production, which could be also of great importance in poorly fertile soils containing high levels of phytopathogens.

Acknowledgements The financial support was provided by the project PRODER 54971 (Portugal). Ricardo Soares and Eva Arcos were granted, respectively, by PRODER 54971 (Portugal) and by FUNDECYT (Spain).

References

Aguilar OM, López MV, Riccillo PM (2001) The diversity of rhizobia nodulating beans in Northwest Argentina as a source of more efficient inoculant strains. J Biotechnol 91(2–3):181–188

Bergersen FJ, Brockwell J, Gibson AH et al (1971) Studies of natural populations and mutants of *Rhizobium* in the improvement of legume inoculants. Plant Soil 35(1):3–16

Beringer JE (1974) R factor transfer in *Rhizobium leguminosarum*. J Gen Microbiol 84(1):188–198

Brasier C (1996) *Phytophthora cinnamomi* and oak decline in southern Europe. Environmental constraints including climate change. Ann For Sci 53(2–3):347–358

Brockwell J, Gault R, Zorin M et al (1982) Effects of environmental variables on the competition between inoculum strains and naturalized populations of *Rhizobium trifolii* for nodulation of *Trifolium subterraneum* L. and on rhizobia persistence in the soil. Aust J Agric Res 33(5):803

Chatel DL, Shipton WA, Parker CA (1973) Establishment and persistence of *Rhizobium trifolii* in western Australian soils. Soil Biol Biochem 5(6):815–824

Collins MT, Thies JE, Abbott LK (2002) Diversity and symbiotic effectiveness of *Rhizobium leguminosarum* bv. *trifolii* isolates from pasture soils in south western Australia. Aust J Soil Res 40(8):1319

Crespo DG (2006) The role of pasture improvement on the rehabilitation of the montado/dehesa system and in developing its traditional products. In: Ramalho Ribeiro JMC, Horta AEM, Mosconi C et al (eds) Animal products from the Mediterranean area. The Netherlands Academic Publishers, Wageningen, pp 185–197

Drew EA, Charman N, Dingemanse R et al (2011) Symbiotic performance of Mediterranean *Trifolium* spp. with naturalised soil rhizobia. Crop Pasture Sci 62(10):903

Ezawa T, Cavagnaro TR, Smith SE et al (2004) Rapid accumulation of polyphosphate in extraradical hyphae of an arbuscular mycorrhizal fungus as revealed by histochemistry and a polyphosphate kinase/luciferase system. New Phytol 161(2):387–392

Felsenstein J (1985) Confidence limits on phylogenies: an approach using the bootstrap. Evolution 39(4):783–791

Ferguson BJ, Indrasumunar A, Hayashi S et al (2010) Molecular analysis of legume nodule development and autoregulation. J Integr Plant Biol 52(1):61–76

Ferreira EM, Castro IV (2011) Fijación biológica de nitrógeno y productividade de pastos naturales en sistemas agrícolas del mediterráneo. In: Megías M, Rivilla R, Soto MJ et al (eds) Fundamentos y Aplicaciones agroambientales de las interacciones beneficiosas plantas-microorganismos. Sociedad Española de Fijación de Nitrógeno (SEFIN), Spain, pp 403–416

Ferreira EM, Marques JF (1992) Selection of Portuguese *Rhizobium leguminosarum* bv. *trifolii* strains for production of legume inoculants. Plant Soil 147(1):151–158

Ferreira EM, Simões N, Castro IV et al (2010) Relationships of selected soil parameters and natural pastures yield in the Montado ecosystem of the Mediterranean area using multivariate analysis. Silva Lus 18(2):151–166

Gaunt MW, Turner SL, Rigottier-Gois L et al (2001) Phylogenies of *atp*D and *rec*A support the small subunit rRNA-based classification of rhizobia. Int J Syst Evol Microbiol 51(Pt 6):2037–2048
Gibson AH, Curnow BC, Bergersen FJ et al (1975) Studies of field populations of *Rhizobium*: effectiveness of strains of *Rhizobium trifolii* associated with *Trifolium subterraneum* L. pastures in south-eastern Australia. Soil Biol Biochem 7(2):95–102
Glick BR (1995) The enhancement of plant growth by free-living bacteria. Can J Microbiol 41 (2):109–117
He ZL, Bian W, Zhu J (2007) Screening and identification of microorganisms capable of utilizing phosphate adsorbed by goethite. Commun Soil Sci Plant Anal 33(5–6):647–663
Howieson JG, Yates RJ, O'Hara GW et al (2005) The interactions of *Rhizobium leguminosarum* biovar *trifolii* in nodulation of annual and perennial *Trifolium* spp. from diverse centres of origin. Aust J Exp Agric 45(3):199
Huo YX, Wernick DG, Liao JC (2012) Toward nitrogen neutral biofuel production. Curr Opin Plant Biol 23(3):406–413
Jensen HL (1941) Nitrogen fixation in leguminous plants. I. General characters of root-nodule bacteria isolated from species of *Medicago* and *Trifolium* in Australia. Proc Linn Soc NSW 67:98–108
Kamboj D, Kumar R, Kumari A et al (2008) Rhizobia, nod factors and nodulation—a review. Agric Rev 29(3):200–206
Kloepper JW, Lifshitz R, Zablotowicz RM (1989) Free-living bacterial inocula for enhancing crop productivity. Trends Biotechnol 7(2):39–44
Lemanceau P, Alabouvette C (1993) Suppression of fusarium wilts by fluorescent pseudomonads: mechanisms and applications. Biocontrol Sci Tech 3(3):219–234
Okon Y, Labandera-Gonzalez CA (1994) Agronomic applications of *Azospirillum*: an evaluation of 20 years worldwide field inoculation. Soil Biol Biochem 26(12):1591–1601
Peix A, Rivas-Boyero AA, Mateos PF et al (2001) Growth promotion of chickpea and barley by a phosphate solubilizing strain of *Mesorhizobium mediterraneum* under growth chamber conditions. Soil Biol Biochem 33(1):103–110
Pérez-Miranda S, Cabirol N, George-Téllez R et al (2007) O-CAS, a fast and universal method for siderophore detection. J Microbiol Methods 70(1):127–131
Rogel MA, Ormeño-Orrillo E, Martinez Romero E (2011) Symbiovars in rhizobia reflect bacterial adaptation to legumes. Syst Appl Microbiol 34(2):96–104
Saitou N, Nei M (1987) The neighbor-joining method: a new method for reconstructing phylogenetic trees. Mol Biol Evol 4(4):406–425
Santi C, Bogusz D, Franche C (2013) Biological nitrogen fixation in non-legume plants. Ann Bot 111(5):743–767
Slattery JF, Coventry DR (1995) Acid-tolerance and symbiotic effectiveness of *Rhizobium leguminosarum* bv. *trifolii* isolated from subterranean clover growing in permanent pastures. Soil Biol Biochem 27(1):111–115
Tamura K, Peterson D, Peterson N et al (2011) MEGA5: molecular evolutionary genetics analysis using maximum likelihood, evolutionary distance, and maximum parsimony methods. Mol Biol Evol 28(10):2731–2739
Tóth G, Guicharnauda R-A, Tóth B et al (2014) Phosphorus levels in croplands of the European Union with implications for P fertilizer use. Eur J Agron 55:42–52
van Rhijn P, Vanderleyden J (1995) The *Rhizobium*-plant symbiosis. Microbiol Rev 59(1):124–142
Vincent JM (1954) The root-nodule bacteria as factors in clover establishment in the red basaltic soils of the Lismore district, NSW. III. Field inoculation trials. Aust J Agric Res 5:61
Vincent JM (1970) A manual for the practical study of root nodule bacteria. IBP Handbook No. 15. Blackwell Scientific Publications, Oxford
Weisburg WG, Barns SM, Pelletier DA et al (1991) 16S ribosomal DNA amplification for phylogenetic study. J Bacteriol 173(2):697–703

Chapter 16
Improving Legume–Rhizobium Symbiosis for Copper Phytostabilization Through Genetic Manipulation of Both Symbionts

Eloísa Pajuelo, Patricia Pérez-Palacios, Asunción Romero-Aguilar, Julián Delgadillo, Bouchra Doukkali, Ignacio D. Rodríguez-Llorente and Miguel A. Caviedes

Abstract The presence of excess copper (Cu) in soils represents an environmental and health problem, due to the risk of groundwater pollution. Besides, it affects plant development and yield. Phytoremediation has consolidated as a low-cost and ecological technique for metal remediation. In this particular, legume–rhizobium symbioses have risen as an attractive biotechnological tool for metal phytostabilization. For this technique to be suitable, metal-tolerant symbionts are needed, which can be generated through genetic engineering. In this work, the genetic manipulation of both symbiotic partners for Cu phytostabilization was described. Concerning the plant, composite *Medicago truncatula* plants expressing the metallothionein gene *mt4a* from *Arabidopsis thaliana* in roots were generated, in an attempt to increase the plant tolerance towards Cu. Concerning the rhizobial strain, an *Ensifer medicae* strain was genetically engineered by expressing the copper resistance genes *copAB* from *Pseudomonas fluorescens*. Our results indicate the following: (a) the expression of *mt4a* in composite plants increases tolerance towards Cu and reduces oxidative stress caused by this pollutant. Lower levels of reactive oxygen species (ROS)-scavenging enzymes were found in *mt4a*-expressing plants; (b) the expression of *mt4a* in composite plants improves nodulation, whereas inoculation with the genetically modified *Ensifer* has a synergistic effect; and (c) The double symbiotic system enhances Cu accumulation in roots, without increasing metal translocation to shoots. We conclude that the genetically modified symbiosis is a suitable tool for Cu rhizo–phytostabilization.

E. Pajuelo · P. Pérez-Palacios · A. Romero-Aguilar · B. Doukkali · I.D. Rodríguez-Llorente · M.A. Caviedes (✉)
Departamento de Microbiología y Parasitología, Facultad de Farmacia, Universidad de Sevilla, c/Profesor García González, 2, 41012 Seville, Spain
e-mail: caviedes@us.es

J. Delgadillo
Área de Microbiología, Colegio de Post-Graduados, Campus de Montecillo, Carretera Federal México-Texcoco, 56230 Montecillo, Mexico

© Springer International Publishing Switzerland 2016

F. González-Andrés and E. James (eds.), *Biological Nitrogen Fixation and Beneficial Plant–Microbe Interactions*, DOI 10.1007/978-3-319-32528-6_16

Keywords *Medicago truncatula* · *Ensifer* (*Sinorhizobium*) *medicae* · Phytoremediation · Heavy metals · Metallothionein *mt4a* · Copper resistance genes *copAB*

16.1 Introduction

The presence of copper in soils can be due to natural biogeological processes, although most copper reaches the environment as a consequence of human activities, such as mining, industrial manufacturing of copper wires, batteries and electronic microchips, together with pesticides used in agriculture (Shrivastava 2009).

Copper has a dual role in plants, being a micronutrient or a toxic depending on the concentration. As a micronutrient, Cu is involved in several physiological processes such as photosynthesis, mitochondrial reactions, cell wall metabolism, pollen tube development and ethylene perception (Yruela 2005; Festa and Thiele 2011). However, Cu excess in plants inhibits germination, reduces growth, inhibits photosynthesis by affecting the photosynthetic apparatus and causes a strong oxidative stress (Cambrollé et al. 2008; Panda et al. 2008; Posmyk et al. 2009; Wang et al. 2011). Besides these effects, in legumes, even low copper concentrations inhibit nodulation, decrease nodule number and nodule mass and affect root hair development (Kopittke et al. 2007; Ahmad et al. 2012).

In the last decade, legumes in association with metal-tolerant rhizobia have emerged as an attractive tool for metal phytostabilization in soils (Dary et al. 2010; Pajuelo et al. 2011; Teng et al. 2015). For this technique to be suitable, both symbiotic partners must be tolerant towards metals. In general, legumes show a moderate metal tolerance (Ahmad et al. 2012; Pajuelo et al., 2011). Moreover, they accumulate metals in roots, with low translocation factors (TF) to shoots (Pajuelo et al. 2007). The stress generated by metal excess in legumes could be attenuated by the presence of metal-resistant rhizosphere microorganisms, particularly rhizobia (Teng et al. 2015; Ullah et al. 2015).

When legumes and/or rhizobia are not sufficient tolerant, genetic engineering is a tool to generate metal-resistant plants and microorganisms (Kotrba et al. 2011; Macek et al. 2008). Regarding Cu, it is known that the expression of metallothioneins in transgenic plants leads to enhanced copper resistance and/or accumulation (Guo et al. 2008; Rodríguez-Llorente et al. 2010). Besides copper binding, metallothioneins seem to have a role in ROS scavenging and detoxification (Hassinen et al. 2011). On their side, genetically modified rhizobia have been obtained with increased resistance towards to cadmium (Sriprang et al. 2002; Ike et al. 2007) and copper (Delgadillo et al. 2015).

In this work, we have advanced one step forward, since we have obtained a double genetically modified symbiotic system with the aim of Cu phytostabilization: composite plants of *Medicago truncatula*, expressing a metallothionein gene *mt4a* from *Arabidopsis thaliana,* and a genetically modified *Ensifer medicae*, expressing copper resistance genes *copAB* from *Pseudomonas fluorescens*. Plant growth, nodulation, oxidative stress and Cu accumulation were evaluated.

16.2 Materials and Methods

16.2.1 Generation of Genetically Modified Symbionts

Composite *M. truncatula* plants were generated by infection with *Agrobacterium rhizogenes* ARquaA1 (Boisson-Dernier et al. 2001) carrying the pMF2 binary vector containing the gene *mt4a* from *A. thaliana* and GUS as a reporter gene (Rodríguez-Llorente et al. 2010). Transgenic roots were selected on kanamycin (25 μg mL^{-1}). Plants transformed with the empty vector were used as control. The genetically modified rhizobial strain *E. medicae* MA11 expressing the *P. fluorescens copAB* genes was previously reported (Delgadillo et al. 2015).

16.2.2 Plant Growth Conditions and Inoculation

After selection with kanamycin, composite *Medicago* plants were grown on sand: vermiculite (3:1) in sterile transparent containers (Duchefa®)—provided with gas exchange filters in the lids. Plants were watered initially with 50 mL of BNM (Buffered Nodulation Medium, without any nitrogen source) and supplemented with increasing Cu concentrations (0, 100 and 200 μM). Plants were inoculated with *E. medicae* MA11 or MA11-*copAB*. The containers were placed in a growth chamber at 22 ± 2 °C and 12 h light:dark cycle. After 15 days, another 25 mL of BNM medium (without additional Cu) were added under sterile conditions in order to keep the humidity inside the pots. The total growth period was one month.

16.2.3 Analysis of Plant Growth Parameters and Nodulation

Plants were collected and number of nodules was determined. Shoots and roots were cut apart and biomass and length measured (20–24 plants). Half of the plants were dried in the oven (80 °C) for the determination of dry weight and Cu accumulation. The other half was homogenized in liquid nitrogen and stored at −80 °C until use.

16.2.4 Determination of Antioxidant Enzymes

The biochemical determination of antioxidant enzymes superoxide dismutase (SOD), total peroxidases (PX), catalase (CAT) and ascorbate peroxidase (APX) in crude extracts of shoots of plants was performed according to Sosa-Alderete et al. (2009). Protein was determined according to Bradford (1976). Results of specific activities are means of three independent determinations.

16.2.5 Analysis of Copper Accumulation in Plant Tissues

Plants were harvested, exhaustively washed with distilled water and dried in an oven at 80 °C until constant dry weight. Cu content in plant tissues was estimated by ICP-OES.

16.2.6 Statistical Analysis

Data of growth and nodulation are means ± standard deviations of 20–25 individual plants. Data of Cu accumulation are means ± standard deviations of 3 determinations of 8 plants pooled together. Data of enzymatic activities are means ± standard deviations of 3 determinations. Statistical analysis was performed by one-way ANOVA, followed by DMS post hoc test at $p < 0.05$.

16.3 Results

16.3.1 Growth of the Double-Modified Symbiotic System in the Presence of Copper

The presence of Cu reduced plant growth particularly at 200 μM (Fig. 16.1a). Root length and biomass decreased, as well as shoot biomass. The maximum decrease was observed in the wild type partnership, whereas the expression of *mt4a* was the most determining factor in protecting root growth inhibition by Cu (Fig. 16.1b). Inoculation with the genetically modified *Ensifer* had only a minor effect on root and plant growth.

16.3.2 Effect on Nodulation

The number of nodules decreased 80 % in the wild-type symbiosis when exposed to 200 μM Cu (Table 16.1). The expression of *mt4a* protected the decrease of nodulation, since in composite plants, nodulation decreased 60 %. There was a synergistic effect of inoculation with the modified *Ensifer*, so that in the double

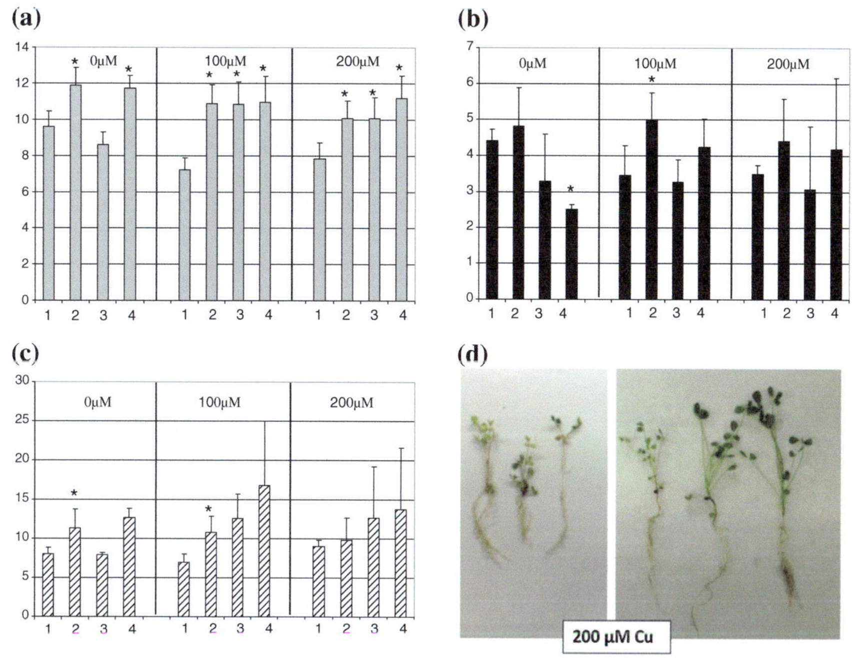

Fig. 16.1 Effect of increasing Cu concentrations of growth parameters for the different symbiotic partnerships (*1*: WT/WT; *2*: MT4a/WT; *3*: WT/*copAB*; *4*: MT4a/*copAB*). **a** Root length (cm). **b** Root biomass (mg/plant). **c** Shoot biomass (mg/plant). **d** Comparison between the wild-type symbiosis (*left*) and the double symbiotic system (*right*) under the presence of 200 μM Cu

Table 16.1 Number of nodules in *M. truncatula* control composite plants (WT) and plants expressing *mt4a* (MT4a) inoculated with either *E. medicae* MA11 (WT) or *E. medicae* MA11-*copAB* (*copAB*) and grown at different Cu concentrations

Plant/inoculant	No external Cu	100 μM Cu	200 μM Cu
WT/WT	6.08 ± 2.15	2.76 ± 1.85	1.23 ± 0.85
WT/*copAB*	5.31 ± 3.02	3.24 ± 1.47	1.55 ± 0.72
MT4a/WT	8.54 ± 2.71*	5.36 ± 2.63*	3.75 ± 1.25*
MT4a/*copAB*	8.53 ± 4.86*	5.55 ± 3.14*	6.11 ± 3.70*

For each metal concentration, significant differences with regard to the control at $p < 0.05$ are indicated by *

symbiotic system, nodulation only decreased 20 % at 200 μM Cu. Moreover, the expression of the reporter gene *lacZ* was confirmed in the presence of X-gal (Fig. 16.2). The blue colour indicated that the nodules were elicited by the modified microsymbiont. Moreover, since the genes *copAB* were located in a distal position respect to the promoter, the expression of *lacZ* was indicative of the correct expression of *copAB,* situated in a proximal position respect to the promoter (Delgadillo et al. 2015).

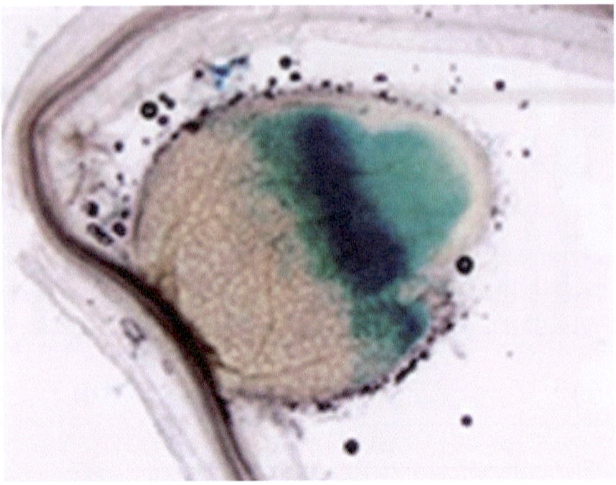

Fig. 16.2 Nodule elicited by the genetically modified *Ensifer medicae* MA11-*copAB*, expressing the copper resistance genes from *Pseudomonas fluorescens* Az13. The blue colour due to β-galactosidase activity is indicative of the expression of *copAB* genes located in a distal position with regard to the promoter

16.3.3 Oxidative Stress in the Presence of Copper

In order to evaluate the oxidative stress of the different symbioses, antioxidant enzymes were determined. Results can be seen in Fig. 16.3. For catalase activity (Fig. 16.3a), the activity increased fivefold in the wild-type symbiosis in the presence of 100–200 μM Cu. Compared to the double-modified symbiosis, there was a synergistic effect between the expression of *mt4a* and inoculation with the

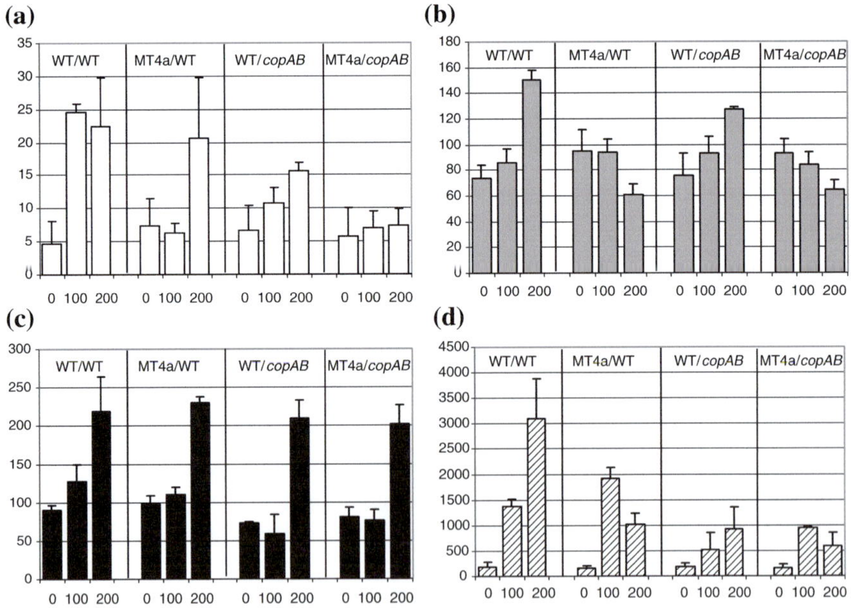

Fig. 16.3 Levels of antioxidant enzymes (in mU/mg of protein) in roots of composite *Medicago* plants (wild-type WT or transgenic plants expressing *mt4a*) inoculated with the two *Ensifer* strains in the presence of Cu. **a** catalase; **b** total peroxidases; **c** ascorbate peroxidase; **d** superoxide dismutase

modified rhizobial strain in diminishing the levels of this enzyme. Similar results were observed for total peroxidase activity (Fig. 16.3b) and for superoxide dismutase activity (Fig. 16.3d). The combined partnerships including the wild-type plant inoculated with the genetically modified rhizobial strain, and the transgenic root inoculated with the wild-type rhizobial strain, showed intermediate enzyme levels. By contrast, ascorbate peroxidase showed a different pattern; neither the expression of *mt4a* nor the inoculation with the genetically modified rhizobial strain led to significant changes in this activity with regard to the wild-type partnership. In general, the results showed that the wild-type symbiosis displayed the highest level of all enzymes, particularly at 200 μM. These results could be interpreted in terms of the double-modified symbiosis having the lowest level of oxidative stress, whereas the combinations of wild-type plants and genetically modified rhizobium or wild-type rhizobium and transgenic roots led to intermediate levels of stress.

16.3.4 Copper Accumulation

Cu determination in plant tissues (Table 16.2) showed that the concentration of Cu in shoots slightly increased at 100 and 200 μM Cu. However, in roots, inoculation with the genetically modified *Ensifer* strongly enhanced Cu accumulation in roots, being this factor the most determining in affecting Cu root accumulation. Expression of *mt4a* had only minor but synergistic effect (Table 16.2). When the TF was calculated (Table 16.3), data showed the lowest values in roots inoculated with the transgenic *Ensifer*, suggesting the usefulness of this transgenic approach in copper phytostabilization.

Table 16.2 Cu accumulation (μg of Cu per mg of plant) in *M. truncatula* control composite plants (WT) and plants expressing *mt4a* (MT4a) inoculated with either *E. medicae* MA11 (WT) or *E. medicae* MA11-*copAB* (*copAB*) and grown at different Cu concentrations

Plant/inoculant	No external Cu	100 μM Cu	200 μM Cu
Shoot			
WT/WT	17 ± 2.1	36 ± 2.5	37 ± 5.7
WT/*copAB*	11 ± 1.4*	43 ± 1.4*	41 ± 3.2
MT4a/WT	17 ± 2.6	30 ± 7.5	42 ± 6.3
MT4a/*copAB*	10 ± 3.1*	31 ± 6.4	48 ± 5.1*
Root			
WT/WT	32 ± 3.5	61 ± 11.4	nd
WT/*copAB*	28 ± 5.4	162 ± 15.7*	147 ± 17.8
MT4a/WT	38 ± 6.4	81 ± 9.7	nd
MT4a/*copAB*	29 ± 9.8	185 ± 17.4*	155 ± 18.7

* Indicate significant differences ($p < 0.05$), (*nd* = not determined, due to little biomass)

Table 16.3 Cu translocation factor in *M. truncatula* control composite plants (WT) and plants expressing *mt4a* (MT4a) inoculated with either *E. medicae* MA11 (WT) or *E. medicae* MA11-*copAB* (*copAB*) and grown at different Cu concentrations

Plant/inoculant	No external Cu	100 μM Cu	200 μM Cu
WT/WT	0.53	0.59	nd
WT/*copAB*	0.28	0.18	0.28
MT4a/WT	0.60	0.53	nd
MT4a/*copAB*	0.34	0.16	0.30

nd = not determined

16.4 Discussion

Copper excess in soils has deleterious effects on the legume–rhizobium symbiotic interaction (Kopittke et al. 2007). Moreover, the accumulation of Cu in the aerial parts of plants could be a way for this element entering the food chain. In this particular, phytostabilization is an adequate technique to immobilize metals in soils, thus preventing accumulation in plant shoots (Méndez and Maier 2008). Genetic engineering allows the generation of more metal-tolerant plants and microorganisms (Kotrba et al. 2011).

In this work, we have approached the manipulation of both symbiotic partners, the legume and the rhizobial strain, for improved Cu phytostabilization. Our data suggest that the expression of *mt4a* in roots of *M. truncatula* improves growth and tolerance towards Cu. This effect could be due to the direct copper-binding activity of metallothioneins, which, together with organic acids, are the main chelators of Cu for its accumulation inside the vacuoles or in the cell walls (Cobbet and Goldsbrough 2002; Guo et al. 2008).

Nodulation has been other parameter highly beneficed in our double-modified symbiotic system. Nodule number strongly decreased (80 %) in the wild-type symbiosis in the presence of Cu, according to what it was previously reported (Kopittke et al. 2007). The decrease in nodulation, and hence in nitrogen fixation, also affects plant health (Rifat et al. 2010). The expression of *mt4a* protected nodulation in the presence of Cu (60 % decrease in nodulation), and inoculation with the genetically modified *Ensifer* had a synergistic effect (only 20 % decrease in nodulation in the presence of 200 μM Cu).

Cu is one of the metals that cause a strongest oxidative stress (Tripathi et al. 2006; Wang et al. 2011). Plants have evolved a complex network of enzymatic and non-enzymatic mechanisms in order to detoxify ROS, which includes enzymes such SOD, APX, PX and CAT. These enzyme activities are widely accepted as markers of oxidative stress (Posmyk et al. 2009; Zhou et al. 2007). According to our data, the double-modified symbiosis showed the lowest levels of antioxidant enzymes, except for ascorbate peroxidase, for which neither the expression of *mt4a* nor inoculation with the genetically modified rhizobial strain led to lower levels in the presence of Cu (200 μM). Besides their Cu-binding properties, metallothioneins are involved in ROS scavenging (Hassinen et al. 2011), which could be another mechanism explaining the higher tolerance towards Cu in the composite plants. Again, inoculation with the engineered rhizobial strain led to further protection of the nodulation in the presence of Cu. Our results are in accordance with some recent reports, in which Cu-resistant rhizobia improve legume tolerance towards this metal, and reduce Cu stress in the plant (Li et al. 2014; Hao et al. 2015).

Regarding Cu accumulation, the main advantage of this double-modified symbiosis is the enhanced accumulation (significantly different) of Cu en roots (two- to threefold with regard to the wild-type symbiosis), without significantly increasing the translocation to shoot. The lowest TFs are achieved in the double-engineered symbiosis. We conclude that the double genetically modified symbiosis is a suitable tool for Cu phyto–rhizostabilization. This is an extra advantage of our proposed system, thus avoiding the entrance of this toxic element in the trophic chain. In fact, all the data of Cu accumulation in shoots are below the threshold recommended for herbivores consumption (Méndez and Maier 2008). This system promotes the accumulation of Cu in roots without additional translocation to shoots and it is a suitable strategy to obtain legumes with improved growth and low Cu concentration in shoots when cultivated in polluted soils.

Acknowledgments Research financed by MINECO (Spain) and FEDER (project reference BIO-2009-7766). P.P.P. acknowledges grants from University of Seville (Spain) and AUIP. J.D. acknowledges a grant from CONACYT (Mexico).

References

Ahmad E, Zaidi A, Khan MS, Oves M (2012) Heavy metal toxicity to symbiotic nitrogen fixing microorganisms and host legumes. In: Zaidi A et al (eds) Toxicity of heavy metals to legumes and bioremediation. Springer, Vienna, pp 29–44

Boisson-Dernier A, Chabaud M, García F, Bécard G, Rosenberg C, Barker DG (2001) *Agrobacterium rhizogenes*-transformed roots of *Medicago truncatula* for the study of nitrogen-fixing and endomycorrhizal symbiotic associations. Mol Plant Microbe Interact 14 (6):695–700

Bradford MM (1976) A rapid and sensitive method for the quantification of microgram quantities of protein utilizing the principle of protein-dye binding. Anal Biochem 72:248–254

Cambrollé J, Redondo-Gómez S, Mateos Naranjo E, Figueroa ME (2008) Comparison of the role of two *Spartina* species in terms of phytostabilization and bioaccumulation of metals in the estuarine sediment. Mar Pollut Bull 56:2037–2042

Cobbet C, Goldsbrough P (2002) Phytochelatins and metallothioneins: roles in heavy metal detoxification and homeostasis. Annu Rev Plant Biol 53:159–182

Dary M, Chamber-Pérez MA, Palomares AJ, Pajuelo E (2010) "In situ" phytostabilisation of heavy metal polluted soils using *Lupinus luteus* inoculated with metal resistant plant growth promoting rhizobacteria. J Hazard Mater 177(1):323–330

Delgadillo J, Lafuente A, Doukkali B, Redondo-Gómez S, Mateos-Naranjo E, Caviedes MA, Pajuelo E, Rodríguez-Llorente ID (2015) Improving legume nodulation and Cu rhizostabilization using a genetically modified rhizobia. Environ Technol 36:1237–1245

Festa RA, Thiele DJ (2011) Copper: an essential metal in biology. Curr Biol 21:877–883

Guo WJ, Meetam M, Goldsbrough PB (2008) Examining the specific contributions of individual *Arabidopsis* metallothioneins to copper distribution and metal tolerance. Plant Physiol 146:1697–1706

Hao X, Xie P, Zhu YG, Taghavi S, Wei G. Rensing C (2015) Copper tolerance mechanisms of *Mesorhizobium amorphae* and its role in aiding phytostabilization by *Robinia pseudoacacia* in copper contaminated soil. Environ Sci Technol 49:2328– 2340

Hassinen VH, Tervahauta AI, Schat H, Kärenlampi SO (2011) Plant metallothioneins—metal chelators with ROS scavenging activity? Plant Biol 13:225–232

Ike A, Sriprang R, Ono H, Murooka Y, Yamashita M (2007) Bioremediation of cadmium contaminated soil using symbiosis between leguminous plant and recombinant rhizobia with the MTL4 and the PCS genes. Chemosphere 66(9):1670–1676

Kopittke P, Dart PJ, Menzie NW (2007) Toxic effects of low concentrations of Cu on nodulation of cowpea (*Vigna unguiculata*). Environ Pollut 145:309–315

Kotrba P, Mackova M, Macek T (2011) Transgenic approaches to enhance phytoremediation of heavy metal polluted soils. In: Khan MS, Zaidi A, Goel R, Musarrat J (eds) Biomanagement of metal contaminated soils. Environmental pollution, vol 20. Springer, Berlin, pp 239–271

Li Z, Ma Z, Hao X, Rensing C, Wei G (2014) Genes conferring copper resistance in *Sinorhizobium meliloti* CCNWSX0020 also promote the growth of *Medicago lupulina* in copper-contaminated soil. Appl Environ Microbiol 80(6):1961–1971

Macek T, Kotrba P, Svatos A, Novakova M, Demnerova K, Mackova M (2008) Novel roles for genetically modified plants in environmental protection. Trends Biotechnol 26:146–152

Méndez MO, Maier RM (2008) Phytostabilisation of mine tailings in arid and semiarid environments: an emerging remediation technology. Environ Health Perspect 116:278–283

Pajuelo E, Carrasco JA, Romero LC, Chamber MA, Gotor C (2007) Evaluation of the metal phytoextraction potential of crop legumes. Regulation of the expression of O-acetylserine (thiol) lyase under metal stress. Plant Biol 9:672–681

Pajuelo E, Rodríguez-LLorente ID, Lafuente A, Caviedes MA (2011) Legume-rhizobium symbioses as a tool for bioremediation of heavy metal polluted soils. In: Khan MS, Zaidi A, Goel R, Musarrat J (eds) Biomanagement of metal contaminated soils, environmental pollution, vol 20. Springer, Germany, pp 95–123

Panda SK (2008) Impact of copper on reactive oxygen species, lipid peroxidation and antioxidants in *Lemna minor*. Biol Plant 52:561–564

Posmyk MM, Kontek R, Janas KM (2009) Antioxidant enzymes activity and phenolic compounds content in red cabbage seedlings exposed to copper stress. Ecotoxicol Environ Saf 72:596–602

Rifat H, Safdar A, Ummay A, Rabia K, Iftikhar A (2010) Soil beneficial bacteria and their role in plant growth promotion: a review. Ann Microbiol 60(4):579–598

Rodríguez-Llorente ID, Perez-Palacios P, Doukkali B, Caviedes MA, Pajuelo E (2010) Expression of the seed-specific metallothionein *mt4a* in plant vegetative tissues increases Cu and Zn tolerance. Plant Sci 178:327–332

Shrivastava AK (2009) A review on copper pollution and its removal from water bodies by pollution control technologies. Ind J Environ Protect 29:552–560

Sosa-Alderete LG, Agostini E, Medina MI (2009) Antioxidant response of tobacco (*Nicotiana tabacum*) hairy roots after phenol treatment. Plant Physiol Biochem 49:1020–1028

Sriprang R, Hayashi M, Yamashita M, Ono H, Saeki K, Murooka Y (2002) A novel bioremediation system for heavy metals using the symbiosis between leguminous plant and genetically engineered rhizobia. J Biotechnol 99(3):279–293

Teng Y, Wang X, Li L, Li Z, Luo Y (2015) Rhizobia and their bio-partners as novel drivers for functional remediation in contaminated soils. Front Plant Sci 6:32

Tripathi BN, Mehta S, Amar A, Gaur JP (2006) Oxidative stress in *Scenedesmus* sp. during short- and long-term exposure to Cu^{2+} and Zn^{2+}. Chemosphere 62:538–544

Ullah A, Heng S, Munis MFH, Fahad S, Yang X (2015) Phytoremediation of heavy metals assisted by plant growth promoting (PGP) bacteria: a review. Environ Exp Bot 117:28–40

Wang SH, Zhang H, Zhang Q, Jin GM, Jiang SJ, Jiang D, He QY, Li ZP (2011) Copper-induced oxidative stress and responses of the antioxidant system in roots of *Medicago sativa*. J Agron Crop Sci 197:418–429

Yruela I (2005) Copper in plants. Braz J Plant Physiol 17(1):145–156

Zhou ZS, Wang SJ, Yang ZM (2007) Biological detection and analysis of mercury toxicity to alfalfa (*Medicago sativa*) plants. Chemosphere 70:1500–1509

Chapter 17
Symbiotic and Agronomic Characterization of Bradyrhizobial Strains Nodulating Cowpea in Northern Peru

Renzo Alfredo Valdez, Bertha Soriano, Gardenia Prado, David Zavaleta, Minoru Matsubara, Doris Zúñiga, Patricie Dion, Ángel Valladolid, Asdrúbal Ruesta, Antonio Castellano-Hinojosa, David Correa-Galeote and Eulogio J. Bedmar

Abstract Eighty bacterial strains were isolated from root nodules of cowpea (*Vigna unguiculata*) plants agriculturally grown in Northern Peru. After isolation, fifty of them reinfected cowpea and eight showed better symbiotic performance than control plants treated with nitrate both under greenhouse and field. Repetitive extragenic palindromic polymerase chain reaction (REP-PCR) clustered the strains into 3 REP-PCR groups. The partial sequence of the 16S rRNA gene from a representative strain of each REP-PCR pattern showed that the strains were closely

R.A. Valdez · A. Castellano-Hinojosa (✉) · D. Correa-Galeote · E.J. Bedmar
Department of Soil Microbiology and Symbiotic Systems,
Estación Experimental Del Zaidín, CSIC, Granada, Spain
e-mail: ach@ugr.es

B. Soriano
Department of Microbiology and Parasitology, National University of Trujillo,
Trujillo, Peru

G. Prado · D. Zavaleta
César Vallejo University, Piura, Peru

M. Matsubara · D. Zúñiga
Laboratory of Microbial Ecology and Biotechnology, La Molina Land
National University, Lima, Peru

P. Dion
Laval University, Quebec, Canada

Á. Valladolid
PROMENESTRAS TEX Association, Chiclayo, Peru

A. Ruesta
National Institute of Land Innovation, Experimental Station of Vista Florida,
Lambayeque, Peru

© Springer International Publishing Switzerland 2016

F. González-Andrés and E. James (eds.), *Biological Nitrogen Fixation and Beneficial Plant–Microbe Interactions*, DOI 10.1007/978-3-319-32528-6_17

related to members of genus *Bradyrhizobium* of the Alphaproteobacteria, but affiliation at the species level was not clear. Pairwise alignments among globally aligned sequences of the 16S rRNA genes, the sequencing of the housekeeping genes *atpD*, *glnII,* and *recA*, and their concatenated phylogenetic analysis showed that strains Rc-391-01 and Rc-458-01 grouped with *B. yuanmingense* B071^T and that strain Rc-352-01 clustered with *Bradyrhizobium liaoningense* 2281^T. Examination of the three representative strains to behave as plant growth-promoting rhizobacteria (PGPRs) showed that none of them solubilized calcium triphosphate and that they all three produced siderophores and IAA-related compounds and had ACC-deaminase activity.

Keywords *Bradyrhizobium* · Cowpea · Symbiosis · 16S rRNA · Housekeeping genes · PGPRs

17.1 Introduction

Legumes play an important ecological role, with representatives in nearly every type of plant group on Earth (Lewis and Forest 2005). They are best characterized by their ability to establish N_2-fixing symbiotic associations with Alphaproteobacteria of the genus *Azorhizobium*, *Bradyrhizobium*, *Mesorhizobium*, *Rhizobium*, and *Ensifer* (formerly *Sinorhizobium*), collectively known as to rhizobia, as well as with Betaproteobacteria of the genus *Burkholderia* and *Cupriavidus* (Peix et al. 2015). Due to this ability, legumes can grow in arid, nitrogen-deficient soils, acting as pioneer plants for soil stabilization and colonization, enhancing their fertility and, consequently, preventing erosion and desertification.

Cowpea (*Vigna unguiculata* (L.) Walp.) is one of the most important, versatile, and nutritious grain legume in semiarid regions of the tropics, covering 14×10^6 ha across Asia, Africa, and South America (Singh et al. 2002). In addition, cowpeas provide a rich source of proteins and calories, as well as minerals and vitamins. A cowpea seed can consist of up to 25 % protein, and its content in anti-nutritional factors is low, which helps the nutritional balance of low-income populations (Rangel et al. 2004). Moreover, cowpea presents high tolerance to edaphic-climatic stresses such as acidity, dryness, and high temperatures (Zilli et al. 2004).

Cowpea establishes symbiotic associations with members of genus *Bradyrhizobium*, among them the species *Bradyrhizobium liaoningense* (Pule-Meulenberg et al. 2010), *B. elkanii* (Zhang et al. 2011), *B. japonicum* (Bejarano et al. 2014), *B. yuanmingense*, *B. arachidis* and *B. daqingense* (Krasova-Wade et al. Krasova-Wade et al. 2003), *B. canariense*, *B. cytisi* and *B. pachyrhizi* (Bejarano et al. 2014), and *B. manausense* (Venancio-Silva et al. 2014).

In addition to symbiotic nitrogen fixation, rhizobial bacteria may act as plant growth-promoting rhizobacteria (PGPR) by solubilization of inorganic P through liberation of organic acids (Marra et al. 2012; Glick 2012), uptake of Fe through siderophores production (Carson et al. 2000), modulation of plant growth by indol acetic acid production (Boeiro et al. 2007), and inhibition of ethylene synthesis by 1-aminocyclopropane-1-carboxylic acid (ACC) deaminase activity (Glick 2014). Together, expression of all these features can improve the efficiency of the rhizobia–legume symbiosis (Naveed et al. 2015).

In the Peruvian agriculture, cowpea is the most important pulse legume with about 25,700 ha producing approximately 59,400 t of grain valued 69 million dollars in period 2011–2014 (PROMPERU 2015). Currently, it is cultivated in the coastal region of Peru including the departments of Piura, Lambayeque, La Libertad, San Martín, and Lima and in the jungle region of Ucayali. In the coastal region, cowpea is cultivated in low-fertility, P- and N-deficient sandy soils where production requires high doses of chemical fertilizers, which, in turns, originates environmental pollution. In the jungle region, specifically Ucayali, cowpea cultivation is performed in "playas," the lowlands formed by the Ucayali River's trailing sediments, and "restingas," the ground floodplain produced by the low caudal of the river during the summer season (Labarta et al. 2007). Chemical fertilization of cowpea crops in the Ucayali region is scarce, and yet productivity is maintained over times.

Because data on the rhizobial species that can be found in nodules of cowpea grown in the Ucayali region of Peru have not been reported, the primary objective of this research was to isolate and identify the rhizobial endosymbionts in nodules of the legume plant. Partial sequence analysis of the core gene 16S rRNA and the housekeeping genes, *atpD*, *glnII*, and *recA,* indicated all the isolates belong to genus *Bradyrhizobium*. We also assessed the symbiotic performance of selected strains in association with *V. unguiculata* grown in soils from different representative regions of cowpea production in Peru. Finally, we have evaluated the plant growth-promoting properties of those selected strains.

17.2 Materials and Methods

17.2.1 Isolation of Bacteria from Root Nodules and Culture Conditions

Nodules (15/plant) were collected from roots of healthy cowpea plants (*V. unguiculata*) (12 plants/location) cultivated in "playa" soil near to the city of Callería (8° 23′ S, 74° 31′ W) and "restinga" soils near to the cities of Yarinacocha (8° 23′ S, 74° 31′ W) and San Juan (8° 19′ S, 74° 36′ W) from the department of Ucayali in the jungle region of Peru. The Callería's soil had a sandy-loam texture and the following characteristics: 85 % sand, 4 % silt, and 11 % clay; pH (in water), 7.9; 0.03 % of total N; 16 ppm of total P; and 0.69 % of organic matter. The Yarinacocha's soil had a silty-clay texture and the following characteristics: 3 % sand, 48 % silt, and 49 % clay; pH (in water), 7.2; 0,09 % of total N; 24.7 ppm of total P; and 2.05 % of organic matter. The San Juan's soil had a silty-clay texture and the following characteristics: 3 % sand, 48 % silt, and 49 % clay; pH (in water), 7.1; 0.08 % of total N; 24.1 ppm of total P; and 2.01 % of organic matter.

Nodules were surface-sterilized according to Somasegaran and Hoben (1994) and then placed independently in Petri dishes and crushed in a drop of sterile water with a sterile glass rod. The resulting suspension was streaked onto Petri dishes containing yeast extract–mannitol (YEM) medium (Vincent 1970) supplemented with Congo red (25 μg/mL). After incubation of the plates at 30 °C for 10 d, colony-forming units, which represented all of the colony types that could be distinguished by microscopic observation of living cells, were chosen.

17.2.2 Seeds Sterilization and Plant Inoculation and Growth

Seeds of cowpea var. CAU-5 were surface-sterilized with 70 % ethanol for 1 min, washed with 5 % sodium hypochlorite for 5 min, rinsed thoroughly with sterile distilled water, embedded in sterile distilled water for 1 h, and then allowed to germinate at 30 °C in the dark. For experiments carried out under greenhouse conditions, seedlings (4/jar) were planted in Leonard jars assemblies containing coarse sand (1 mm) in their upper part and 1/4 strength Jensen's N-free mineral solution (Somasegaran and Hoben 1994) in the lower part of the jar. In parallel experiments, Jensen's mineral solution was supplemented with 2.15 M KNO_3. Seedlings were separately inoculated at planting with 1 mL (about 10^8 cells) of each one of the strains isolated from the cowpea nodules, placed in a greenhouse, and harvested at 10 % flowering. The experiments were conducted in a randomized block design using 2 plants/jar and 3 jars/strain.

For experiments developed under field conditions, two experimental fields representing major cowpea-planting regions in Peru were used. They are owned by Instituto Nacional de Innovación Agraria (INIA) and are located in Paiján (7° 43′ S, 79° 18′ W) (La Libertad region, Peru) and Picsi (6° 43′ S, 79° 46′ W). Soil from Paiján has a sandy-loam texture and the following characteristics: 72 % sand, 15 % silt, and 13 % clay; pH (in water), 7.4; 0.11 % of total N; 7.5 ppm of total P; and 2 % of organic matter. Soils from Picsi (6° 43′ S; 79° 46′ W) has loamy texture and the following characteristics: 42 % sand, 35 % silt, and 23 % clay; pH (in water), 7.20; 0.18 % of total N; 5.7 ppm of total P; and 2.75 % of organic matter. The experiments were conducted in a randomized block design where the block size was 7 m length × 4 m width, 80 cm between rows, and 1 m between plots (for 4 replicates). Seeds of cowpea var. CAU-5 were coated separately with bacterial suspension ($\sim 10^8$ cells/mL) using sucrose (10 % w/v) as an adhesive agent for 2 h before planting. Uninoculated plants were used as a control. At soil preparation, all blocks received P (360 kg P_2O_5/ha as super phosphate) and K (120 kg K_2SO_4/ha). Twenty-five days after planting, the soil received N (60 kg N/ha as NH_4NO_3). The plants were grown under until fruit maturity (about 110 d from planting).

17.2.3 Physiological Assays

Dry weights of the plants (PDW) were obtained by drying the fresh samples in an oven at 65 °C. Complete drying was indicated by reaching a constant weight. Efficiency (EFI) and effectiveness (EFV) of the strains were calculated according to Muniz et al. (2012) using the equations: EFI = (PDW of inoculated plants/PDW of plants not treated with combined N) × 100, and EFV = (PDW of inoculated plants/PDW of plants treated with combined N) × 100. Only strains that showed higher EFV than the control plants under greenhouse conditions were used for taxonomic affiliation and field trials.

17.2.4 DNA Extraction and PCR Amplifications

For DNA extraction and PCR amplifications, genomic DNA was isolated from bacterial cells using the RealPure Genomic DNA Extraction kit (Durviz, Spain), according to the manufacturer's instructions. The quantity of DNA was determined by using a NanoDrop spectrophotometer (NanoDrop ND1000). Repetitive extragenic palindromic polymerase chain reactions (REP-PCR) were performed using primers REPIR-I and REP2-I, according to de Bruijn (1992). The genetic diversity of the strains was analyzed by the presence or absence of polymorphic bands in the gel. The data were grouped by the unweighted pair group method with arithmetic mean (UPGMA) algorithm and Pearson coefficient using the software program InfoQuest FP version 4.5 (Biorad). PCR amplifications of 16S rRNA gene fragments were carried out using the two opposing primers fD1 and rD1, as previously reported (Weisburg et al. 1991). Partial sequences of *atpD*, *recA*, and *glnII* genes were obtained using the pair primers atpD 255F/atpD 782R, glnII 12 F/glnII 689R, and recA 41 F/recA 640R described by Vinuesa et al. (2005), respectively. Three forward primers, nodCF, nodCF2, and nodCFn, and the reverse primer nodCI, were used for amplification and sequencing of the *nodC* gene (Laguerre et al. 2001). Amplification products were purified using the Qiagen PCR product purification system, and they were subjected to cycle sequencing using the same primers as for PCR amplification, with ABI Prism Dye Chemistry, and analyzed with a 3130xl automatic sequencer at the sequencing facilities of Estación Experimental del Zaidín, Granada, Spain.

The 16S rRNA gene sequences were compared to those deposited in EzTaxon-e (Kim et al. 2012), and the gene sequences of the housekeeping genes were compared to homologous sequences in GenBank using the BLASTN program (Altschul et al. 1990). Phylogenetic analyses were performed with the Geneious package version 8.1.3 (Kearse et al. 2012), inferred using the neighbor-joining algorithm (Saitou and Nei 1987), and visualized with MEGA5 (Tamura et al. 2011). DNA sequences were aligned using the Phydit software (Chun 2001).

17.2.5 Plant Nodulation Tests

Seeds of *V. unguiculata* var. CAU-5 were surface-sterilized and allowed to germinate at 30 °C in the dark. Seedlings (3/jar) were planted in Leonard jars assemblies containing vermiculite and inoculated separately (about 10^8 cells/mL) with each one of the 3 representative strains selected after taxonomic affiliation. Plants were grown under greenhouse conditions, fed with ¼ strength Jensen's N-free mineral solution, and harvested at 30 % flowering.

17.2.6 Plant Growth-Promoting Traits of Selected Bradyrhizobial Species

Indole acetic acid (IAA) production, siderophore production, phosphate solubilization, and ACC-deaminase activity were assayed as indicated previously (Castellano-Hinojosa et al. 2015). Briefly, production of IAA and IAA-related compounds was evaluated after growth of the cells in Bergersen liquid minimal medium (Bergersen 1977) to reach about 10^8 cells/mL. After centrifugation at 13,000 rpm for 3 min, the supernatant was mixed with the Salkowsky reagent (1:1 v/v). Color intensity was measured spectrophotometrically at 535 nm after 30 min. Auxin levels were estimated as IAA equivalents using standard curves prepared with pure IAA. The effect on tryptophan on IAA production was also evaluated after addition of 600 μg/mL of *L*-tryptophan to the growth media. *Brevibacillus brevis* BEA1 (Moreno et al. 2009) was used as a positive control.

Production of siderophores was quantified using the liquid chrome azurol S (CAS)-shuttle assay (Payne 1994). For that purpose, bacteria were grown in Fe-free SM minimal medium (Brown and Dilworth 1975) at 30 °C to reach about 10^8 cells/mL. After centrifugation at 10,000 rpm for 15 min in a microfuge, the supernatant was mixed (1:1, v:v) with CAS solution, allowed to stand for 20 min, and used to assay the intensity of blue color at 630 nm. SM minimal medium was used as blank. Siderophore production was estimated following the formula: % siderophore units = [(Ar-As)/Ar] × 100, where Ar = absorbance of reference (minimal media + CAS assay solution) and As = absorbance of sample (Sayyed et al. 2005).

Estimation of phosphate solubilization was carried out as described by Peix et al. (2001) using YED-P medium supplemented with 0.1 % tricalcium phosphate. Bacteria were first grown in YEM medium up to reach a density of about 10^8 cells/mL. Then, 20 μL aliquots of each culture were inoculated in YED-P medium and incubated at 30 °C for 20 d. The diameter of the solubilization halo was measured at the beginning and at the end of the incubation period. From those measurements, a solubilization index (SI) was calculated according to the formula SI = halo diameter (mm)/colony diameter (mm) (Marra et al. 2012).

Ability of the strains to hydrolyze 1-aminocyclopropane-1-carboxylic acid (ACC) was assayed according to Ma et al. (2003). The bacterial cells were grown in

5 mL YEM medium at 30 °C until they reach about 10^8 cells/mL. The cells were collected by centrifugation at 6000 rpm, for 10 min at 4 °C, washed twice with 0.1 M Tris-HCl (pH 7.5), suspended in 2 mL of modified M9 minimal medium (Miller 1972) supplemented with 5 mM ACC, and then incubated at 25 °C with shaking (170 rpm) for 72 h. The induced bacterial cells were harvested and washed as described above, and resuspended in 200 μL of 0.1 M Tris-HCl (pH 8.5). The cells were solubilized using 5 % toluene (v/v). The amount of α-ketobutyrate in the supernatant was determined after centrifugation at 3000 rpm for 10 min and determination of the absorbance of the samples at 540 nm. Standard curves were prepared with pure α-ketobutyrate.

17.2.7 Protein Determination

Protein concentration in bacterial samples was estimated according to Bradford (1976).

17.2.8 Accession Numbers

Accession numbers of the nucleotide sequences of the bradyrhizobial species used in this study are shown in the figure trees.

17.2.9 Statistical Analysis

Measured variables in this study were first explored using the Shapiro–Wilk test to check whether they meet the normality assumptions. We used the Tukey test, Duncan test, and Kruskal–Wallis and Conover-Iman combined tests for the comparisons of the results. All the statistical analyses were carried out by the XLSTAT version 2013.1 software (Addinsoft).

17.3 Results

17.3.1 Isolation of Bacteria and Symbiotic Tests

Eighty bacterial strains were isolated from extracts of root nodules of cowpea plants growing in soils of the Ucayali region, 40 from Callería and 20 from each Yarinacocha and San Juan. Out of the 80 bacterial strains, 50 formed nodules on

Table 17.1 Plant dry weight (PDW), efficiency (EFI), and effectiveness (EFV) of selected bradyrhizobial strains used for inoculation of cowpea var CAU-5

Strain/treatment	PDW (g)	EFI (%)	EFV (%)
Rc-391-01	2.66a	277	122
Rc-353-02	2.59ab	271	120
Rc-458-01	2.57ab	268	119
Rc-402-01	2.52b	262	116
Rc-450-01	2.35bc	244	108
Rc-352-01	2.31bc	240	106
Rc-455-02	2.22c	231	102
Rc-355-05	2.17d	226	100
C + N	2.17d	226	100
C	0.96e	100	44

Plants were grown in Leonard jars assemblies under greenhouse conditions until 10 % flowering. Numbers in a column followed by the same letter are not significantly different according to the Tukey test ($n = 5$; $P \leq 0.05$). Controls included plants without inoculation grown in the absence (C) and in the presence of 2.15 M mineral nutrient solution (C + N)

cowpea CAR-5 planted in Leonard jar assemblies that were grown under greenhouse conditions, 20 from Callería, 17 from Yarinacocha, and 13 from San Juan. However, only 8 strains, 6 from Callería (Rc-352-01, Rc-353-02, Rc-391-01, Rc-450-01, Rc-455-02, and Rc-458-01), 1 from Yarinacocha (Rc-402-01), and 1 from San Juan (Rc-355-05) had equal or greater EFI and EFV than the control plants, ranging from 2 to 22 % higher PDW values (Table 17.1). The remaining 42 strains showed lower EFI and EFV values than those of the control plants (data not shown). Non-inoculated controls produced no nodules, indicating no cross-contamination.

17.3.2 Field Trials

The eight selected strains were used to inoculate cowpea CAR-5 plants in experimental fields located at Paiján and Picsi. Regardless of the experimental site, inoculation with whatever strain produced higher crop yield than those of the control plants (Table 17.2). Grain yield in Paiján varied with the strain used for inoculation with values ranging from 1710 to 2215 kg/ha; again, strain Rc-352-01 showed superior performance than the remaining strains with a 29.5 % grain yield increase in relation to the control plants. Different results were found in Picsi's experimental field as statistical differences were not found among treatments.

Table 17.2 Grain yield of cowpea var. CAU-5 plants inoculated with selected bradyrhizobial strains

Strain/treatment	Paiján	Picsi
	Grain yield (kg/ha)	
Rc-352-01	2215a	1761a
Rc-450-01	2145b	1682a
Rc-455-02	2128b	1875a
Rc-355-05	2087b	1751a
Rc-458-01	2049b	1786a
Rc-353-02	1903bc	1817a
Rc-402-01	1868bc	1892a
Rc-391-01	1816bc	1631a
60 kg N/ha	2083b	1745a
Control	1710c	1745a

Plants were grown at Paijan and Picsi (Peru) experimental fields during the growing season. Numbers in a column followed by the same letter are not significantly different according to the Tukey test ($n = 5$; $P \leq 0.05$)

17.3.3 REP-PCR and 16S RNA Gene Phylogenetic Analyses

The eight selected strains were represented by 3 different REP-PCR patterns (Table 17.3). The partial sequence of the 16S rRNA gene from a representative strain of each REP pattern showed that the 3 strains were members of genus *Bradyrhizobium* of the family Rhizobiaceae of the Alphaproteobacteria. An EzTaxon-e analysis inferred from the 16S rRNA gene partial sequences indicated that representative strains had similarity values higher than 99.7 % with several *Bradyrhizobium* type species, so that a clear affiliation at the species levels could not be established. A phylogenetic tree (Fig. 17.1) based on the 16S rRNA gene partial sequences indicated that representative strains Rc-458-01 and Rc-391-01 grouped with *B. yuanmingense* B071^{T} and that representative strain Rc-352-01 clustered with *B. liaoningense* 2281^{T}.

Table 17.3 Identification of selected bacteria isolated from root nodules of cowpea var. CAR-5 plants

Strains[a]	REP-PCR pattern	Closest related genus on basis of 16S rRNA gene sequence
Rc-352-01, Rc-355-05, Rc-402-01	I	*Bradyrhizobium*
Rc-391-01, Rc-353-02	II	*Bradyrhizobium*
Rc-458-01, Rc-455-02, Rc-450-01	III	*Bradyrhizobium*

[a]Strains in bold were chosen as representative of those with identical 16S rRNA gene sequence

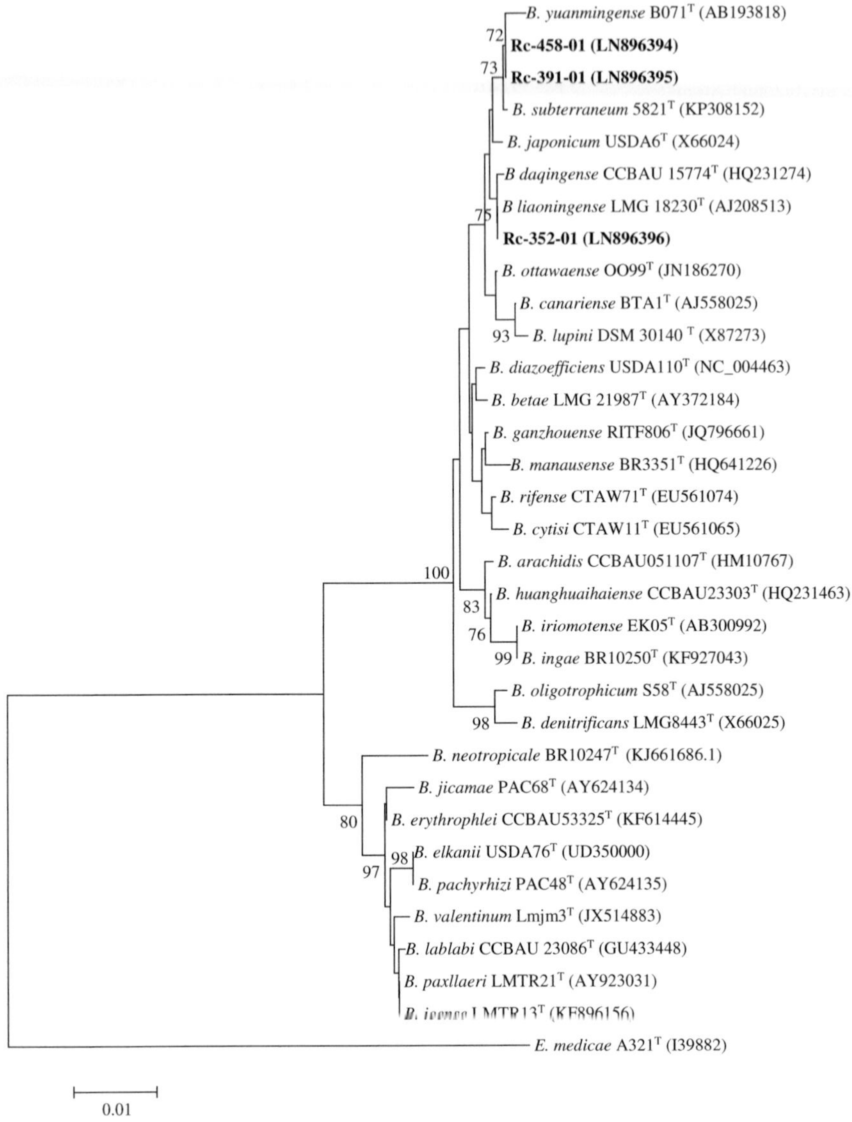

Fig. 17.1 Neighbor-joining phylogenetic tree based on partial 16S rRNA sequences of strains from nodules of *V. unguiculata* and phylogenetically related species within genus *Bradyrhizobium*. Bootstraps values are indicated as percentages derived from 1000 replications. Values lower than 70 are not shown. Bar 1 nucleotide substitution per 100 nucleotides. The tree is rooted on *Ensifer medicae* A321^{T}

17.3.4 *atpD, glnII, and recA* Genes Phylogenetic Analyses

PCR amplifications of the *atpD*, *glnII*, and *recA* genes from each of the 3 representative strains yielded single bands of approximately 0.5, 0.6, and 0.4 kb, respectively (data not shown), and their respective DNA sequences showed similarity with those of the *Bradyrhizobium* type strains in the public databases. A phylogenetic tree based on each of the *atpD*, *glnII*, and *recA* gene partial sequences indicated that strains Rc-391-01 and Rc-458-01 grouped with *B. yuanmingense* B071^T and that strain Rc-352-01 clustered with *B. liaoningense* 2281^T (data not shown). A concatenated phylogenetic tree based on *atpD* + *glnII* + *recA* sequences (Fig. 17.2) gave similar results to those obtained for the single

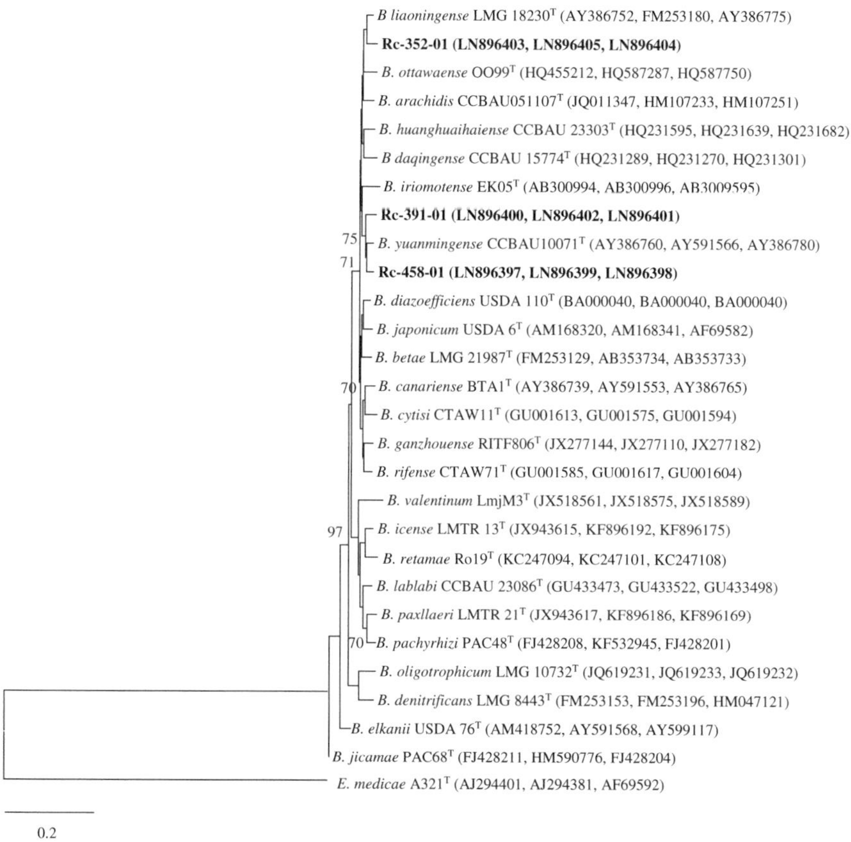

Fig. 17.2 Neighbor-joining phylogenetic tree based on combined partial *atpD* + *recA* + *glnII* sequences of strains from nodules of *V. unguiculata* and phylogenetically related species within genus *Bradyrhizobium*. Bootstraps values are indicated as percentages derived from 1000 replications. Values lower than 70 are not shown. Bar, 2 nucleotide substitution per 100 nucleotides. The tree is rooted on *Ensifer medicae* A321^T

phylogenetic trees of the *atpD*, *glnII*, and *recA* genes and confirmed affiliation of strain Rc-352-01 with *B liaoningense* and strains Rc- Rc-391-01 and 458-01 with *B. yuanmingense*.

17.3.5 Plant Growth-Promoting Traits of Selected Bradyrhizobial Species

Despite our repeated attempts, none of the 3 strains Rc-352-01, Rc-391-01, and Rc-458-01 solubilized tricalcium phosphate, but they produced IAA-related compounds and had ACC-deaminase activity (Table 17.4). Production of IAA-related compounds was detected after growth in Bergersen minimal medium and addition of *L*-tryptophan increased IAA production. The three strains were better IAA producers than the control strain *B. brevis* BEA1 used in this study (Table 17.4).

The quantitative assay used for determination of siderophores production revealed that values of activity were 5.02, 27.00, and 59.52 siderophores units μg/protein for strains Rc-458-01, Rc-352-01, and Rc-391-01, respectively. Also, strains Rc-352-01, Rc-391-01, and Rc-458-01 expressed ACC-deaminase activity with values of 17.39, 24.32, and 51.84 μmol α-ketobutyrate/mg protein x h, respectively.

Table 17.4 Indol acetic acid (IAA) and siderophore production and ACC-deaminase activity of selected *Bradyrhizobium* strains from root nodules of cowpea var. CAU-5

Trait	Strain			
	Rc-458-01	Rc-352-01	Rc-391-01	Control strain
IAA production (μg/μg protein)	8.98 ± 3.31 Ba	10.00 ± 1.77 Aa	9.45 ± 1.15 Ac	5.25 ± 0.10 Cb
	17.85 ± 0.62 Ab	21.60 ± 0.47 Aa	12.51 ± 0.72 Ac	9.46 ± 0.21 Ad
Siderophore production (% US μg/protein)	5.02 ± 0.86 A	27.00 ± 4.39 A	59.52 ± 0.62 A	NA*
ACC-deaminase activity (nmol α-ketobutyrate/mg x h)	51.84 ± 15.27 a	17.39 ± 0.95 c	24.32 ± 6.00 b	NA*

Cells were grown in Bergersen minimal medium supplemented or not with 600 μg/ml *L*-tryptophan. Numbers in a row followed by the same lowercase letter and numbers in a column followed by the same uppercase letter are not significantly different according to Kruskal–Wallis test ($n = 3$; $P \leq 0.05$). *B. brevis* BEA1 was used as a control in IAA production

*NA: not applicable

17.4 Discussion

In the jungle region of Peru, cowpea is grown traditionally in "playas" and "restingas," mostly without chemical fertilization and no irrigation. In these areas, soils are of low fertility and yet they maintain a sustainable cowpea production for years. It is well established that, in addition to mineral fertilizers, N inputs to agriculture can be obtained from symbiotic nitrogen fixation by rhizobial bacteria inside root nodules of leguminous plants (Sprent 2009) and that rhizobia associated with feed/fodder legumes contribute a substantial proportion of this fixed nitrogen (Herridge et al. 2008). Because it is more likely that a rhizobia *V. unguiculata* symbiotic association could be the main N source for growth of cowpea plants in "playas" and "restingas," the aim of this study was to identify the microsymbiont from nodules of cowpea plants. Only 50 (62.5 %) rhizobial strains from a total of 80 bacterial strains isolated from cowpea nodules were able to nodulate again cowpea plants grown in Leonard jar assemblies under greenhouse conditions. This result agrees with that of 75 % previously reported by Martins et al. (1997) during isolation of rhizobial bacteria from nodules of cowpea grown in soils from Pernambuco (Brazil). Using a similar soil, however, Leite et al. (2009) found that only a 25 % of the strains were true rhizobial strains. More likely, the remaining bacterial strains found within the nodules could be endophytes (da Costa et al. 2013).

Selection of cowpea-isolated rhizobial strains based on the dry weight values of plants being N_2-fixation dependent or treated with combined nitrogen has been used as a criterium to select strains with superior symbiotic performance (Fernandes et al. 2003, Lima et al. 2005, Florentino et al. 2010). In our study, eight strains were selected and used to inoculate plants of cowpea, producing higher dry weight than control plants treated with combined nitrogen when they were grown under greenhouse conditions (Table 17.1). A higher effectiveness, estimated as grain yield, of the selected strains was also obtained when the plants grew under field conditions. At each Paiján and Picsi experimental fields, grain yield (Table 17.2) was higher than the Peruvian national average of 1290 kg/ha (FAO 2015). These results contrast with those that indicate that inoculation of a regional cultivar of cowpea with strains TAL193 and TAL658 of *Bradyrhizobium* sp. did not result in any significant increase in grain yield as compared with the control plants in soils of Yurimaguas (Peru) (Piha and Nicholaides 1983). The 22 % increase in grain yield due to inoculation of the plants with strain Rc-352-01 is, however, similar to those of 32 and 22 % reported by Almeida et al. (2010) and Costa et al. (2011), respectively, when the cowpea cultivar BR17 was inoculated with *B. japonicum* strain INPA 0311B in soils of Piauí (Brazil).

In Paiján experimental field, inoculation of plants with each of the eight selected strains resulted in significant differences in grain yield among them as compared with the control plants; however, differences were not found when the experiment was carried out in Picsi experimental field (Table 17.2). Lack of significant differences in grain yield of cowpea plants inoculated with different rhizobial strains has been attributed to differences in the number and efficiency of soil native rhizobial populations to establish effective symbiosis with the host plants (Almeida et al. 2010; Gualter et al. 2011; Ferreira et al. 2013).

REP-PCR fingerprinting of the 8 selected strains showed they grouped in 3 different patterns (data not shown) and allowed selection of a representative strain from each group for further 16S rRNA gene sequencing (Table 17.3). This technique has been extensively used to cluster bacteria at the subspecies or strain level (de Bruijn 1992; Versalovic et al. 1994) and has proved to be a powerful tool for studies on microbial ecology and evolution (Ishii and Sadowsky 2009). The partial sequence of the 16S rRNA gene from each one of the representative strains Rc-352-01, Rc-391-01, and Rc-458-01 (REP-PCR groups I to III, respectively) showed that they could be classified into the genus *Bradyrhizobium* (Table 17.3). Ez-taxon-e analysis and a phylogenetic tree inferred from 16S rRNA gene sequences (Fig. 17.1) indicated that these strains clustered with the type strains *B. yuanmingense* B071^T and *B. liaoningense* 2281^T. Because housekeeping gene analysis complements that of the 16S rRNA gene in taxonomic studies at species level (Tindall et al. 2010), the *atpD*, *glnII*, and *recA* genes were used to assess identification of the bradyrhizobial strains nodulating cowpea plants. These genes have been previously used for taxonomic identification of *Bradyrhizobium* species and gave the best relative performance when used as genetic markers to identify other rhizobial species (Vinuesa et al. 2005). A concatenated phylogenetic tree based on *atpD* + *glnII* + *recA* partial sequences confirmed affiliation of strain Rc-352-01 to *B. liaoningense* 2281^{T}and strains Rc-391-01 and Rc-458-01 to *B. yuanmingense* B071^T (Fig. 17.2). According to the results in this study, *B. yuanmingense* is the most abundant endosymbiont isolated from cowpea nodules. This species represents 99 % of the strains isolated from cowpea in India (Appunu et al. 2009), 94 % in Japan (Sarr et al. 2009), 40 % in China (Zhang et al. 2011), and 35 % in Botswana (Steenkamp et al. 2008). It is also widely distributed in Asia, Africa, and Latin America (Sarr et al. 2009). In addition to cowpea, *B. yuanmingense* has been reported to form nodules in *Phaseolus lunatus*, a native legume of Peru (Ormeño-Orrillo et al. 2006).

None of the *Bradyrhizobium* strains isolated from cowpea nodules solubilized tricalcium phosphate, which is in agreement with the previous results from other authors (Boeiro et al. 2007; Marra et al. 2012). The 3 strains produced IAA-related compounds when cultured in minimal medium, a property previously described among members of *Bradyrhizobium* (Boeiro et al. 2007; Camerini et al. 2008), which has been shown to increase nitrogen fixation capability as well as tolerance to various stresses (Camerini et al. 2008; Karmakar et al. 2015). Our bradyrhizobial strains grew well when cultured in liquid CAS, and quantification of siderophore production showed wide variations among the 3 strains (Table 17.4). Similar results have been published for other bradyrhizobial species (van Rossum et al. 1994; Khandelwal 2002).

Generally, rhizobial strains that express ACC-deaminase are up to 40 % more efficient at forming nitrogen-fixing nodules than strains that lack this activity (Ma et al. 2003, 2004). However, generally, level of ACC-deaminase activity in rhizobia is low compared with free-living plant growth-promoting bacteria (Glick 2014).

Nevertheless, high expression of activity has been found in *B. japonicum* strain 110spc4 (Murset et al. 2012) and in rhizobia species from a collection of 233 strains from Canada (Duan et al. 2009). The 3 strains characterized in this study show levels of activity similar to those found by Duan et al. (2009). It has been reported that there is a high probability of finding strains with ACC-deaminase activity in strains selected for its high symbiotic performance (Glick 2014). Thus, it is possible that relatively high levels of ACC-deaminase activity in our strains are due to their selection by their high effectiveness when inoculated to cowpea plants both at greenhouse and field conditions.

Acknowledgments This work was supported by MINAGRI-financed INCAGRO project (Peru) and the ERDF-cofinanced project P12-AGR-1968 from Junta de Andalucía (Spain)

References

Almeida ALG, Alcántara RMCM, Nóbrega RSA, Nóbrega JCA, Leite LFC, Silva JAL (2010) Yield and nodulation of *Vigna unguiculata* (L.) Walp. inoculated with rhizobia strains in Bom Jesus, PI. Rev Bras Sci Agr 5:364–369

Altschul SF, Gish W, Miller W, Myers EW, Lipman DJ (1990) Basic local alignment search tool. J Mol Biol 215:403–410

Appunu C, N'zoue A, Moulin L, Depret G, Laguerre G (2009) *Vigna mungo*, *V. radiata* and *V. unguiculata* plants sampled in different agronomical-ecological–climatic regions of India are nodulated by *Bradyrhizobium yuanmingense*. Syst Appl Microbiol 32:460–470

Bejarano A, Ramírez-Bahena MH, Velázquez E, Peix A (2014) *Vigna unguiculata* is nodulated in Spain by endosymbionts of Genisteae legumes and by a new symbiovar (vignae) of the genus *Bradyrhizobium*. Syst Appl Microbiol 37:533–540

Bergersen FJ (1977) A treatise on dinitrogen fixation. In: Hardy RWF, Silver WS (eds) Biology, section III. Wiley, New York, pp 519–556

Boeiro L, Perrig D, Masciarelli O, Penna C, Cassan F, Luna V (2007) Phytohormone production by three strains of *Bradyrhizobium japonicum* and possible physiological and technological implications. J Appl Microbiol Biotechnol 74:874–880

Bradford MM (1976) A rapid and sensitive method for the quantitation of microgram quantities of protein utilizing the principle of protein-dye binding. Anal Biochem 72:248–254

Brown CM, Dilworth MJ (1975) Ammonia assimilation by *Rhizobium* cultures and bacteroids. J Gen Microbiol 86:39–48

Camerini S, Senatore B, Lonardo E, Imperlini E, Bianco C, Moschetti G, Rotino GL, Campion B, Defez R (2008) Introduction of a novel pathway for IAA biosynthesis to rhizobia alters vetch root nodule development. Arch Microbiol 190:67–77

Carson KC, Meyer J, Dilworth MJ (2000) Hydroxamate siderophores of root nodule bacteria. Soil Biol Biochem 32:11–21

Castellano-Hinojosa A, Correa-Galeote D, Palau J, Bedmar EJ (2015) Isolation of N_2-fixing rhizobacteria from *Lolium perenne* and evaluating their plant growth promoting traits. J Basic Microbiol 55:1–7

Chun J (2001) PHYDIT version 31. Available at http://plaza.snu.ac.kr/~jchun/phydit/

Costa EM, Nóbrega RSA, Martins LV, Amaral FHC, Moreira FMS (2011) Yield and nodulation of *Vigna unguiculata* (L.) Walp. inoculated with rhizobia strains in Bom Jesus. PI. Ciênc Agron 42:1–7 de Bruijn FJ (1992) Use of repetitive sequences and the polymerase chain reaction to fingerprint the genomes of *Rhizobium meliloti* isolates and other soil bacteria. Appl Environ Microbiol 58:2180–2187

da Costa EM, Nóbrega RSA, Carvalho F, Trochmann A, Ferreira LFM, Moreira FMS (2013) Plant growth promotion and genetic diversity of bacteria isolated from cowpea nodules. Pesq Agrop Bras 48:1275–1284

Duan J, Müller K, Charles T, Vesely S, Glick B (2009) 1-Aminocyclopropane-1-carboxylate (ACC) deaminase genes in rhizobia from Southern Saskatchewan. Microb Ecol 57:423–436

Fernandes MF, Fernandes RPM, Hungria M (2003) Seleção de rizóbios nativos para guandu, caupi e feijão-de-porco nos tabuleiros costeiros de Sergipe. Pesq Agropec Bras 38:835–842

Ferreira L, de Nobrega VM, Nóbrega RSA, Nóbrega JCA, Aguir FL, de Moreira FM, Pacheco LP (2013) Biological nitrogen fixation in production of *Vigna unguiculata* (L.) Walp, family farming in Piauí, Brazil. J Agr Sci 5:153–160

Florentino LA, Sousa PM, Silva JS, Silva KB (2010) Diversity and efficiency of *Bradyrhizobium* strains isolated from soil samples collected from around *Sesbania virgata* roots using cowpea as trap species. Rev Bras Ciên Solo 34:1113–1123

Glick BR (2012) Plant growth-promoting bacteria: mechanisms and applications. Hindawi Publishing Corporation, Scientifica, pp 1–15

Glick BR (2014) Bacteria with ACC deaminase can promote plant growth and help to feed the world. Microbiol Res 169:30–39

Gualter RMR, Boddey RM, Ruamjanel NG, Freitas ACRm de Xavier GR (2011) Eficiência agronômica de estirpes de rizóbio em feijão-caupi cultivado na região da Pré-Amazônia maranhense. Pesq Agropec Bras 46:303–308

Herridge DF, Peoples M, Boddey RM (2008) Global inputs of biological nitrogen fixation in agricultural systems. Plant Soil 311:1–18

Ishii S, Sadowsky MJ (2009) Applications of the rep-PCR DNA fingerprinting technique to study microbial diversity, ecology and evolution. Environ Microbiol 11:733–740

Khandelwal KR (2002) Practical pharmacognosy techniques and experiments, 9th edn, Nirali Prakashan, pp 156–161

Karmakar K, Rana A, Rajwar A, Sahgal M, Johri BN (2015) Legume-rhizobia symbiosis under stress. In: Kumar Arora N (ed) Plant microbes symbiosis: applied facets. Springer, Berlin, pp 209–239

Kearse M, Moir R, Wilson A, Stones-Havas S, Cheung M, Sturrock S, Buxton S, Cooper A, Markowitz S, Duran C, Thierer T, Ashton B, Mentjies P, Drummond A (2012) Geneious basic: an integrated and extendable desktop software platform for the organization and analysis of sequence data. Bioinformatics 28:1647–1649

Kim OS, Cho YJ, Lee K, Yoon SH, Kim M, Na H, Park SC, Jeon YS, Lee JH, Yi H, Won S, Chun J (2012) Introducing EzTaxon-e: a prokaryotic 16S rRNA gene sequence database with phylotypes that represent uncultured species. Int J Syst Evol Microbiol 62:716–721

Krasova-Wade T, Ndoye I, Braconnier S, Sarr B, Lajudie P, Neyra M (2003) Diversity of indigenous bradyrhizobia associated with three cowpea cultivars (*Vigna unguiculata* (L). Walp.) grown under limited and favorable water conditions in Senegal (West Africa). Afr J Biotech 2:13–22

Labarta A, White D, Leguía E, Guzmán W, Soto J (2007) La agricultura en la Amazonia ribereña del río Ucayali: ¿una zona productiva pero poco rentable? Acta Amaz 37:177–186

Laguerre G, Nour SM, Macheret V, Sanjuan J, Drouin P, Amarger N (2001) Classification of rhizobia based on *nodC* and *nifH* gene analysis reveals a close phylogenetic relationship among *Phaseolus vulgaris* symbionts. Microbiol 147:981–993

Lewis GP, Forest F (2005) Cercideae. In: Lewis G, Schrire B, Mackinder B, Lock M (eds) Legumes of the world. Royal Botanic Gardens, Kew, Richmond, UK, pp 57–67

Leite J, Seido SL, Passos SR, Xavier GR, Rumjanek NG, Martins LMV (2009) Biodiversity of rhizobia associated with cowpea cultivars in soils of the lower half of the São Francisco river valley. Rev Bras Ciên Solo 33:1215–1226

Lima AS, Pereira JPAR, Moreira FMS (2005) Diversidade fenotípica e eficiência simbiótica de estirpes de *Bradyrhizobium* sp. de solos da Amazônia. Pesq Agropec Bras 40:1095–1104

Ma W, Sebestianova S, Sebestian J, Burd GI, Guinel F, Glick BR (2003) Prevalence of 1-aminocyclopropaqne-1-carboxylate in deaminase in *Rhizobium* spp. Anton van Leeuwenhoek 83:285–291

Ma W, Charles TC, Glick BR (2004) Expression of an exogenous 1-aminocyclopropane-1-carboxylate deaminase gene in *Sinorhizobium meliloti* increases its ability to nodulate alfalfa. Appl Environ Microbiol 70:5891–5897

Marra LM, Soares CRFS, Oliveira SM, Ferreira PAA, Soares BL, Carvalho RF, Lima JM, Moreira FMS (2012) Biological nitrogen fixation and phosphate solubilization by bacteria isolated from tropical soils. Plant Soil 357:289–307

Martins LMV, Neves MCP, Rumjanek NG (1997) Growth characteristics and symbiotic efficiency of rhizobia isolated from cowpea nodules of the northeast region of Brazil. Soil Biol Biochem 5:1005–1010

Miller J (1972) Experiments in molecular genetics. Cold Spring Harbor Laboratory, Cold Spring Harbor

Moreno B, Vivas A, Nogales R, Benitez E (2009) Solvent tolerance acquired by *Brevibacillus brevis* during an olive-waste vermicomposting process. Ecotoxicol Environ Saf 72:2109–2114

Muniz AW, Costa MD, Wolff CL, Oliveira WAF, de Sá ES (2012) Selecao de estirpes rizóbio para Adesmia latifolia (Spreng) Vogel. Biotemas 25:177–180

Murset V, Hennecke H, Pessi G (2012) Disparate role of rhizobial ACC deaminase in root-nodule symbioses. Symbiosis 57:43–50

Naveed M, Mehbood I, Hussain MB, Zahit A (2015) Perspectives of rhizobial inoculation for sustainable crop production. In: Kumar Arora N (ed) Plant microbes symbiosis: applied facets. Springer, Berlin, pp 209–239

Ormeño-Orrillo E, Vinuesa P, Zúñiga-Dávila D, Martinez Romero E (2006) Molecular diversity of native bradyrhizobia isolated from Lima bean. Syst Appl Microbiol 29:253–262

Payne SM (1994) Detection, isolation and characterization of siderophores. Methods Enzymol 235:329–344

Peix A, Rivas-Boyero AA, Mateos PF, Rodríguez-Barrueco C, Martínez-Molina E, Velázquez E (2001) Growth promotion of chickpea and barley by a phosphate solubilizing strain of *Mesorhizobium mediterraneum* under growth chamber conditions. Soil Biol Biochem 33:103–110

Peix A, Ramírez-Bahena MH, Velázquez E, Bedmar EJ (2015) Bacterial associations with legumes. Crit Rev Plant Sci 34:17–42

Piha MI, Nicholaides JJ (1983) Cowpea (*Vigna unguiculata*) agronomic-economic research on soils of the tropics, Technical Report, Soils Science Department, North Carolina State University, Ralaigh, North Carolina, USA, pp 43–45

Pule-Meulenberg F, Belane AK, Krasova-Wade T, Dakora F (2010) Symbiotic functioning and bradyrhizobial biodiversity of cowpea (*Vigna unguiculata* L. Walp) in Africa. BMC Microbiol 10:1–12

Rangel A, Saraiva K, Schwengber P, Narciso MS, Domont GB et al (2004) Biological evaluation of a protein isolate from cowpea (*Vigna unguiculata*) seeds. Food Chem 87:491–499

Saitou N, Nei MA (1987) Neighbour-joining method: a new method for reconstructing phylogenetic trees. Mol Biol Evol 44:406–425

Sarr PS, Yamakawa T, Fujimoto S, Saeki Y, Thao HT, Myint AK (2009) Phylogenetic diversity and symbiotic effectiveness of root-nodulating bacteria associated with cowpea in the south-west area of Japan. Microb Environ 24:105–112

Sayyed RZ, Badguja MD, Sonwane HM, Mhaske MM, Chincholkar SB (2005) Production of microbial iron chelators (siderophores) by fluorescent pseudomonads. J Ind Biotechnol 4:486–490

Singh B, Ehlers JD, Sharma B, Freire-Filho FR (2002) Recent progress in cowpea breeding. In: Fatokun CA, Tarawali SA, Singh BB, Kormawa OM, Tamo M (eds) Challenges and opportunities for enhancing sustainable cowpea production. Ibadan, IITA, pp 22–40

Somasegaran P, Hoben HJ (1994) Handbook for rhizobia. Methods in Legume-*Rhizobium* Technology. Springer, New York, pp 240–258

Sprent JI (2009) Legume nodulation: a global perspective. Wiley-Blackwell, Oxford

Steenkamp ET, Stepkowski T, Przymusiak-Botha WJ, Law IJ (2008) Cowpea and peanut in southern Africa are nodulated by diverse *Bradyrhizobium* strains harboring nodulation genes that belong to the large pantropical clade common in Africa. Mol Phylo Evol 48:1131–1144

Tamura K, Peterson D, Peterson N, Stecher G, Nei M, Kumar S (2011) MEGA5: molecular evolutionary genetics analysis using maximum likelihood, evolutionary distance, and maximum parsimony methods. Mol Biol Evol 28:2731–2739

Tindall BJ, Rosselló-Móra R, Busse HJ, Ludwig W, Kämpfer P (2010) Notes on the characterization of prokaryote strains for taxonomic purposes. Int J Syst Evol Microbiol 60:249–266

van Rossum D, Muyotcha A, De Hope BM, van Verseveld HW, Stouthamer AH, Boogerd FC (1994) Soil acidity in relation to groundnut-*Bradyrhizobium* symbiotic performance. Plant Soil 163:165–175

Venancio-Silva F, de Meyer SE, Simoes de Araujo JL, da Costa Barbe T, Xavier GR, O'Hara G, Ardley J, Rumjanek NG, Willems A, Zilli JE (2014) *Bradyrhizobium manausense* sp. nov., isolated from effective nodules of *Vigna unguiculata* grown in Brazilian Amazon rainforest soils. Int J Syst Evol Microbiol 64:2358–2363

Versalovic J, Schneider M, Bruijn FJ, Lupski JP (1994) Genomic fingerprinting of bacteria using repetitive sequence-based polymerase chain reaction. Meth Mol Cell Biol 5:25–40

Vincent J (1970) A manual for the practical study of the root nodule bacteria International Biological Programme Handbook, vol 15. Blackwell, Oxford

Vinuesa P, León-Barrios M, Silva C, Willems A, Jarabo-Lorenzo A, Pérez-Galdona R, Werner D, Martínez-Romero E (2005) *Bradyrhizobium canariense* sp. nov., an acid-tolerant endosymbiont that nodulates endemic genistoid legumes (*Papilionoideae: Genisteae*) from the Canary Islands, along with *Bradyrhizobium japonicum* bv. genistearum, *Bradyrhizobium genospecies* alpha and *Bradyrhizobium genospecies* beta. Int J Syst Evol Microbiol 55:569–575

Weisburg WG, Barns SM, Pelletier DA, Lane DJ (1991) 16S ribosomal DNA amplification for phylogenetic study. J Bacteriol 173:697–703

Zhang S, Xie F, Yang J, Li Y (2011) Phylogenetic of bradyrhizobia from Chinese cowpea miscellany inferred from 16S rRNA, *atpD*, *glnII*, and 16S-23S intergenic spacer sequences. Can J Microbiol 57:316–327

Zilli JE, Vallishesky RB, Filho FRF, Neves MCP, Rumjanek NG (2004) Assessment of cowpea *Rhizobium* diversity in Cerrado areas of northeastern Brazil. Braz J Microbiol 35:281–287

Chapter 18
Rhizobium as Potential Biofertilizer of *Eruca Sativa*

Alba Rubio-Canalejas, Lorena Celador-Lera, Xavier Cruz-González, Esther Menéndez and Raúl Rivas

Abstract *Eruca sativa*, commonly known as rocket salad, is one of the leading leafy vegetables, belonging to the Brassicaceae family that include other species of agricultural importance, such as broccoli, cabbage, cauliflower, and mustard, amongst others. It is widely cultivated and consumed, being its popularity due to its peculiar flavour and nutritional value. On the other hand, to maintain agricultural production, the indiscriminate use of chemical fertilizers, pesticides, and herbicides results in damage to soil and biodiversity reduction and may result in adverse health effects. An effective alternative is the use of biofertilizers based on plant growth-promoting rhizobacteria (PGPR). In this study, we evaluated the use of a *Rhizobium* sp. strain as biofertilizer for *E. sativa*, performing different assays to demonstrate its potential. According to the obtained results, the partial sequencing of 16S rRNA gene classified this strain into the genus *Rhizobium*. This strain is not able to solubilize phosphate under in vitro conditions. However, this strain was an excellent producer of indole-3-acetic acid (IAA) and its precursors. Moreover, this strain produced siderophores in a low concentration. In vitro inoculation of *E. sativa* plants with this strain resulted in a significative increase of the number of secondary roots at 6 and 8 days post-inoculation compared to the uninoculated treatment. Therefore, our results support the possible inclusion of this strain of *Rhizobium* sp. in formulations as potential biofertilizer for *E. sativa* crops.

Keywords *Eruca sativa* · Plant growth promotion · *Rhizobium* · Biofertilizer · Indole-3-acetic acid

---

A. Rubio-Canalejas · L. Celador-Lera · X. Cruz-González (✉) · E. Menéndez · R. Rivas
Department of Microbiology and Genetics, University of Salamanca, Salamanca, Spain
e-mail: xavier.cruz@usal.es

R. Rivas
Associated Unit I+D University of Salamanca—CSIC 'Plant-Microbe Interactions Group', Salamanca, Spain

© Springer International Publishing Switzerland 2016

F. González-Andrés and E. James (eds.), *Biological Nitrogen Fixation and Beneficial Plant–Microbe Interactions*, DOI 10.1007/978-3-319-32528-6_18

18.1 Introduction

According to the data from the Population Division of United Nations, in the last 50 years, the world population has shown a vertiginous growth and is predicted to reach 9 billion in the next 50 years. Then, population growth will lead to an increase in demand on resources, both land and water resources. Furthermore, the thickness of that population growth will be in developing countries, which has added problems.

An alternative to increase crop production is the use of chemical fertilizers, pesticides, and herbicides. However, its indiscriminate use causes land damage, reduces biodiversity, and could have adverse health effects (Selma et al. 2010). Nonetheless, there is an effective and greener alternative that does not affect the environment, which is the employment of biofertilizers. These biofertilizers may contain rhizobacteria, which promote plant growth and affect positively in the uptake of soil nutrients. In other words, these plant growth-promoting rhizobacteria (PGPR) colonize the rhizosphere, promoting an increase in the concentration of available minerals; in addition, some of them are able to biosynthesize phytohormones and/or inhibit plant pathogenic organisms. All of these features, known as plant growth-promoting mechanisms, may result in an increase in germination rate, plant weight, and tolerance to biotic and abiotic stresses, amongst other benefits, when those rhizobacteria are in contact with the plant root system (Chou 2013). In addition to these advantages, some of those PGPRs are low-energy consumer bacteria with a fast growth rate, which may be turned into a decrease in the environmental pollution and also become an economic option in comparison with the formulation of chemical fertilizers, which are expensive and contribute to increase pollution. One of the most prominent PGPR groups is collectively called Rhizobia, which are able to symbiotically fix atmospheric nitrogen within legume nodules. Recent studies have shown that the use of strains belonging to the genus *Rhizobium,* which were able to enhance nutrient uptake and synthesize phytohormones, improves growth and development of vegetable crops, such as *Lactuca sativa, Daucus carota, Solanum lycopersicum,* and *Capsicum annuum* (García-Fraile et al. 2012, Flores-Félix et al. 2013).

One of the leafy vegetable crops that are acquiring relevance is *Eruca sativa* Mill., commonly called rocket, a perennial herb native of the Mediterranean coast (Miyazawa et al. 2002), which is widely grown and consumed worldwide. The main producers are Italy, Greece, and Turkey but also its production has increased in Argentina (Barlas et al. 2011; Zapata et al. 2005). Moreover, during the last decade, its production has been expanded in north European markets (Omirou et al. 2012). This particular crop belongs to the family Brassicaceae, as well as other important vegetable crop, such as cabbage, broccoli, cauliflower, mustard, turnip, radish, and watercress, amongst others (Chou 2013). One of the reasons why its notoriety and economic potential has increased is due to its distinctive flavour and the nutritional value of its leaves, which are a good source of essential nutrients and bioactive compounds (Barillary et al. 2005; Barlas et al. 2011; Omirou et al. 2012).

Up to date, there are few studies reporting the inoculation of *E. sativa* crops with PGP micro-organisms. It is already reported the inoculation of *E. sativa* plants with a strain of *Pseudomonas putida*, which improves plant growth under cadmium (Cd) stress (Kamran et al. 2015). Similarly, plants inoculated with the beneficial fungi *Trichoderma harzianum* and exposed to different concentrations of copper (Cu) and zinc (Zn) reduced the toxic effect of these metals (Al-Rajhi 2013). Nonetheless, there is a lack of studies reporting the use of rhizobial strains as biofertilizers for this crop. Thus, the aim of this work is to evaluate the use of a rhizobial strain, isolated as root endophyte in *Zea mays*, as a potential biofertilizer for *E. sativa*.

18.2 Materials and Methods

18.2.1 Endophyte Isolation

The isolate was obtained from *Zea mays* roots grown in a soil of NW Spain. Briefly, collected maize roots were surface-sterilized with 70 % ethanol for 2 min and 5 % NaClO for 2 min. Then, roots were washed with sterile distilled water and macerated. Afterwards, roots were crushed and cultured in Yeast Mannitol Agar (YMA) at 28 °C and monitorized daily for endophyte collection.

18.2.2 DNA Extraction and 16S rRNA Gene Sequencing

DNA was obtained from bacterial suspensions, which were centrifuged at 9000 rpm for 2 min. The supernatant was removed, and the pellet was washed with 200 μl of 0.1 % Sarkosyl and 100 μl of 0.05 M NaOH. Sample was included for 4 min at 100 °C, and the 200 μl of sterile ultrapure water was added.

The 16S rRNA gene was amplified using the following PCR reaction: 2.5 μl of GeneAmp 10× buffer, 1 μl BSA 0.1 %, 6 μl of 25 mM $MgCl_2$, 0.25 μl of 20 mM of dNTPs, 0.2 μl of 5U AmpliTaq Gold DNA polymerase (PerkinElmer Biosystems), 2.5 μl of 2 μM of 27F (5′ GAGGGTGGCGGTTCT 3′) and 1522R (5′ AAGGAGGTGATCCANCCRCA 3′), and 3 μl of DNA template. The thermocycler parameters were 95 °C for 9 min; 35 cycles of 95 °C for 1 min, 59 °C for 1 min, and 75 °C for 2 min; and a final extension at 72 °C for 7 min. The amplification product was extracted from the agarose gel (1 %) using the QIAquick PCR Purification Kit (Qiagen Ltd). The amplification product was sequenced in the Sequencing Service of NUCLEUS, University of Salamanca, obtaining a sequence of 1417 bp, which was compared with sequences of type strains held in the EzTaxon-e database (Kim et al. 2012).

18.2.3 PGPR Tests

To determine whether the strain is capable of solubilizing the phosphate, the strain was placed in a YED-P medium (Peix et al. 2003). Siderophore production was determined with the CAS-M9-Agar medium (Alexander and Zuberer 1991). The indole-3-acetic acid production was tested in JMM medium as described by O'Hara et al. (1989), using Salkowsky agent for quantification by spectrophotometry.

18.2.4 Seedling Assays

E. sativa seeds were sterilized using 0.1 % SDS (w/v) for 10 min, followed by three washes in sterile distilled water of 30 min each (Moussa 2006). Seeds were pregerminated on 1 % agar plates at 28 °C for 24–48 h in darkness. Then, seedlings were placed on 1.5 % agar plates and the roots inoculated with 250 µl/plant of a rhizobial suspension with an optical density (600 nm) of 0.6. Seedlings were covered with sterile filter paper. Plates were placed in a growth chamber, monitorizing their growth and development at 6 and 8 days post-inoculation.

18.3 Results and Discussion

18.3.1 16S rRNA Gene Sequencing

According to the partial 16S rRNA gene sequencing, the strain isolated as maize endophyte is classified as *Rhizobium* sp. The genus *Rhizobium* is commonly found in association with legumes, due to its role in nitrogen-fixing symbiosis (Laranjo et al. 2014), but recently, several studies reported its role as endophytes in various crops, such as cereals and horticultural crops (Chabot et al. 1996; Mishra et al. 2006, García-Fraile et al. 2012; Flores Félix et al. 2013). Moreover, species of this genus, such as *Rhizobium etli,* have been found as an endophyte in maize roots, which was able to recolonize the roots of other maize varieties and increase maize dry weight between 10 and 100 %, compared with non-inoculated plants (Gutiérrez-Zamora and Martínez-Romero, 2001). Therefore, our rhizobial strain may be able to colonize the roots of this non-leguminous plant efficiently.

18.3.2 Phosphate Solubilization

To determine the potential of this rhizobial strain as plant growth promoter, several PGPR mechanisms were evaluated qualitatively. Phosphate solubilization was tested in YED-P medium, showing that this strain is not able to solubilize $Ca_3(PO_4)_2$, under the established in vitro conditions. No clear halo was observed around its colonies. However, this result is not conclusive due to the lack of information of its growth in other phosphate-containing media and under greenhouse or field conditions. The used concentration of tricalcium phosphate might be inappropriate for this strain, as suggested by previous studies, which proposed an alternative using a combination of two or three metal P for testing phosphate-solubilizing bacteria (Bashan et al. 2013). Nevertheless, there are a considerable number of *Rhizobium* strains that are able to solubilize inorganic phosphate sources under in vitro conditions (Halder and Chakrabartty 1993; Chabot et al. 1996; Flores-Felix et al. 2013).

18.3.3 Siderophores Production

In the case of siderophores production, this strain produced a clear halo around its colonies when grown in CAS-M9-Agar medium, revealing the ability of this strain for producing siderophores. In other words, the strain is capable of reducing Fe^{3+} to Fe^{2+}, which is captured for metabolic reactions, such as oxygen reduction for ATP synthesis, ribonucleotide precursors reduction, or *haem* group formation, amongst others (Verma et al. 2012). Our results are in agreement with previous studies reporting strains of *Rhizobium* as siderophore producers (Duhan et al. 1998; Garcia-Fraile et al. 2012; Flores-Félix et al. 2013).

18.3.4 Indole-3-Acetic Acid (IAA)

Concerning the ability to produce phytohormones, specifically indole-3-acetic acid (IAA), this strain produced 170 mg l^{-1}. This concentration is higher than the previously reported for other *Rhizobium* strains, such as *Rhizobium leguminosarum* PEPV16, which produced 77 mg l^{-1} (Flores-Félix et al. 2013). This phytohormone is involved in the regulation of various plant physiological processes. Bacterial IAA increases the length and number of primary and secondary roots (Castillo et al. 2005; Duca et al. 2014). Moreover, IAA is considered a signalling molecule in bacteria with a direct effect on bacterial physiology (Spaepen et al. 2007). Several strains of the genera *Rhizobium*, *Pseudomonas*, *Azospirillum*, *Azotobacter,* or *Streptomyces* are reported as indole-3-acetic acid producers (Duca et al. 2014; García-Fraile et al. 2012). However, bacterial IAA concentration has a considerable effect on plant development; a deficiency causes a delay in plant growth, and high concentrations inhibit primary root growth (Duca et al. 2014).

Table 18.1 Shoot length, root growth, and secondary root number at 6 and 8 days post-inoculation

Days post-inoculation	Treatment	Shoot length (cm)	Root length (cm)	Secondary root number
6	Control	1.98 ± 0.08^{a}	6.52 ± 0.803^{a}	1.364 ± 0.203^{a}
	Rhizobium sp.	2.04 ± 0.119^{a}	6.56 ± 0.712^{a}	2.333 ± 0.441^{b}
8	Control	2.16 ± 0.116^{a}	8.52 ± 0.936^{a}	1.50 ± 0.261^{a}
	Rhizobium sp.	2.3 ± 0.126^{a}	8.62 ± 0.696^{a}	3.875 ± 0.295^{b}

[a] Values followed by the same letter are not significantly different at $P < 0.05$

18.3.5 Seedlings Assays

To test the potential of this rhizobial strain in plant growth at early stages of plant development, we performed in vitro inoculation assays on *E. sativa* seedlings. Our results showed no significant differences in shoot and root length; however, secondary root number was significatively increased in plants inoculated with *Rhizobium* sp. in comparison with non-inoculated plants (Table 18.1). Duca et al. (2014) reported similar effects in root number stimulation due to IAA levels. The obtained data in plant development at early stages may be related to the levels of IAA produced by this strain.

Our results support the possible inclusion of this strain of *Rhizobium* sp. in formulations as potential biofertilizer for *E. sativa* crops, due to the positive effects observed, especially the production of IAA, which promoted the increase in the number of roots of *E. sativa*. This radicular system increment will provide a better uptake of soil nutrients to the plant and therefore better growth and development. The inclusion of this strain in biofertilization schemes or its use as bioinoculant may contribute to resolve the problematic of chemical fertilizers added to this crop, not only due to its positive effects but also to the safeness and eco-friendliness of this strain.

18.4 Conclusions

The *Rhizobium* sp. strain, isolated as maize endophyte, is able to interact with *E. sativa* plants, promoting its growth and development by increasing the root system surfaces, which may result in a better survival in stressful situations, such as water or nutrient deficiency, and might lead to a reduction in the use of chemical fertilizers in this crop.

Acknowledgements This work was supported by the Junta de Castilla y León (JCyL SA 169 U 14 project). XCG acknowledges a fellowship from the Fundación Kinesis.

References

Alexander B, Zuberer D (1991) Use of chrome azurol S reagents to evaluate siderophore production by rhizosphere bacteria. Biol Fertil Soils 12:39–45

Al-Rajhi A (2013) Impact of biofertilizer *Trichoderma harzianum* Rifai and the biomarker changes in *Eruca Sativa* L. plant grown in metal-polluted soils. World Appl Sci J 22(2):171–180

Barillari J, Canistro D, Paolini M, Ferroni F, Pedulli G, Iori R, Valgimigli L (2005) Direct antioxidant activity of purified glucoerucin, the dietary secondary metabolite contained in rocket (*Eruca sativa* Mill.) seeds and sprouts. J Agric Food Chem 53:2475–2482

Barlas N, Irget M, Tepecil M (2011) Mineral content of the rocket plant (*Eruca sativa*). Afr J Biotechnol 10(64):14080–14082

Bashan Y, Kamnev A, de-Bashan L (2013) Tricalcium phosphate is inappropriate as a universal selection factor for isolating and testing phosphate-solubilizing bacteria that enhance plant growth: a proposal for an alternative procedure. Biol Fertil Soils 49:465–479

Castillo G, Altuna B, Michelena G, Sánchez-Bravo J, Acosta M (2005) Cuantificación del contenido de ácido indolacético (AIA) en un caldo de fermentación microbiana. Anales de Biología 27:137–142

Chabot R, Antoun H, Cescas M (1996) Growth promotion of maize and lettuce by phosphate-solubilizing *Rhizobium leguminosarum* biovar phaseoli. Plant Soil 184:311–321

Chou M (2013): Soil bacterium *Bacillus subtilis* (GB03) augments plant growth and volatile emissions in *Eruca sativa* (Arugula). Unpublished master's thesis, Texas Tech University

Duca D, Lorv J, Patten C, Rose D, Glick B (2014) Indole-3-acetic acid in plant–microbe interactions. A Van Leeuw J Microb 106:85–125

Flores-Felix J-D, Menéndez E, Rivera L, Marcos-García M, Martínez-Hidalgo P, Mateos P, Martínez-Molina E, Velázquez M, García-Fraile P, Rivas R (2013) Use of *Rhizobium leguminosarum* as a potential biofertilizer for *Lactuca sativa* and *Daucus carota* crops. Plant Nutr Soil Sci 176:876–882

Duhan JS, Dudeja SS, Khurana AL (1998) Siderophore production in relation to N_2 fixation and iron uptake in pigeon pea-*Rhizobium* symbiosis. Folia Microbiol 43(4):421–426

García-Fraile P, Carro L, Robledo M, Ramírez-Bahena M-H, Flores-Félix J-D et al (2012) *Rhizobium* promotes non-legumes growth and quality in several production steps: towards a biofertilization of edible raw vegetables healthy for humans. PLoS One 7(5)

Gutiérrez-Zamora M, Martínez-Romero E (2001) Natural endophytic association between *Rhizobium etli* and maize (*Zea mays* L.). J Biotechnol 91:117–126

Halder A, Chakrabartty P (1993) Solubilization of inorganic phosphate by *Rhizobium*. Folia Microbiol 38(4):325–330

Kamran M, Syed J, Eqani S, Munis M, Chaudhary H (2015) Effect of plant growth-promoting rhizobacteria inoculation on cadmium (Cd) uptake by *Eruca sativa*. Environ Sci Pollut R 22 (12):9275–9283

Kim OS, Cho YJ, Lee K, Yoon SH, Kim M, Na H, Park SC, Jeon YS, Lee JH, Yi H, Won S, Chun J (2012) Introducing EzTaxon-e: a prokaryotic 16S rRNA gene sequence database with phylotypes that represent uncultured species. Int J Syst Evol Microbiol 62:716–721

Laranjo M, Alexandre A, Oliveira S (2014) Legume growth-promoting rhizobia: an overview on the *Mesorhizobium* genus. Microbiol Res 169(1):2–17

Mishra R, Singh R, Jaiswal H, Kumar V, Maurya S (2006) *Rhizobium* mediated induction of phenolics and plant growth promotion in rice (*Oryza sativa* L.). Curr Microbiol 52:383–389

Miyazawa M, Maehara T, Kurose K (2002) Composition of the essential oil from the leaves of *Eruca sativa*. Flavour Fragr J 17:187–190

Moussa H (2006) Gamma irradiation regulation of nitrate level in rocket (*Eruca vesicaria* subsp. sativa) plants. J New Seeds 8(1):91–100

O'Hara G, Goss T, Dilworth M, Glenn A (1989) Maintenance of intracellular pH and acid tolerance in *Rhizobium meliloti*. Appl Environ Microbiol 55(8):1870–1876

Omirou M, Papastefanou C, Katsarou D, Papastylianou I, Passam H, Ehaliotis C, Papadopoulou K (2012) Relationships between nitrogen, dry matter accumulation and glucosinolates in *Eruca sativa* Mills. The applicability of the critical NO3-N levels approach. Plant Soil 354(1): 347–358

Peix A, Rivas R, Mateos P, Martínez-Molina E, Rodríguez-Barrueco C, Velázquez E (2003) *Pseudomonas rhizosphaerae* sp. nov., a novel species that actively solubilizes phosphate *in vitro*. Int J Syst Evol Micr 53:2067–2072

Selma M, Martínez-Sanchez A, Allende A, Ros M, Hernández M, Gil M (2010) Impact of organic soil amendments on phytochemicals and microbial quality of rocket leaves (*Eruca sativa*). J Agr Food Chem 58:8331–8337

Spaepen S, Vanderleyden J, Remans R (2007) Indole-3-acetic acid in microbial and microorganism-plant signaling. FEMS Microbiol Rev 31(4):425–448

Verma V, Joshi K, Mazumdar B (2012) Study of siderophore formation in nodule-forming bacterial species. Res J Chem Sci 2(11):26–29

Zapata R, Romero A, Maseda P (2005) First report of white rust of arugula caused by *Albugo candida* in Argentina. Plant Dis 89(2):207

Chapter 19
Biocontrol of *Fusarium oxysporum* f.sp. *phaseoli* and *Phytophthora capsici* with Autochthonous Endophytes in Common Bean and Pepper in Castilla y León (Spain)

Marcia Barquero, Arsenio Terrón, Encarna Velázquez and Fernando González-Andrés

Abstract *Fusarium oxysporum* f.sp. *phaseoli* and *Phytophthora capsici* are soil pathogens which cause diseases to common bean (*Phaseolus vulgaris*) and pepper (*Capsicum annuum*), respectively. Some microorganisms associated with plants show different biocontrol mechanisms against pathogens. The agricultural use of these mechanisms is a chief strategy as an alternative to chemicals. In a previous work, we isolated and taxonomically identified 68 endophytic bacteria from common bean and 54 from pepper. The isolations were carried out in farms located in regions protected by a PGI: "Alubia de La Bañeza—León" in the case of common bean, and "Pimiento de Fresno-Benavente" in the case of pepper. The aim of the present work was the evaluation of such bacteria and AM fungus, as potential biocontrollers of *F. oxysporum* f.sp. *phaseoli* and *P. capsici*. In total 18 out of the 122 isolates controlled in vitro the growth either of *F. oxysporum* or *P. capsici*, or both pathogens. Such controller bacteria were tested in the plants in which they were isolated (common bean or pepper), which were grown in hydroponic conditions and inoculated with *F. oxysporum* f.sp. *phaseoli* and *P. capsici*, respectively. Two isolates belonging to the species *Pseudomonas brassicacearum* and *Bacillus siamensis*, respectively, were the best controllers of Fusarium root rot in common bean. Moreover, two isolates belonging to *Bacillus pumilus* were the best controllers of *P. capsici* in pepper plants. In trials carried out in microcosm conditions, common bean plants inoculated with the pathogen *F. oxysporum*, plus an autochthonous *Rhizobium leguminosarum* strain and the *B. siamensis* isolate, showed a

M. Barquero · E. Velázquez
Departamento de Microbiología y Genética, Universidad de Salamanca, Salamanca, Spain

A. Terrón · F. González-Andrés (✉)
Instituto de Medio Ambiente, Recursos Naturales y Biodiversidad, Universidad de León, León, Spain
e-mail: fgona@unileon.es

© Springer International Publishing Switzerland 2016

F. González-Andrés and E. James (eds.), *Biological Nitrogen Fixation and Beneficial Plant–Microbe Interactions*, DOI 10.1007/978-3-319-32528-6_19

very weak pathogen attack. Regarding the pepper microcosm assay, the co-inoculation with one of the *B. pumilus* isolates plus an autochthonous mycorrhiza controlled *P. capsici* in the plants inoculated with the pathogen, even at the same level than the chemical fungicide. This opens the door for the development of biocontrollers, alternative to chemicals, in these two crops.

Keywords Biocontrol · *Phaseolus vulgaris* · *Capsicum annuum* · *Fusarium oxysporum* f.sp. *phaseoli* · *Phytophthora capsici* · *Bacillus siamensis* · *Pseudomonas brassicacearum*

19.1 Introduction

The control of *Fusarium oxysporum* f.sp. *phaseoli* (*Fop*) and *Phytophthora capsici* in common bean (*Phaseolus vulgaris*) and pepper (*Capsicum annuum*), respectively, is commonly based on the use of chemicals, which may entail several negative consequences for the environment, such as soil and water contamination, and also the development of resistances. In addition to this, the increasing willingness of the consumers to buy products free from pesticides prioritizes the search for alternatives to chemicals. This is especially important in products with quality protection such as Protected Geographic Indications (PGI), which is usually preferred by consumers concerned with the quality and healthiness of the food products (Aprile et al. 2012).

Microorganisms associated with plants, such as the so-called plant growth promoting rhizobacteria (PGPR) (Haas and Keel 2003), Arbuscular Mycorrhizal (AM) fungi (Azcón-Aguilar et al. 2002), and rhizobia (Bardin et al. 2004), have developed different biocontrol mechanisms against pathogens. These mechanisms can be used to develop biocontrol strategies; consequently, their agricultural use means a chief alternative strategy with respect to chemicals (Gerhardson 2002).

PGPR exert biocontrol through several mechanisms including antibiotics, lytic enzymes, siderophores and aminocyclopropane-1-carboxylate (ACC) deaminase production, competence, induced systemic resistance (ISR), and growth induction (Compant et al. 2005; Glick 2012). PGPR can use one or several mechanisms. For instance, certain strains of fluorescent pseudomonads present aggressive root colonization, induction of systemic resistance in the plant, and production of diffusible or volatile antifungal antibiotics (Haas and Keel 2003) and *Bacillus* lipopeptides are involved in antagonistic activity, root colonization, and stimulation of host defense mechanisms (Ongena and Jacques 2008).

Likewise, in addition to the improvement of plant nutrition, AM fungi can also influence the pathogens' control across the action of different mechanisms such as competence and activation of defense mechanisms in the plant. Several studies have demonstrated AM fungi activity against nematodes (Calvet et al. 2001), oomycetes, and soil fungus such as *Phytophthora* (Vigo et al. 2000), *Pythium* (Larsen et al. 2012), *Fusarium* (Akköprü and Demir 2005), *Rhizoctonia* (Yao et al. 2002), and

Armillaria (Nogales et al. 2009). In the same way as PGPR and AM fungi, rhizobia group is also able to control pathogens. Arora et al. (2001) and Bardin et al. (2004) have obtained interesting results with *Macrophomina* in peanuts and *Pythium* in peas and sugar beet.

Co-inoculation studies, based on the microbial natural interactions occurrence in the soil, have shown exciting results. Co-inoculation of PGPR such as *Pseudomonas* and *Bacillus* has been achieved to control plant pathogens in different crops. In example, Saravanakumar et al. (2009) used *Pseudomonas* mixtures against the sheath rot pathogen *Sarocladium oryzae* in rice, enhancing the disease resistance and Recep et al. (2009) tested 17 strains of PGPR, among them *Bacillus amyloliquefaciens* and *Bacillus subtilils*, against *Fusarium* spp., causing dry rot disease in potato. Co-inoculation of *Pseudomonas* and *Rhizobium* induced systemic resistance against Fusarium wilt (Dutta et al. 2008), and Sayeed Akhtar and Siddiqui (2008) observed significant effects of co-inoculation of *Glomus intraradices*, *Rhizobium* sp., and *Pseudomonas striata* on the root rot disease complex control of chickpea caused by *Meloidogyne incognita* and *Macrophomina phaseolina*.

The starting point for this work was a collection of endophytic bacteria fully identified on the basis of the 16S rRNA sequencing (Barquero 2014, unpublished PhD. Thesis), composed by 68 endophytic bacteria from common bean and 54 from pepper. Moreover, six AMF spiral morphotypes were also isolated and identified as *Glomus mosseae* (Barquero 2014, unpublished PhD. Thesis). The aim of the present work was the evaluation of such bacteria and AM fungus, as potential biocontrollers of *F. oxysporum* f.sp. *phaseoli* and *P. capsici*.

19.2 Materials and Methods

19.2.1 Soil Microorganisms Isolation

Endophytic bacteria and AM fungi were isolated and identified on the basis of DNA sequencing, by Barquero (2014, unpublished PhD. Thesis) from roots of pepper and common bean crops, collected from farms located in the regions protected by the PGI "Alubia de La Bañeza—León" in the case of common bean and by the PGI "Pimiento de Fresno-Benavente" in the case of pepper.

19.2.2 Assessment of the Antibiotic Effect of the Isolated Bacteria Against the Pathogenic Fungi, by Dual Culture Assay

Endophytic bacteria were tested for their ability to inhibit the growth of *Fop* and *P. capsici* on PDA plates. The inhibition halo was measured in three replications

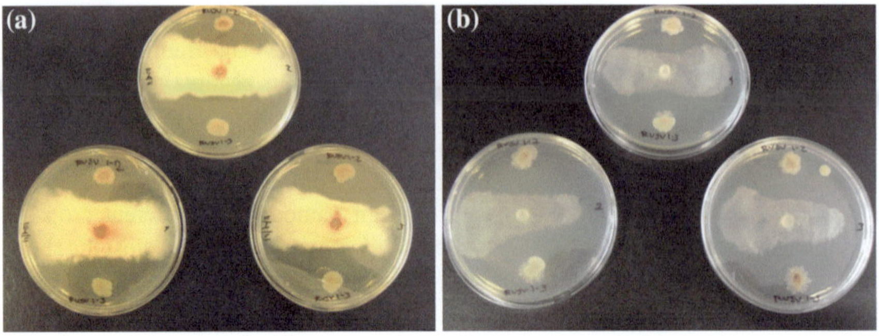

Fig. 19.1 In vitro growth inhibition of **a** *F. oxysporum* f.sp. *phaseoli* and **b** *P. capsici* by endophytic rhizobacteria

per bacterial isolation, arranged in three Petri dishes (Fig. 19.1). Mycelial plug of *Fop* or *P. capsici* takes it from the growth zone of 7-day-old colonies was placed at the center of the plate. Two different bacterial strains from 8-day-old pure cultures on TSA medium were placed at 1 cm of the edge of the plate. Plates were incubated for 1 week at 28 °C. Inhibition zone between the fungus and the bacterial strain was measured (mm).

19.2.3 In Vivo Assay

The endophytic bacteria which produced a clear inhibition halo to *Fop* in the dual culture were tested in common bean plants inoculated with the pathogen. Likewise, the bacteria which produced a clear inhibition halo to *P. capsici* were tested in pepper plants, with the aim of evaluating its potential activity as biocontrollers. Plants were grown on 1 l pots with washed sterile exfoliated vermiculite (WSEV), on a growth chamber under controlled conditions (16 h in light at 24 °C and 8 h in darkness at 18 °C, and relative humidity 65 % ± 5 %). Plants were irrigated with Rigaud and Puppo (1975) nutritive solution with N added as NH_4NO_3 to a final concentration of 1.6 g l^{-1}. The statistical design for the experiment was complete random design with 8 replicates per treatment. Bacterial suspensions were obtained from cultures in TSA after 3 days of incubation at 28 °C. Appropriate controls inoculated only with the pathogen, and non-inoculated at all, were used. Aerial and root biomass were dried in oven at 70 °C during 72 h. Incidence and severity of the disease and plant growth parameters were estimated in the plants after 35 days. Analysis of variance was performed with the statistical package SPSS v. 17.0.

19.2.4 In Vivo Assay in Common Bean Plants

The treatments consisted on the inoculation of the common bean plants with 15 different endophytic bacterial strains. Common bean seeds were surfaced sterilized (ethanol 70 % 1 min, NaClO 6 %, 6 min and three times washed with sterile water) and 20 seeds per bacterial strain were seeded on a recipient with WSEV. Each seed was inoculated with 2 ml of a suspension of 9×10^8 UFC ml^{-1}. Controls were inoculated with a 0.9 % saline solution (2 ml). After 1 week, seedlings were transplanted to 1 l pots with WSEV and were infected with the pathogen. Roots of seedlings were washed on running tap water and deep during 5 min in spore suspension of 1×10^6 spores ml^{-1} of *Fop*. Once infected, seedlings (8/endophytic bacterial strain) were planted on the pots and were inoculated a second time with the endophytic bacterial strain (2 ml of a suspension of 9×10^8 UFC ml^{-1}). Incidence and severity was recorded 1 week after inoculation using the Centro Internacional de Agricultura Tropical (CIAT) scale (Pastor-Corrales 1987). Dry and fresh aerial mass weight was obtained.

19.2.5 In Vivo Assay in Pepper Plants

Thirteen endophytic bacterial strains were tested on pepper plants. Pepper seeds were surface sterilized (ethanol 70 % 1 min, NaClO 2 % 5 min, and three times washed with sterile water) and seeded on water agar (WA). After 4 days, seedlings were transplanted on seedling trays (25 × 28 × 45 mm) with WSEV and 3 weeks later plants were transplanted to 1 l pots with WSEV. Each plant was inoculated with 25 ml of a suspension of 1.5×10^8 UFC ml^{-1} at the bottom of the plant, avoiding the direct contact of the suspension with the plant. After 1 week, plants were infected with *P. capsici*. Each plant was inoculated with 1 ml of a zoospore suspension of 1×10^5 zoospores ml^{-1} of *P. capsici*. Zoospores were obtained following the procedure previously described by Kim et al. (2008). Incidence and severity was recorded 2 weeks after inoculation using the Kingkun et al. (1989) scale. Dry and fresh aerial mass weight and dry root weigh were also obtained.

19.2.6 Microcosm Assays

Co-inoculations with rhizosphere microorganisms were carried out in microcosm trials with real soil as substratum, with the purpose of studying its biocontroller effect against *Fop* and *P. capsici*. Dual combination of two selected endophytic bacteria (the best according to in vivo assay) plus either an autochthonous AM fungus or a rhizobia inoculum (*Rhizobium leguminosarum* LET1910) previously isolated from the PGI "Alubia de La Bañeza—León" (Mulas et al. 2011) were co-inoculated together with the pathogen. Controls inoculated only with the pathogen, and non-inoculated at all, were used. In the case of pepper, also control inoculated with the pathogen and treated with the fungicide Previcur® Energy (propamocarb 53 % and fosetil 31 %) at the recommended dose was used. Experimental design was a Randomized Complete Block Design (RCBD) with 3 repetitions. Each experimental unit consisted of four plants.

Trials were established in greenhouse. Inoculated common bean and pepper plants were grown in 4 l pots. Substrate was composed by a mixture 4:1 (v:v) of soil and exfoliated vermiculite. Soil was taken from common bean and pepper farms located within the PGI demarcation. Plants were irrigated and fertilized properly according to hydric and nutrient requirements. The incidence of the disease and the plant growth parameters were estimated in the plants after 2 months of growth in the glasshouse. Experimental data were analyzed with the statistical package SPSS v. 17.0.

19.2.7 Microcosm Assay in Common Bean Plants

Common bean seeds were surface sterilized (as described above) and seeded on a recipient with WSEV. Each seed was inoculated with 2 ml of a suspension of 9×10^8 UFC ml^{-1} of the corresponding bacterial strain according to the treatment. In the case of treatments composed by endophytic bacteria and rhizobia, seeds were inoculated with 1 ml of the suspension of each one. After 1-week growth on growth chamber, seedlings were transplanted on the 4 l pots and were infected with *Fop* as indicated for the in vivo assay. Infected seedlings were planted on pots and were inoculated again with the endophytic bacterial strain just at transplanting (8 ml of a suspension of 9×10^8 UFC ml^{-1} directly onto the roots, and in the case of treatments composed by endophytic bacteria and rhizobia 4 ml/each). In the treatments inoculated with the AM fungus, 10 g of AM inoculum (AM spores and mycelia mixed with sterilized sand as a carrier) were put on the hole before to put the seedling, following the inoculation with the endophytic bacteria. Controls were inoculated with saline solution. Incidence and severity was recorded 20 days after inoculation using the Centro Internacional de Agricultura Tropical (CIAT) scale (Pastor-Corrales 1987). Dry and fresh aerial mass weight was also obtained.

19.2.8 Microcosm Assay in Pepper Plants

Pepper seedlings obtained as described for the in vivo assay were transplanted to 4 l pots and inoculated with the corresponding microorganisms according to the treatment just at transplanting. Each plant was inoculated with 12.5 ml of a suspension of 6×10^8 UFC ml^{-1} directly in the root. Treatments composed by endophytic bacteria and rhizobia were inoculated with 6.25 ml of the suspension of each one and treatments composed by AM fungus were also inoculated as described for common bean inoculation. A second inoculation (at the same dose used on the first inoculation) with the endophytic bacteria and the rhizobia was done one week later of transplanting and one week later, the plants were infected with *P. capsici*. Each plant was inoculated with a 5 ml zoospore suspension of 1×10^4 zoospores ml^{-1} of *P. capsici.* Incidence and severity was recorded 4 weeks after infection using the Kingkun et al. (1989) scale. Dry and fresh aerial mass weight was obtained.

19.3 Results and Discussion

19.3.1 Dual Culture Assay

Results of dual culture assay showed that 18 endophytic bacteria out of the 122 tested, produced an inhibition halo (Fig. 19.1). Of them, eight only controlled the growth of *F. oxysporum*, and four only controlled the growth of *P. capsici*. Other six isolates controlled the growth of the two pathogens. Four of the isolates that inhibited the two pathogens also produced the largest inhibition halo and belonged to the genus *Pseudomonas*, specifically 3 strains of *Pseudomonas brassicacearum* (strains SV1, SV2, PB1), and one of *P. thivervalensis* (strain SV3) (Barquero 2014, unpublished PhD. Thesis). Inhibition halo size recorded in dual culture assay varied between 12.67 and 1 mm. Those isolates producing an inhibition zone on pathogens growth in different degrees belonged to the genera *Pseudomonas* and *Bacillus*, microorganisms commonly used as soil diseases controllers (Lugtenberg et al. 2013)

19.3.2 In Vivo Assay in Common Bean Plants

The isolates PB1 (*P. brassicacearum*) and FA1 (*Bacillus siamensis*) (Barquero 2014, unpublished PhD. Thesis) (Table 19.1) showed the best Fusarium root rot control and moreover the best plant growth rate (dry and fresh aerial biomass) in common bean plants, with a similar growth compared to the non-inoculated control. The plants inoculated with *B. siamensis* plus the pathogen showed an aerial growth rate even higher than the control, non-inoculated neither with the PGPR nor with the pathogen; however, this difference was not statistically significant. While *B. siamensis*, species recently described, have not been tested against *Fusarium*, Jeong et al. (2012) found a significant mycelial growth inhibition of the plant pathogens *Rhizoctonia solani* and *Botrytis cinerea*. In the other hand, *P. brassicacearum* is a common specie associated with plant roots and one of the major root colonizing

Table 19.1 Disease severity index and biomass produced by common bean plants grown in hydroponic conditions, inoculated with the indicated treatments based in autochthonous rhizobacteria and the pathogen *Fop*

Treatment	Fresh aerial mass (g plant^{-1})		Dry aerial mass (g plant^{-1})		Severity	
Pseudomonas thivervalensis (strain SV1)	1.60	a	0.215	a	6.50	de
P. thivervalensis (strain CC1)	1.58	a	0.216	a	7.25	e
Control inoculated with *Fusarium*	1.85	a	0.244	a	4.75	abcde
P. thivervalensis (strain SV3)	2.32	ab	0.266	ab	6.00	cde
P. brassicacearum subsp. *neoaurantiaca* (strain CV1)	2.83	ab	0.368	abc	5.25	abcde
Bacillus siamensis (strain TA2)	2.41	ab	0.373	abc	5.75	bcde
B. siamensis (strain PA1)	2.88	ab	0.374	abc	5.25	abcde
B. safensis (strain TA1)	3.07	ab	0.39	abc	5.00	abcde
Stenotrophomonas maltophilia (strain CE1)	2.75	ab	0.393	abc	4.75	abcde
B. siamensis (strain MM1)	3.37	ab	0.443	abc	3.5	abcd
P. thivervalensis (Strain SV2)	3.78	ab	0.448	abc	2.5	abc
B. siamensis (strain FB1)	3.68	ab	0.551	abc	5.25	abcde
B. siamensis (strain FB2)	4.06	ab	0.554	abc	2.25	ab
B. siamensis (strain FB3)	4.59	ab	0.668	abc	1.75	a
P. brassicacearum subsp. *neoaurantiaca* (strain PB1)	4.81	ab	0.701	abc	2.75	abc
Non-inoculated control	5.21	ab	0.808	bc	Not affected	
B. siamensis (strain FA1)	5.83	b	0.906	c	1.75	a
Mean square	12.694		0.338		25.438	
F	2.523**		3.444***		6.300***	

The mean values followed by the same letter did not significantly differ ($p < 0.01$) in the ANOVA. Significance level: *** $p<0.001$; ** $0.001<p<0.01$; *$p<0.05$

species of the plants *Arabidopsis thaliana* and *Brassica napus* (Ortet et al. 2011). Likewise, conducted studies have demonstrated its action against *Microdochium nivale* and *Fusarium* spp., causing snow mold in wheat (Levenfors et al. 2008), and as well against the soil-borne pathogen *Gaeumannomyces graminis* var. *tritici*, causing the "take all" of wheat (Ross et al. 2000).

Plants belonging to treatment inoculated with the strain of *B. siamensis* (strain FA1) plus *Fop* showed a significantly higher dry matter production, compared to the treatment inoculated only with *Fop*. Jeong et al. (2012) found an increase on seedling growth in *A. thaliana* without physical contact, suggesting that bacterial volatiles promoted plant growth. The results obtained in this study demonstrated a clear growth promotion action besides of *Fop* control, by the isolate of *B. siamensis*.

Also, it is important to note that there is no correlation between the results obtained in dual culture assay and the in vivo assay results, indicating that more than one mechanism of action are involved in biocontrol process.

Fig. 19.2 In vivo (**a**) and microcosm (**b**) biological control assays by endophytic autochthonous rhizobacteria. **a** Biocontrol of *P. capsici* in pepper by *B. pumilus* (from *left* to *right* control inoculated with *P. capsici*; treatment inoculated with the pathogen plus the strain of *B. pumilus*; non-inoculated control). **b** Biocontrol of *F. oxysporum* f.sp. *phaseoli* in common bean by *R. leguminosarum* and *B. siamensis* (from *left* to *right* non-inoculated control; control inoculated with *Fop*; treatment inoculated with the pathogen plus the strain of *B. pumilus* plus *Rhizobium leguminosarum* strain LET1910)

19.3.3 In Vivo Assay in Pepper Plant

In the case of pepper plants, the best results in the control of *P. capsici* were achieved in the treatments inoculated with the strains FC2 (Fig. 19.2a) and SA1, both belonging to *Bacillus pumilus* (Barquero 2014, unpublished PhD. Thesis) (Table 19.2); the growth parameters (dry and fresh aerial biomass) were significantly higher in the treatments ($p < 0.001$) compared to the control inoculated with *P. capsici*. However for other five bacterial strains, the growth parameters (specifically the fresh aerial biomass) were also significantly higher than in the control inoculated with the pathogen. In addition to this, the plant growth rate in the plants inoculated with the strain FC2 of *B. pumilus* plus the pathogen was even higher than in the control, non-inoculated neither with the PGPR nor with the pathogen. *B. pumilus* has been commercialized as biofungicide in fruit crops, vegetables, and soybean against foliar pathogens such as rust, powdery mildew, and botrytis (Pérez-García et al. 2011).

19.3.4 Microcosm Assays

The application of mixtures of microorganisms to control plant diseases is an important approach to the real conditions occurring naturally in the soil, where the pathogen control is achieved due to the action of several microorganisms instead of the action of high amounts of a single microorganism (Whipps 2001; Berendsen et al. 2012).

Table 19.2 Disease severity index and biomass produced by pepper plants grown in hydroponic conditions, inoculated with the indicated treatments based in autochthonous rhizobacteria and the pathogen *P. capsici*

Treatment	Fresh aerial mass (g plant^{-1})		Dry aerial mass (g plant^{-1})		Severity	
					Mean	Standard deviation
Control inoculated with *Phytophthora*	1.44	a	0.55	a	4.75	0.71
B. anthracis (strain FC1)	3.17	ab	0.73	ab	4.38	1.19
P. thivervalensis (strain CC1)	5.27	abc	0.99	abc	3.00	2.51
P. thivervalensis (strain SV1)	5.94	abc	1.01	abc	2.75	1.98
P. brassicacearum subsp. *neoaurantiaca* (strain CV1)	6.59	abc	1.02	abc	2.38	1.85
P. thivervalensis (strain SV2)	6.59	abc	1.05	abc	2.75	1.28
B. toyonensis (strain VA1)	7.08	abcd	1.07	abc	1.75	1.75
P. brassicacearum subsp. *neoaurantiaca* (strain PB1)	8.04	bcd	1.15	abc	1.63	0.52
B. pumilus (strain MM1)	8.23	bcd	1.21	cd	2.13	0.64
B. pumilus (strain VA2)	8.32	bcd	1.13	cd	1.63	1.06
B. pumilus (strain LQ1)	8.55	bcd	1.27	cd	1.75	1.16
P. thivervalensis (strain SV3)	8.80	bcd	1.23	cd	1.63	1.41
B. pumilus (strain SA1)	9.44	cd	1.33	cd	1.38	0.52
Non-inoculated control	10.12	cd	1.25	cd	Not affected	Not affected
B. pumilus (strain FC2)	11.21	d	1.48	d	0.50	0.93
Mean square	50.416		0.41			
F	4.772 ***		3.234***			

In the parameters analyzed with ANOVA, the mean values followed by the same letter did not significantly differ ($p < 0.01$). Significance level: *** p<0.001; ** 0.001<p<0.01; *p<0.05

19.3.5 Microcosm Assay in Common Bean Plants

Co-inoculation of common bean plants with the strain FA1 of the endophytic bacteria *B. siamensis* plus *R. leguminosarum* LET1910 showed the best control of Fusarium root rot in common bean (Fig. 19.2b and Table 19.3). The aerial biomass produced by the plants inoculated with the mentioned cocktail of microorganisms plus the pathogen (*F. oxysporum*) was significantly higher ($p < 0.001$) compared to the control inoculated with the pathogen. Furthermore, the differences in the disease severity were highly significant compared to the control inoculated with *Fusarium*. Antagonistic action of *B. siamensis* exhibited in the in vivo assay added to *R. leguminosarum* action was the best combination to control *Fop*. Besides to the well-known action of rhizobia as nitrogen fixing in legumes, it is assumed that direct modes of action against the pathogens are performed by the rhizobia. Dar

Table 19.3 Disease severity index and biomass produced by common bean plants inoculated with mixtures of autochthonous rhizobacteria and the pathogen *Fop*, in microcosm conditions

Treatment	Fresh aerial mass (g plant^{-1})		Dry aerial mass (g plant^{-1})		Severity	
Glomus moseae (genotype F) + *B. siamensis* (strain FA1)	2.60	a	0.60	a	7.5	c
G. moseae (genotype F) + *P. brassicacearum* subsp. *neoaurantiaca* (strain PB1)	3.33	a	0.79	ab	7.8	c
G. moseae (genotype A) + *B. siamensis* (strain FA1)	3.67	a	0.56	a	7.3	c
Control inoculated with *Fusarium*	3.84	a	0.73	a	7.6	c
G. moseae (genotype A) + *Rhizobium leguminosarum* LET1910	4.35	a	0.76	a	6.0	bc
R. leguminosarum LET1910 + *P. brassicacearum* subsp. *neoaurantiaca* (strain PB1)	5.23	ab	0.80	abc	5.8	bc
G. moseae (genotype F) + *R. leguminosarum* LET1910	6.86	abc	0.98	abc	5.7	bc
G. moseae (genotype A) + *P. brassicacearum* subsp. *neoaurantiaca* (strain PB1)	7.51	abc	0.94	abc	5.7	bc
R. leguminosarum LET1910 + *B. siamensis* (strain FA1)	10.78	bc	1.43	bc	3.8	ab
Non-inoculated control	11.17	c	1.46	c	1.0	a
Mean square	112.802		1.177		52.712	
F	5.624 ***		4.617***		9.034***	

The mean values followed by the same letter did not significantly differ in the ANOVA test ($p < 0.01$). Significance level: *** p<0.001; ** 0.001<p<0.01; *p<0.05

et al. (1997) evaluated the control of *Fusarium* in common bean achieved by the AM fungus *G. mosseae* and *R. leguminosarum*, finding a significant reduction on disease incidence. Studies have reported rhizobitoxin production, responsible for the restriction of tissue invasion of parasitic fungus (Chakraborty and Chakraborty 1989), also has been observed mycelial growth inhibition through secondary metabolites production (Ehteshamul-Haque and Ghaffar 1993; Bardin et al. 2004) and production of bacteriocine, an antibiotic produced by strains of *Rhizobium. trifolii* and *Rhizobium lupine* (Roslycky 1967).

19.3.6 Microcosm Assay in Pepper Plants

The best biocontrol of *P. capsici* in pepper plants was achieved co-inoculating the plants with the AM fungus *Glomus mosseae* genotype A and the strain SA1 of *B. pumilus* (Table 19.4). Plants co-inoculated with such cocktail plus the pathogen showed a very low level of disease severity, similar to the control treated with the

Table 19.4 Disease severity index and biomass produced by pepper plants inoculated with mixtures of autochthonous rhizobacteria and the pathogen *P. capsici*, in microcosm conditions

Treatment	Fresh aerial mass (g plant^{-1})		Dry aerial mass (g plant^{-1})		Severity	
Control with *Phytophthora*	0.109	a	0.025	a	4.6	e
G. moseae (genotype F) + *B. pumilus* (strain SA1)	0.395	ab	0.068	ab	3.9	de
G. moseae (genotype F) + *B. pumilus* (strain FC2)	0.531	ab	0.121	ab	3.8	de
R. leguminosarum LET1910 + *B. pumilus* (strain FC2)	0.542	ab	0.116	ab	3.2	cde
G. moseae (genotype F) + *R. leguminosarum* LET1910	0.67	ab	0.095	ab	2.0	abcd
G. moseae (genotype A) + *R. leguminosarum* LET1910	0.738	ab	0.13	ab	3.6	cde
R. leguminosarum LET1910 + *B. pumilus* (strain SA1)	0.932	ab	0.155	ab	1,4	abc
G. moseae (genotype A) + *B. pumilus* (strain FC2)	1.426	ab	0.219	ab	0.7	ab
G. moseae (genotype A) + *B. pumilus* (strain SA1)	1.509	ab	0.224	ab	2.4	bcde
Control with fungicide	1.649	b	0.264	b	0.1	ab
Non-inoculated control	1.664	b	0.232	ab	0.0	a
Mean square	3.613		0.069		32.073	
F	3.109**		2.520***		9.999***	

The mean values followed by the same letter did not significantly differ in the ANOVA test ($p < 0.01$). Significance level: *** p<0.001; ** 0.001<p<0.01; *p<0.05

chemical fungicide. Moreover, the biomass production of the co-inoculated plants and treated with the pathogen did not significantly differ either with the non-inoculated control or with the fungicide-treated control ($p < 0.001$). Antagonism has widely been studied as the key mode of action of the genus *Bacillus* (Pérez-García et al. 2011). Production of cell wall degrading enzymes (Shoda 2000) and lipopeptides showing activity against bacteria, fungi, and oomycetes and stimulating host defense mechanisms are the most studied weapons involved in biocontrol of pathogens (Ongena and Jacques 2008). Regardless the mode of action of *G. mosseae*, Cordier et al. (1998) attributed the disease inhibition observed against *Phytophthora parasitica* on tomato plants, to Induced Systemic Resistance (ISR) mechanisms; also competition for nutrients and root colonization with pathogens are considered modes of action of AM fungi (Azcón-Aguilar et al. 2002).

19.4 Conclusions

The obtained results in this study demonstrate the microorganism ability to control soil-borne pathogens, opening the door for the development of biocontrollers for both common bean and pepper. Although this is only one of the first stages in the development process of a commercial product, the data evidenced a clear effect on diseases control. Despite of numberless studies that have been conducted to search and use microorganism as biocontrol agents, the amount of bioformulations available in the market is limited. On the one hand, issues such as modes of action, host specificity, kind of mixes, and environmental conditions to have the best pathogen adaptability, still need a better understood. On the other hand, hesitation to change to biological products of farmers and the lack of consistency on field results, make of biological control a skill still unexploited. Finally, the condition of living organisms of the biocontroller bacteria makes necessary the search for regionally adapted strains to increase the rate of success, and for this reason works such as the present one gain interest for the commercial development of biocontrollers.

Acknowledgements This work has been financially supported by the research project LE029A10-2 (*Junta de Castilla y León*, Spain). The Spanish Ministry of Foreign Affairs and Cooperation (MAEC) granted M. Barquero

References

Akköprü A, Demir S (2005) Biological control of Fusarium wilt in tomato caused by *Fusarium oxysporum* f. sp. *lycopersici* by AMF *Glomus intraradices* and some rhizobacteria. J Phytopathol 153(9):544–550

Aprile MC, Caputo V, Nayga RM Jr (2012) Consumers' valuation of food quality labels: the case of the European geographic indication and organic farming labels. Int J Consum Stud 36 (2):158–165

Arora N, Kang S, Maheshwari D (2001) Isolation of siderophore-producing strains of *Rhizobium meliloti* and their biocontrol potential against *Macrophomina phaseolina* that causes charcoal rot of groundnut. Curr Sci 81(6):673–677

Azcón-Aguilar C, Jaizme-Vega M, Calvet C (2002) The contribution of arbuscular mycorrhizal fungi to the control of soil-borne plant pathogens. In: Gianinazzi S, Schüepp H, Barea JM, Haselwandter K (eds) Mycorrhiza technology in agriculture: from genes to bioproducts. Birkhaüser Verlag, Basel, Switzerland

Bardin SD, Huang H, Pinto J, Amundsen EJ, Erickson RS (2004) Biological control of *Pythium* damping-off of pea and sugar beet by *Rhizobium leguminosarum* bv. *viceae*. Can J Bot 82 (3):291–296

Barquero M (2014) Caracterización y selección de bacterias y hongos micorrícicos aislados en raíces de alubia y pimiento, en la provincia de León, para el desarrollo de biofertilizantes. Universidad de Salamanca (Unpublished PhD. Thesis)

Berendsen RL, Pieterse CM, Bakker PA (2012) The rhizosphere microbiome and plant health. Trends Plant Sci 17:478–486

Calvet C, Pinochet J, Hernández-Dorrego A, Estaún V, Camprubí A (2001) Field microplot performance of the peach-almond hybrid GF-677 after inoculation with arbuscular mycorrhizal fungi in a replant soil infested with root-knot nematodes. Mycorrhiza 10(6):295–300

Chakraborty U, Chakraborty B (1989) Interaction of *Rhizobium leguminosarum* and *Fusarium solani* f. sp. *pisi* on pea affecting disease development and phytoalexin production. Can J Bot 67(6):1698–1701

Compant S, Duffy B, Nowak J, Clément C, Barka EA (2005) Use of plant growth-promoting bacteria for biocontrol of plant diseases: principles, mechanisms of action, and future prospects. Appl Environ Microbiol 71(9):4951–4959

Cordier C, Pozo M, Barea J, Gianinazzi S, Gianinazzi-Pearson V (1998) Cell defense responses associated with localized and systemic resistance to *Phytophthora parasitica* induced in tomato by an arbuscular mycorrhizal fungus. Mol Plant Microbe Interact 11(10):1017–1028

Dar GH, Zargar M, Beigh G (1997) Biocontrol of Fusarium root rot in the common bean (*Phaseolus vulgaris* L.) by using symbiotic *Glomus mosseae* and *Rhizobium leguminosarum*. Microb Ecol 34(1):74–80

Dutta S, Mishra A, Dileep Kumar B (2008) Induction of systemic resistance against fusarial wilt in pigeon pea through interaction of plant growth promoting rhizobacteria and rhizobia. Soil Biol Biochem 40(2):452–461

Ehteshamul-Haque S, Ghaffar A (1993) Use of rhizobia in the control of root rot diseases of sunflower, okra, soybean and mungbean. J Phytopathol 138(2):157–163

Gerhardson B (2002) Biological substitutes for pesticides. Trends Biotechnol 20(8):338–343

Glick BR (2012) Plant growth-promoting bacteria: mechanisms and applications. Scientifica 2012:1–15

Haas D, Keel C (2003) Regulation of antibiotic production in root-colonizing *Pseudomonas* spp. and relevance for biological control of plant disease. Annu Rev Phytopathol 41(1):117–153

Jeong H, Jeong DE, Kim SH, Song GC, Park SY, Ryu CM, Choi SK (2012) Draft genome sequence of the plant growth-promoting bacterium *Bacillus siamensis* KCTC 13613T. J Bacteriol 194(15):4148–4149

Kim HS, Sang MK, Jeun Y, Hwang BK, Kim KD (2008) Sequential selection and efficacy of antagonistic rhizobacteria for controlling Phytophthora blight of pepper. Crop Prot 27(3):436–443

Kingkun D, Taeyangkun H, Hongsanho J (1989) Expression of age-related resistance in pepper plants infected with *Phytophthora capsici*. Plant Dis 73:745–747

Larsen J, Graham JH, Cubero J, Ravnskov S (2012) Biocontrol traits of plant growth suppressive arbuscular mycorrhizal fungi against root rot in tomato caused by *Pythium aphanidermatum*. Eur J Plant Pathol 133(2):361–369

Levenfors JP, Eberhard TH, Levenfors JJ, Gerhardson B, Hökeberg M (2008) Biological control of snow mould (*Microdochium nivale*) in winter cereals by *Pseudomonas brassicacearum*, MA250. Biocontrol 53(4):651–665

Lugtenberg BJJ, Malfanova N, Kamilova F, Berg G (2013) Microbial control of plant root diseases. In: de Bruijn, FJ (ed) Molecular microbial ecology of the rhizosphere. Wiley, London

Mulas D, García-Fraile P, Carro L, Ramírez-Bahena M, Casquero P, Velázquez E, González-Andrés F (2011) Distribution and efficiency of *Rhizobium leguminosarum* strains nodulating *Phaseolus vulgaris* in northern Spanish soils: selection of native strains that replace conventional N fertilization. Soil Biol Biochem 43(11):2283–2293

Nogales A, Aguirreolea J, Santa María E, Camprubí A, Calvet C (2009) Response of mycorrhizal grapevine to *Armillaria mellea* inoculation: disease development and polyamines. Plant Soil 317(1–2):177–187

Ongena M, Jacques P (2008) *Bacillus lipopeptides*: versatile weapons for plant disease biocontrol. Trends Microbiol 16(3):115–125

Ortet P, Barakat M, Lalaouna D, Fochesato S, Barbe V, Vacherie B, Achouak W (2011) Complete genome sequence of a beneficial plant root-associated bacterium, *Pseudomonas brassicacearum*. J Bacteriol 193(12):3146

Pastor Corrales MA (1987) Reactions of selected bean germplasms to infection by *Fusarium oxysporum* f. sp. *phaseoli*. Plant Dis 71:990–993

Pérez-García A, Romero D, de Vicente A (2011) Plant protection and growth stimulation by microorganisms: biotechnological applications of *Bacilli* in agriculture. Curr Opin Biotechnol 22(2):187–193

Recep K, Fikrettin S, Erkol D, Cafer E (2009) Biological control of the potato dry rot caused by *Fusarium species* using PGPR strains. Biol Control 50(2):194–198

Rigaud J, Puppo A (1975) Indole-3-acetic acid catabolism by soybean bacterioids. J Gen Microbiol 88:223–228

Roslycky E (1967) Bacteriocin production in the rhizobia bacteria. Can J Microbiol 13(4):431–432

Ross IL, Alami Y, Harvey PR, Achouak W, Ryder MH (2000) Genetic diversity and biological control activity of novel species of closely related pseudomonads isolated from wheat field soils in South Australia. Appl Environ Microbiol 66(4):1609–1616

Saravanakumar D, Lavanya N, Muthumeena K, Raguchander T, Samiyappan R (2009) Fluorescent Pseudomonad mixtures mediate disease resistance in rice plants against sheath rot (*Sarocladium oryzae*) disease. Biocontrol 54(2):273–286

Sayeed Akhtar M, Siddiqui ZA (2008) Biocontrol of a root-rot disease complex of chickpea by *Glomus intraradices*, *Rhizobium* sp. and *Pseudomonas straita*. Crop Prot 27(3–5):410–417

Shoda M (2000) Bacterial control of plant diseases. J Biosci Bioeng 89(6):515–521

Vigo C, Norman J, Hooker J (2000) Biocontrol of the pathogen *Phytophthora parasitica* by arbuscular mycorrhizal fungi is a consequence of effects on infection loci. Plant Pathol 49 (4):509–514

Whipps JM (2001) Microbial interactions and biocontrol in the rhizosphere. J Exp Botany 52(Spec Issue):487–511

Yao M, Tweddell R, Desilets H (2002) Effect of two vesicular-arbuscular mycorrhizal fungi on the growth of micropropagated potato plantlets and on the extent of disease caused by *Rhizoctonia solani*. Mycorrhiza 12(5):235–242

Chapter 20
Disease Control and Plant Growth Promotion (PGP) of Selected Bacterial Strains in *Phaseolus vulgaris*

Miriam Memenza, Elvia Mostacero, Félix Camarena and Doris Zúñiga

Abstract The aim of this study was evaluating the effect of *Rhizobium* (RE10 and RM02) and *Bacillus* (B02 and B15) strains to reduce the incidence of disease caused by *Sclerotinia sclerotiorum* as well as to improve the yield of common bean cv. Centenary plants in field trial. The field trial was located in Pachacamac, Lima, Peru, and consisted of 8 treatments arranged in a random block design with three replicates. For the chemical control, different fungicides were applied on seeds and leaves. For each treatment, bacteria were first inoculated on the seeds and thereafter inoculated three times more in the plant neck. The seeds were pelleted with bacterial suspension (10^8 CFU/ml) previously mixed with 1 g of agricultural soil. The plant neck was sprayed with 6 ml of 2×10^5 CFU/ml, in three stages of crop growth. At harvest time, all the inoculated treatments showed the lowest percentage of plants infected by the disease caused by *S. sclerotiorum* with significant difference compared with untreated plants (14 %). The best treatment was RE10 + B02 (2 %), followed by RM02 (3 %) and B02 (4 %) without significant difference compared with the chemical control (2 %). The same strains also showed the highest percentage of germination. Treatment RE10 + B02 also presented the highest yield (16 pods/plant and grain yield of 1717.8 kg/ha) with significant difference compared with untreated plants (11 pods/plant and 883.2 kg/ha). On the other hand, plants inoculated with *Rhizobium* strains presented between six and ten nodules per plant, while untreated plants presented less than four.

Keywords Common bean · Disease incidence · Yield · PGP bacteria

M. Memenza · D. Zúñiga (✉)
Laboratory of Microbial Ecology and Biotechnology, Biology Department, Sciences Faculty, Universidad Nacional Agraria La Molina (UNALM), Lima, Peru
e-mail: dzuniga@lamolina.edu.pe

E. Mostacero · F. Camarena
Legumes and Oil Program, Agronomy Faculty, Universidad Nacional Agraria La Molina, Lima, Peru

© Springer International Publishing Switzerland 2016
F. González-Andrés and E. James (eds.), *Biological Nitrogen Fixation and Beneficial Plant–Microbe Interactions*, DOI 10.1007/978-3-319-32528-6_20

20.1 Introduction

The common bean (*Phaseolus vulgaris* L.) is original from the American continent, and in Peru, it is one of the most consumed legumes due to its high content of proteins, fats, carbohydrates, and minerals. Common beans are the most cultivated legume in Peru (30,800 Ha) (MINAGRI 2014). The crop management is performed by intensive use of chemical pesticides and fertilizers which are prone to cause environmental and social problems, coupled with the effects of global warming. By the other hand worldwide, the overuse of chemicals has increased in the long term, the rate of incidence of crop diseases with crop yield decreasing up to 70 %, threatening the food safety around the country (Camarena et al. 2009). Eighty percent of these diseases are caused by pathogenic fungi able to infect plant tissues in different phenological stages of the crop, highlighting *Sclerotinia sclerotiorum* as responsible of the green parts and pods rot (Camarena et al. 2012; SENASA 2012).

The use of bacteria from the rhizosphere of common bean, as plant growth promoters (PGP) in commercial field, has a high potential for the control of phytopathogens and the improvement of the crop yield (Olalde and Aguilera 1998). The PGP bacteria exert several direct and indirect bacterial mechanisms in the plant rhizosphere, having as a consequence the improvement of the plant growth (Gopalakrishnan et al. 2015). Direct plant growth promotion results from (1) nutrient acquisition by nitrogen fixation and phosphate solubilization and (2) hormonal production of indole-3-acetic (IAA) acid (auxin), cytokinins, gibberellins, and other hormones. In the other hand, several compounds are involved in the suppression of plant pathogens, which is indirectly connected with plant growth. This includes (1) production of antibiotics; (2) secretion of siderophores for iron uptake depriving it from the fungal pathogens; (3) production of low molecular weight metabolites such as hydrocyanic acid (HCN); (4) production of lytic enzymes such as chitinase, b-1,3 glucanase, protease, and lipase, which lyse the pathogenic fungal cell walls; and (5) induced systemic resistance (ISR) processes in plants (Reddy 2013; Kloepper and Schroth 1978). Many of the bacteria with interest as inoculants for crops are from the genera *Rhizobium* and *Bacillus*. Rhizobia are well known as inoculants for nitrogen fixation in legumes and also for phosphate solubilization. Moreover, *Rhizobium leguminosarum*, *Rhizobium meliloti*, *R. leguminosarum* bv. *trifolii*, *R. leguminosarum* bv. *viciae*, *R. meliloti,* and *Rhizobium trifolii* have been reported to secrete antibiotics and cell wall-degrading enzymes to inhibit phytopathogens (Afzal and Bano 2008; Egamberdiyeva et al. 2004; Bardin et al. 2004; Vessey 2003; Ozkoc and Deliveli 2001; Siddiqui and Mahmood 2001). In the case of *Bacillus,* this group presents high genetic diversity, and it is consider one of the main sources of microbial biopesticides. Different mechanisms have been proposed to explain the capacity of *Bacillus* spp. to inhibit the mycelial growth of fungal pathogens, which includes production of antimicrobial metabolites, release of hydrolytic enzymes, competence for nutrients and space, or combination of these mechanisms (Compant et al. 2005). There is a broad variety of antibiotic compounds with structural and functional diversity which control pathogenic microorganisms: oligomycin A, kanosamin, zwittermicin A, and polyketides as macrolactin (Cawoy et al. 2011). Peptides are the most common antibiotic compounds produced by this genus, which includes polymixin, surfactin, iturin, and fengycin, all are characterized to be resistant to hydrolysis by peptidases and proteases (Zhang et al. 2012). *B. brevis* and *B. polymyxa* produce peptide antibiotics that strongly inhibit germination of *Botrytis cinerea* causal agent of gray rot on strawberry fruit in vitro and field tests (Haggag et al. 2013). *Bacillus* strains also produce phytohormones and solubilize phosphates (Richardson et al. 2009); several species of the genus *Bacillus* have induced an increase of the length and the number of secondary roots, which has been attributed to the production of the auxin indole-3-acetic acid (IAA) (Idriss et al. 2002).

The aim of this study was to evaluate the effect of *Rhizobium* and *Bacillus* strains to reduce the incidence of disease caused by *S. sclerotiorum* and to improve the yield of common bean cv. centenary plants in field trial.

20.2 Materials and Methods

20.2.1 Preparation of Inoculum

The PGP bacterial were previously isolated from rhizosphere of different crops. The strains RM2a (*Rhizobium*) and B02 (*Bacillus*) were isolated from common bean plants cultivated in the valley of Ica (Zuñiga 2015); strain RE10 (*Rhizobium*) from common bean plants cultivated in La Molina (Gonzales et al. 2007); and strain B15 (*Bacillus*) from potato plants from the Andean highland (Calvo et al. 2010). In the previous study (data not shown), all *Bacillus* strains showed antagonistic effect against pathogens and plant growth-promoting activity in vitro, and *Rhizobium* strains also in vivo (Zúñiga 2015). For inoculum production, *Rhizobium* (RE10 and RM2a) and *Bacillus* (B02 and B15) strains were cultivated in 25-ml Erlenmeyer flasks containing 10 ml of yeast mannitol broth (YMB) and nutrient broth medium, respectively. All strains were incubated at 28 °C on a rotary shaker at 150 rpm until they reached 10^8 CFU/ml measured in optical density at 600 nm.

20.2.2 Field Experiment

Study was conducted at the Experimental Station of Legumes Program of Universidad Nacional Agraria La Molina (UNALM), located in Pachacamac Lima, Peru (12°16′37″S, 76°52′34″W), at 78 m above sea level. The experiment was performed with seeds of common bean cv. Centenary obtained from Legumes and Oil Program, UNALM. The experimental field presented loam soil with 3.15 % organic matter, phosphorus 4.67 ppm, potassium 150 ppm, electrical conductivity 4.20 dS/m, and pH 8.22. During the experiment, weather conditions were 94.71 % of humidity, average temperatures between 13.67 and 17.09 °C, and total rainfall of 40 mm (SENAMHI 2015). The experiment was arranged in field plots (7.2 m^2) comprised of three rows and 44 common bean cv. Centenary plants per row laid in a random complete block design with three replications. Eight treatments were considered: plants inoculated with only one strain (RE10, RM2a, B02, and B15), double inoculation (RE10 + B02 and RM2a + B02), chemical control, and untreated plants. Bacteria were first inoculated on seeds and then by successive applications on neck of plants in different growth stages: early flowering, pod formation, and pod filling. Three hundred and fifty six grams of seeds (1056 seeds) were pelleted with bacterial suspension (10^8 CFU/ml) previously mixed with 1 g of agricultural soil (Zuñiga 2012). For plant neck treatment, plants were sprayed with 6 ml of 2×10^5 CFU/ml. Fungicides based on the active ingredients thiophanate methyl and thiram were used as chemical control, and they were applied on seeds and leaves following the corresponding instructions.

20.2.3 Growth Parameters Evaluation

To assess the efficacy of *Rhizobium* and *Bacillus* strains to promote growth plant, plants were sampled in different growth stages. At the beginning of experiment, percentage of germination was evaluated in the three rows per plot. At pre-flowering stage, ten plants were uprooted from the central row of each plot and used to measure the plant height (cm), the fresh and dry weight (g), and the number of flower buds. The dry weight of the plant was recorded after drying the plants at 80 °C for three days. At harvest, the number of pods/plant (PP) and the grain yield (kg/ha) were scored.

20.2.4 Disease Evaluation

Percentage of disease incidence (% DI) caused by *S. sclerotiorum* was evaluated at the time of bacteria inoculation using the following formula:

$$\%\,\mathrm{DI} = \frac{\text{Number of infected plants}}{\text{Total number of tested plants}} \times 100$$

20.2.5 Statistical Analysis

The Statgraphics statistical package was used for data analysis. When a significant *F* value was detected, LSD estimates of significant differences were calculated from the ANOVA analysis. The level of statistical significance was set at 0.05.

20.3 Results

20.3.1 *Analysis of Growth Parameters*

The assessment of the plants growth showed that there was a positive response to bacterial inoculation, despite the poor soil conditions, which was alkaline, saline, and deficient in phosphorus. The treatments inoculated with bacterial strains showed higher percentages of germination compared with untreated plants (71 %). The best results were obtained with the strains RE10 + B02 (84 %) and RM2a (82 %), similar to the chemical control (85 %) (Table 20.1). Plants treated with RE10 + B02 also presented higher value of plant height (46 cm), fresh (82.88 g) and dry (23.43 g) weight, and number of flower buds (17), compared to the untreated control. The second best treatment was plants inoculated with *Bacillus* B02 strain with plant height (43.37 cm), fresh (114.64 g) and dry (26.62 g) weight, and flowers bud (17).

The result also showed that all treatments had positive effect on the grain yield with significant differences compared with the untreated control (11 pods/plant and 883.2 kg/ha) (Table 20.1 and Fig. 20.1). The treatment inoculated with RE10 + B02 showed the highest yield (16 pod/plant and 1717.8 kg/ha), following the treatment with B02 (18 pod/plant and 1361.99 kg/ha).

Table 20.1 Effect of PGP bacteria on percentage of seed germination and growth parameters on common bean plants under field conditions

Treatments	% G	PH (cm)	FW (g)	DW (g)	FB	PP	GY (kg/ha)
RE10	74	36.83	86.80	21.67	22	17 b	1162.41 c
RM2a	82	40.43	77.53	21.95	17	16 b	1272.73 c
B15	78	39.07	75.98	20.69	18	16 b	1391.16 c
B02	81	43.37	114.64	26.62	17	18 b	1361.99 c
RE10 + B02	84	46.00	82.28	23.43	17	16 b	1717.82 a
RM2a + B02	71	43.77	63.45	20.44	17	16 b	1361.90 c
Chemical control	85	44.27	82.77	21.11	15	17 b	1430.93 b
Untreated plants	73	40.90	69.89	20.05	16	11 a	883.19 d
Mean square	92.613	27.796	700.772	13.794	14.095	11.518	4431.23
F	0.914 ns	0.887 ns	0.544 ns	0.356 ns	0.252 ns	3.412*	64.433**

The mean values followed by the same letter did not significantly differ in the ANOVA (test LSD)
% G Percentage of germination, *PH* Plant height, *FW* Fresh weight, *DW* Dry weight, *FB* Number of flower buds, *PP* Pod/plant, *GY* Grain yield
Significance values: ns, not significant; $*p \leq 0.001$; $**0.001 < p \leq 0.005$

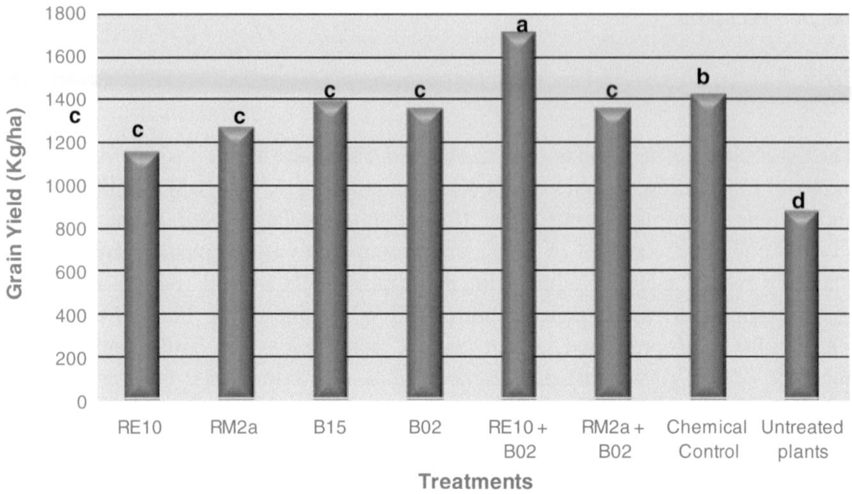

Fig. 20.1 Effect of PGP bacteria on grain yield for each treatment on common bean plants. Different letters mean statistically significant differences ($p \leq 0.05$) in the LSD test

20.3.2 Analysis of Incidence Disease

During the field trial, weather conditions with averages of 94.71 % of humidity and temperatures between 13.67 and 17.09 °C had moderated influence in the disease development. Common bean plants showed rotting of pods and green parts covered by white fungal mycelium of *S. sclerotiorum*. At the harvest time, all the inoculated treatments showed the lowest percentage of disease incidence caused by *S. sclerotiorum* with significant differences compared with the untreated control (14 % DI). The best treatment was RE10 + B02 (2 % DI), followed by RM2a (3 % DI) and B02 (4 % DI) without significant difference compared with chemical control (2 % DI) (Fig. 20.2).

20.4 Discussion

The inoculation with *Rhizobium* sp. or better with a mixture of *Rhizboium* sp. and *Bacillus* sp. has proven to reduce the disease incidence in common bean and to improve the crop yield in field conditions. Therefore, inoculation has been shown as a replacement to chemicals in common bean crop, contributing to the development of a sustainable agriculture. *Rhizobium* spp. and *Bacillus* spp. have been widely reported as fungal diseases controllers and crop productivity improvers (Gopalakrishnan et al. 2015; Cawoy et al. 2011). In our field experiment, all the treatments had a beneficial effect in the reduction of the diseases incidence;

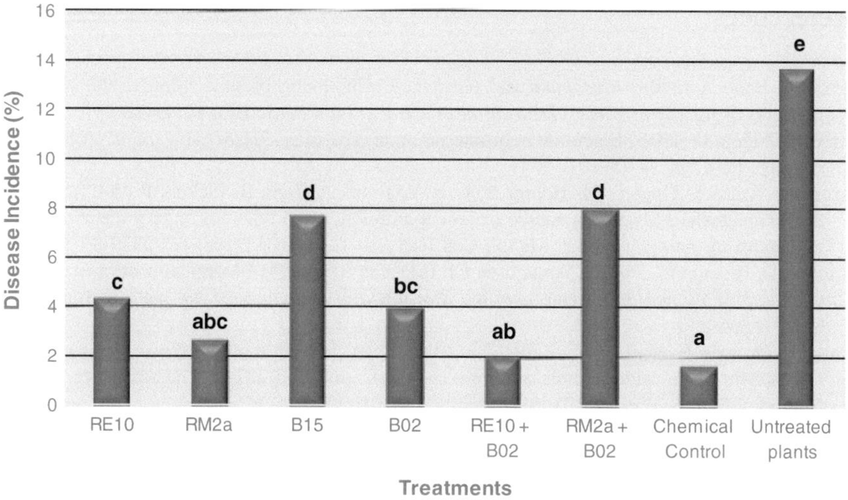

Fig. 20.2 Effect of PGP bacteria on the disease incidence caused by *S. sclerotiorum* between treatments on common bean plants. Different letters mean statistically significant differences ($p \leq 0.05$) in the LSD test

however, the synergistic effect of the combined inoculation with RE10 + BO2 showed better results when compared with single inoculation. In the other hand, the B15 strain isolated from Andean soils reduced in almost 50 % the diseases incidence of common bean (Fig. 20.2). Arcos and Zúñiga (2015) also reported the reduction of the disease caused by *Rhizoctonia solani* using *Bacillus* strains isolated from Andean soils, in potato. Several mechanisms have been proposed to explain the inhibitory effect of *Bacillus* sp. in the mycelial growth of *S. sclerotiorum*, which includes production of antimicrobial metabolites and hydrolytic enzymes, competence for nutrients and/or space, or combination of these mechanisms (Arguelles-Arias et al. 2009; Zhang and Xue 2010). In the other hand, *Rhizobium* strains are not only N-fixers in symbiosis with legumes, but also major promoters of the plant growth by the production of phytohormones and solubilization of phosphates (Gopalakrishnan et al. 2015).

This result displays the potential of inoculants based on PGP bacteria as controllers of fungal diseases in common bean crops, with a direct consequence in the yield improvement.

Acknowledgments This work was supported by IT-UNALM No. 003-2013 and FONDECYT No. 004-2013-CONCYTEC PROJECTS.

References

Afzal A, Bano A (2008) *Rhizobium* and phosphate solubilizing bacteria improve the yield and phosphorus uptake in wheat (*Triticum aestivum* L.). Int J Agric Biol 10:85–88

Arcos J, Zúñiga D (2015) Efecto de rizobacterias en el control de *Rhizoctonia solani* en el cultivo de papa. Ecol apl 14:95–101

Arguelles-Arias A, Ongena M, Halimi B, Lara Y, Brans A, Joris B, Fickers P (2009) *Bacillus amyloliquefaciens* GA1 as a source of potent antibiotics and other secondary metabolites for biocontrol of plant pathogens. Microb Cell Fact 8:1–12. doi:10.1186/1475-2859-8-63

Bardin SD, Huang HC, Pinto J, Amundsen EJ, Erickson RS (2004) Biological control of *Pythium* damping-off of pea and sugar beet by *Rhizobium leguminosarum* bv. *viceae*. Can J Bot 82:291–296

Calvo P, Ormeño E, Martínez E, Zúñiga D (2010) Characterization of *Bacillus* isolates of potato rhizosphere from andean soils of Peru and their potential PGPR characteristics. Braz J Microbiol 41:899–906. doi:10.1590/S1517-83822010000400008

Camarena M, Huaringa J, Mostacero N (2009) Innovación tecnológica para el incremento de la producción de frijol común (*Phaseolus vulagaris* L.). Primera Edición. Universidad Nacional Agraria La Molina-Consejo de Ciencia, Tecnología e Innovación Tecnológica, pp 20–25 (ISBN 978-9972-50-093-0)

Camarena M, Huaringa A, Mostacero E, Patricio J (2012) Tecnologías para el incremento de la producción del frijol vainita (*Phaseolus vulgaris* L.) para la exportación. Primera Edición. Universidad Nacional Agraria La Molina, pp. 13, 72–75 (ISBN 978-612-4147-00-5)

Cawoy H, Bettiol W, Fickers P and Ongena M (2011) *Bacillus*-based biological control of plant diseases. In: Pesticides in the modern world pesticides use and management, pp. 351–401 (ISBN 978-953-307-459-7)

Compant S, Duffy B, Nowak J, Clement C, Barka E (2005) Use of plant growth-promoting bacteria for biocontrol of plant diseases: principles, mechanisms of action, and future prospects. Appl Environ Microbiol 71:4951–4959

Egamberdiyeva D, Juraeva D, Poberejskaya S, Myachina O, Teryuhova P, Seydalieva L, Aliev A (2004) Improvement of wheat and cotton growth and nutrient uptake by phosphate solubilizing bacteria. In: Proceeding of 26th annual conservation tillage conference for sustainable agriculture, Auburn, pp 58–65

Gonzales E, Flores W, Matsubara M, Ormeño-Orrillo E, Martínez-Romero E, Zuñiga D (2007) Pérfiles REP-PCR y bioquímicos de cepas de *Rhizobium* aislados de tres variedades de frijol común (*Phaseolus vulgaris* L.). In: XXIII Reunión de la Asociación Latinoamericana de Rizobiología. Córdoba, Argentina. 25–28 de Marzo, pp 88

Gopalakrishnan S, Sathya A, Vijayabharathi R, Varshney R, Gowda C, Krishnamurthy L (2015) Plant growth promoting rhizobia: challenges and opportunities. 3. Biotech 5:355–377. doi:10.1007/s13205-014-0241-x

Haggag W, Abd-El-Kareem F, Abou-Hussein S (2013) Bioprocessing of *Brevibacillus Brevis* and *Bacillus Polymyxa*: a potential biocontrol agent of gray mould disease of strawberry fruits. IJEIT 3:509–518

Hashidoko Y, Nakayama T, Homma Y, Tahara S (1999) Structure elucidation of xanthovaccin A, a new antibiotci produced from *Stenotrophomonas* sp. stran SB-K88. Tetrahedron Lett 40:2957–2960

Idriss E, Makarewicz O, Farouk A, Rosner K, Greiner R, Bochow H, Richter T, Borriss R (2002) Extracellular phytase activity of *Bacillus amyloliquefaciens* FZB45 contributes to its plant growth-promoting effect. Microbiology 148:2097–2109

Kloepper J, Schroth M (1978) Plant growth-promoting rhizobacteria on radishes. In: Proceedings of the 4th international conference on plant pathogenic bacteria. Station de Pathologie Ve´ge´tale et de Phytobacte´riologie, INRA, Angers, France, pp 879–882

Olalde P, Aguilera L (1998) Microorganismos y Biodiversidad. Terra 16:289–292

Ozkoc I, Deliveli MH (2001) In vitro inhibition of the mycelial growth of some root rot fungi by *Rhizobium leguminosarum* biovar *phaseoli* isolates. Turk J Biol 25:435–445

Reddy PP (2013) Plant growth promoting rhizobacteria (PGPR). Recent advances in crop protection. Springer, India, pp 131–145

SENAMHI (2015) Available at http://www.senamhi.gob.pe/

Siddiqui ZA, Mahmood I (2001) Effects of rhizobacteria and root symbionts on the reproduction of *Meloidogyne javanica* and growth of chickpea. Bioresour Technol 79:41–45. doi:10.1016/S0960-8524(01)00036-0

Vessey KJ (2003) Plant growth promoting rhizobacteria as biofertilizers. Plant Soil 255:571–586. doi:10.1023/A:1026037216893

Zhang JX, Xue AG (2010) Biocontrol of sclerotinia stem rot (*Sclerotinia sclerotiorum*) of soybean using novel *Bacillus subtilis* strain SB24 under control conditions. Plant Pathol 59:382–391. doi:10.1111/j.1365-3059.2009.02227.x

Zhang SM, Wang YX, Meng LQ, Li J, Zhao XY, Cao X, Chen XL, Wang AX, Li JF (2012) Isolation and characterization of antifungal lipopéptidos produced by endophytic *Bacillus amyloliquefaciens* TF28. Afr J Microbiol Res 6:1747–1755. doi:10.5897/AJMR11.1025

Zúñiga D (2012) Manual de Microbiología Agrícola: *Rhizobium*, PGPRs, Indicadores de Fertilidad e Inocuidad. In: Ed. Olaya M. Universidad Nacional Agraria La Molina. Lima-Perú, pp. 53–55 (ISBN 978-612-4147-04-3)

Zuñiga D (2015) Estudio de las Interacciones *Phaseolus vulgaris*-PGPR-Micorrizas como Herramienta Biotecnológica para el Control de Fitopatógenos del Cultivo de Frijol. Informe Técnico. Proyecto IT-UNALM N°003-2013

Index

© Springer International Publishing Switzerland 2016
F. González-Andrés and E. James (eds.), *Biological Nitrogen Fixation and Beneficial Plant–Microbe Interactions*, DOI 10.1007/978-3-319-32528-6

MIX
Papier aus verantwortungsvollen Quellen
Paper from responsible sources
FSC® C105338

If you have any concerns about our products,
you can contact us on
ProductSafety@springernature.com

In case Publisher is established outside the EU,
the EU authorized representative is:
Springer Nature Customer Service Center GmbH
Europaplatz 3, 69115 Heidelberg, Germany

Printed by Libri Plureos GmbH
in Hamburg, Germany